EFCE Event No. 350

Process Optimisation

A three-day Symposium organised by the Midlands Branch of The Institution of Chemical Engineers and held at the University of Nottingham, 7–9 April 1987.

Organising Committee

J.K. Walters (Chairman)	University of Nottingham
P.K. Andow	Loughborough University of Technology
A.V. Bridgwater	University of Aston
N. Emery	University of Birmingham
J.B. Guy	The Boots Co plc
W.E. Jones	University of Nottingham
U. Joshi	British Gypsum
J.S. Parkinson	Courtaulds Ltd
K.E. Porter	University of Aston
J.I.T. Stenhouse	Loughborough University of Technology
R.B. Wilcockson	Loughborough University of Technology
J.A Wilson	University of Nottingham
A. Wint	University of Nottingham

THE INSTITUTION OF CHEMICAL ENGINEERS SYMPOSIUM SERIES No. 100

ISBN 0 85295 205 8

PUBLISHED BY THE INSTITUTION OF CHEMICAL ENGINEERS

First edition 1987 – ISBN 0 85295 205 8

MEMBERS OF THE INSTITUTION OF CHEMICAL ENGINEERS (Worldwide) SHOULD ORDER DIRECT FROM THE INSTITUTION

Geo. E Davis Building, 165–171 Railway Terrace, Rugby, Warks CV21 3HQ.

Australian orders to:

R M Wood, School of Chemical Engineering and Industrial Chemistry, University of New South Wales, PO Box 1, Kensington, NSW, Australia 2033.

Distributed throughout the world (excluding Australia) by Pergamon Press Ltd, except to IChemE members.

U.K.	Pergamon Press Ltd., Headington Hill Hall Oxford OX3 0BW, England.
U.S A.	Pergamon Press Inc., Maxwell House, Fairview Park, Elmsford, New York 10523, U.S.A.
CANADA	Pergamon Press Canada Ltd., Suite 104, 150 Consumers Road, Willowdale, Ontario M2J 1P9, Canada
FEDERAL REPUBLIC OF GERMANY	Pergamon Press GmbH, 6242 Kronberg-Taunus, Hammerweg 6, Federal Republic of Germany
JAPAN	Pergamon Press, 8th Floor, Matsuoka Central Building, 1–7–1 Nishishinjuku, Shinjuku-ku, Tokyo 160, Japan
BRAZIL	Pergamon Editora Ltda, Rua Eça de Quciros, 346, CEP 04011, Sào Paulo, Brazil
PEOPLES REPUBLIC OF CHINA	Pergamon Press, Qianmca Hotel, Beijing, People's Republic of China

---

BRITISH LIBRARY CATALOGUING IN PUBLICATION DATA

Process optimisation: a three-day symposium organised by the Midlands Branch of the Institution of Chemical Engineers and held at the University of Nottingham, 7–9 April 1987. – (EFCE event; no. 350). – (The Institution of Chemical Engineers Symposium Series; no. 100).

1. Chemical processes

I. Institution of Chemical Engineers *Midlands Branch* II. Series III. Series

660.2'8 TP155.7

ISBN 0-08-034818-1 ISBN 0-85295-205-8 Institution of Chemical Engineers

LIBRARY OF CONGRESS CATALOGING IN PUBLICATION DATA

Process optimization.

(The Institution of Chemical Engineers symposium series; no. 100)

"EFCE event no. 350."

1. Chemical process control – Congresses.

I. Walters, J. K. II. Institution of Chemical Engineers (Great Britain). Midlands Branch. III. Series: Symposium series (Institution of Chemical Engineers (Great Britain)); no. 100.

TP155.75.P764 1987 660.2'81 87-14790 ISBN 0-08-034818-1

Dedication

To the memory of Dr. J.C. Mecklenburgh

Foreword

The initiative for this Symposium on Process Optimisation came from Dr. J.C. Mecklenburgh, a Senior Lecturer at the University of Nottingham, who was Chairman of the Organising Committee from its inception until his sudden death in November 1986.

John Mecklenburgh was a scholarly man, an acknowledged authority on Plant Layout and Safety and a man very active in all aspects of the Institution's affairs. His death represents a great loss to the profession of Chemical Engineering.

The Symposium splits logically into three parts and follows the development of Process Plant from conception to operation. On the first day Project Selection is discussed, with financial implications, resource availabilty and energy matters being emphasised. This is followed by a day devoted to Process Design where detailed optimisation possibilities are considered including safety and hazard assessment. On the third day Plant Operation is discussed with condition monitoring, revamp and computing and control systems all considered for optimisation.

Improvement is always possible at all stages of a project and we have in this Symposium presentations on the latest ideas for Process Optimisation.

J.K. Walters

Contents

Project Selection

Process Design

Detailed Design

Plant Operation

Part I: Case Studies

Part II

* These papers are printed out of page sequence.

OPTIMISATION OF GAS SEPARATION SCHEMES FOR THE ENSOL PROCESS

D.G. Arundale*

SYNOPSIS

This paper describes the application to the ENSOL (natural gas to ethanol) process of a method of evaluating alternative flowschemes.

The method was developed as a flexible, computer based model capable of generating material and energy balances together with capital cost estimates. Flowschemes are outlined for production and separation of synthesis gas by the steam reforming, partial oxidation and methanol decomposition routes.

INTRODUCTION

In 1984 BASF, Monsanto and H&G jointly announced the ENSOL process route as a new, efficient and economical method of converting natural gas (or other hydrocarbons) into pure bulk ethanol. In view of the range of distinct processes linked by the ENSOL concept there is considerable scope for integration and optimisation. This paper describes the method developed for screening and evaluating the options available when designing for specific feedstock, site and market requirements.

THE ENSOL PROCESS

It will be sufficient in this paper to offer only a brief description of the essential ENSOL process. Technological, economic and market factors have been discussed in greater detail in various publications - see references 1 and 2, for instance.

The process chemistry and flowscheme are illustrated by Figures 1 and 2, from which it is clear that a series of more or less separate plants are involved. Initially a gaseous feedstock is converted into synthesis gas using H&G's own process technology and this syngas is then separated into three streams:

Stream (1) : Hydrogen
Stream (2) : Carbon Monoxide
Stream (3) : Methanol synthesis gas

* Humphreys & Glasgow Ltd.

Stream (3) passes as make up to the Methanol Loop to produce part or all of the methanol which joins with stream (2) as feed to the Acetic Acid plant. This plant applies Monsanto's well established carbonylation technology.

The key innovation is vapour phase hydrogenation of the acetic acid with stream (1) to give ethanol at very high yield. This reaction is well known but has only become feasible on a commercial scale now that BASF have developed and proved the catalyst.

Ethanol has excellent properties as a gasoline additive and a blending octane number as high or higher than other oxygenates such as MTBE, GTBA and methanol used for this purpose. As well as ethanol's position as a major motor fuel in Brazil, blends containing up to 10% of fermentation ethanol are already marketed in the USA and to a lesser extent in Western Europe. However, special circumstances or subsidies have been needed for economic viability. The new process, particularly when based on remotely located sources of natural or associated gas which would otherwise be flared, opens up the prospect of large scale use of ethanol. Moreover, there is not the restriction on feedstock availability that applies to more complex chemicals based on C_2, C_3 and C_4 hydrocarbons.

PROCESS OPTIONS

The part of the ENSOL process for which H&G have primary responsibility is the front end i.e. gas generation, gas separation and methanol synthesis. The essential products are feed materials at defined battery limit conditions for the acetic acid/ethanol end of the plant. This oversimplification is shown on Figure 3 for the balanced case (although methanol import or export cases are equally possible).

The "ideal" ENSOL material balance can be represented as:

$$2.CO + 4.H_2 = C_2H_6O + H_2O$$

i.e. the Gas Generator output must have a minimum H_2/CO ratio of 2. In practice CO_2 is also present (and is a vital component) and leads to some loss of hydrogen so that actual ratios are in the range 2.2-2.5. Ratios outside this indicate downgrading of CO or H_2 from product to fuel value.

Problems in Gas Separation are mainly caused by the presence of inert components - CH_4, N_2 and Argon. CO_2 is, of course, a useful part of the methanol make up gas. The most difficult operation is reduction of inerts in the CO stream to less than 5% and preferably 2%. Therefore the strategy is to build up a Loop make-up-gas which has both an acceptable stoichiometry and contains most of the inerts. The presence of the Methanol Loop confers a great advantage on the ENSOL process because of its incidental ability to act as an inerts purge.

A wide choice of options is available for both gas generation and gas separation:

Gas Generation Processes

Steam Reforming
Steam Reforming with CO_2 recycle
Partial Oxidation
Partial Oxidation with part purge recycle
Steam Reforming followed by Partial Oxidation
CO Shift (if needed for H_2/CO ratio adjustment)
Methanol Decomposition

Gas Separation Processes

CO_2 Removal
Cryogenic Separation
PSA (Pressure Swing Adsorption by molecular sieves)
Prism (Monsanto's proprietary membrane separator)
CO Extraction (e.g. Cosorb)
Methanol Synthesis

Choice of processes for these front end operations will be influenced not only by the factors mentioned above but also by project variables, e.g. feedstock composition or requirements for integration with other site facilities.

PROCESS FLOWSCHEMES

The methods developed to evaluate four flowschemes based upon the above processes will now be explained and some results presented.

Steam Reforming

The syngas produced by a steam reformer contains an excess of H_2 (typically H_2/CO about 3) but most or all of the surplus is readily used for furnace firing. Figure 4 shows the standard arrangement where CO_2 is removed from the syngas and compressed into the Loop, after which the balance of the gas is treated cryogenically.

Figure 4 also shows the alternative scheme where the CO_2 is recycled to the reformer feed. This reduces the H_2 excess and, as demonstrated later, is likely to give a better overall performance.

Partial Oxidation

The partial oxidation process gives rise to a syngas with H_2/CO ratio less than 2 but by means of the CO Shift reaction this can be raised sufficiently to prevent excessive carbon rejection to the fuel by-product stream.

The scheme shown in Figure 5 takes advantage of the high reaction pressure to purify the CO through two stages of Prism membrane separation. The Methanol Loop and PSA parts of the scheme are similar to those in steam reforming. In some circumstances it may be worth recycling part of the Loop purge to recover hydrogen. References 3 and 4 give information about the principles of PSA and Prism operation.

Methanol Decomposition

Syngas separation would be a much simpler operation if the inert components were not there. To some extent ENSOL already achieves this effect by use of the Methanol Loop and an interesting possibility is to carry this to its logical conclusion by passing all the syngas into the Loop.

A speculative scheme is shown in Figure 6. The conventional large, high efficiency methanol synthesis plant produces enough methanol both for direct ENSOL feed and for the $H_2/CO=2$ product of decomposition:

$$CH_3OH = CO + 2.H_2$$

Although this idea represents a novel application, decomposition itself is an established catalytic process carried out at moderate conditions. Reactor products include unreacted methanol and trace by-products as well as CO and H_2. Methanol and most of the by products are washed from the gas and recycled to the synthesis plant. The resultant syngas contains 1-3% inerts and is relatively easy to separate by a PSA/Prism arrangement into CO and H_2 streams.

METHOD OF EVALUATION

It was apparent from the outset of this work that, faced with so many schemes and options, we needed to develop a systematic and reproducible method of evaluating process alternatives in relation to feedstock consumption, energy balance and capital cost. The method had to be flexible enough for process optimisation over the range of basic raw material and design constraints likely to be encountered.

The approach taken was to reduce the essential balances to modular subroutines with easily adjustable parameters. These modules were incorporated into a computer program which also contained the data correlations and procedures for calculations such as equilibrium conditions, energy and power consumptions.

Process and Energy Balances

Referring to the general scheme of Figure 3, these calculations fall into three parts:

(1) Gas generation

Thermodynamic calculations define the synthesis gas composition together with (by overall energy balance) waste heat in the form of high pressure steam.

(2) Gas separation

The syngas is split into CO, H_2, and a make up gas (MUG) which is converted to methanol. A waste gas is also produced for use as fuel.

(3) Energy System

The electrical or mechanical power requirements of gas generation, separation and compression are met by a boiler/turbo-alternator system utilising all the HP steam and fuel gases, plus additional feed gas as necessary.

Figures 4 to 6 present specific examples of this general scheme for steam reforming, partial oxidation and methanol decomposition cases. Each block represents a modular set of calculation subroutines. Changes in pressure level are indicated by compressors, the power consumption of which is estimated from stream flow and pressure. The Reforming and Partial Oxidation modules carry out conventional Ks and Kr equilibrium and heat balance calculations. Most of the other modules are simple stream splitters into 2 or 3 products, the partition factors being treated as program parameters to be reset as specific design or vendor information becomes available.

Primarily to show module structure, factors typical of those which have been used for the Cryoseparation and Prism cases are given on Figure 7.

Capital Cost Estimate

Gross feedstock consumptions derived from material and energy balances cannot be compared in isolation - it is essential to have some indication of the capital cost impact. The method has therefore been developed to generate approximate costs for the various units and services defined on the flowschemes (e.g. Figures 4 - 6).

At this level of evaluation the requirement is more to highlight broad differences between schemes rather than to produce detailed estimates. A number of in-house estimates for comparable plants have been used to derive unit rates for total installed cost at the 2000 MTD Ethanol scale. These rates are then adjusted by the cost/capacity exponent method for the relatively small differences between 2000 MTD cases. However for services, where the range can be greater, slightly more complex functions are used.

This series of 2000 MTD based factors is given as Figure 8. They are satisfactory when used to compare variations between the ENSOL schemes but will not necessarily be suitable for use in different contexts.

RESULTS AND COMMENTS

The theme of this paper is development of a methodology applicable to most sets of project requirements in order to make a "broad brush" optimisation and selection between alternative ENSOL flowschemes. The intention is not to arrive at a single conclusion, but to present series of results which identify significant variables and demonstrate typical quantities for schemes of the type under discussion. The basic assumptions are as follows:

Feedstock composition	CH_4	89.4% (mol)
	C_2H_6	10.0%
	N_2	0.6%
Product streams	H_2	4135 kg.mol/h
	CO	2080 kg.mol/h
	Methanol	1846 kg.mol/h
Ethanol equivalent	2000 MID	

In addition to total feedstock consumption and capital cost the most important performance index for any scheme is the purity of the CO product stream - because of its impact upon the economics of the (extremely expensive) downstream acetic acid plant.

Steam Reforming Schemes

Effect of reforming conditions:-

Reformer pressure (bara)	12	20	25
temperature (C)	900	870	850
CO Product purity (mol%)	96.7	95.2	93.8
Process feedstock (MMkcal/h)	959	1055	1147
Total Plant Feed (MMkcal/h)	1142	1138	1151
Total Installed Cost (£million)	139.7	141.4	146.7

Effect of CO_2 Recycle:-

Reformer pressure (bara)	12	20	25
temperature (C)	900	870	850
CO Product purity (mol%)	97.3	96.6	96.0
Process feedstock (MMkcal/h)	933	986	1035
Total Plant Feed (MMkcal/h)	1126	1115	1119
Total Installed Cost (£million)	132.3	130.8	133.8

Effect of methanol loop efficiency:-

Carbon conversion (%)	94	85	75
Loop Pressure (bara)	80	70	60
CO Product purity (mol%)	96.6	96.5	96.4
Process feedstock (MMkcal/h)	986	1041	1118
Total Plant Feed (MMkcal/h)	1115	1119	1128
Total Installed Cost (£million)	130.8	129.8	129.1

Partial Oxidation Schemes

Effect of methanol loop efficiency:-

Carbon conversion (%)	94	85	75
Loop Pressure (bara)	80	70	60
CO Product purity (mol%)	96.4	96.9	97.4
Process feedstock (MMkcal/h)	1046	1049	1074
Total Plant Feed (MMkcal/h)	1095	1085	1080
Total Installed Cost (£million)	156.2	153.0	150.7

Flowsheet variations with 85%/70 bar Loop:-

Type of variation	Standard	Steam addition	Partial recycle
CO Product purity (mol%)	96.2	95.3	94.3
Process feedstock (MMkcal/h)	1103	1175	1127
Total Plant Feed (MMkcal/h)	1125	1175	1171
Total Installed Cost (£million)	166.1	173.9	175.1

CONCLUSIONS

At this stage in the development of ENSOL there is no virtue in rejecting any of the schemes. All of them have something in their favour, as may be seen from the following table:

Comparison of best case for each scheme

Scheme	Steam Reforming	Stm.Ref.+ CO_2 Recycle	Partial Oxidation	Methanol Decomp'n
CO purity (mol%)	96.7	97.3	97.4	98.5
Feedstock (MMkcal/h)	959	933	1074	1122
Total Feed (MMkcal/h)	1142	1126	1080	1203
Cost (£million)	139.7	132.3	150.7	173

This evaluation shows that Steam Reforming, particularly with CO_2 recycle, is the process with the lowest capital cost. On the other hand Partial Oxidation gives the highest thermal efficiency. Methanol Decomposition has the potential to give the best level of CO product purity and hence to reduce the cost of the downstream acetic acid plant.

The particular market for which the ENSOL process is considered most suitable is the exploitation of surplus gas at remote locations i.e. low price feedstock and high cost of capital. In those circumstances the steam reforming schemes are clearly to be preferred and can give attractive investment returns (see Reference 1).

REFERENCES

1. C.L. Winter: Make Ethanol via Syngas. Hydrocarbon Processing-April 1986. pp 71-73.

2. J. Griffiths & A. Horton: Synthetic Ethanol - Gas to Liquid Assets. Carbon One Chemical Technology: I.Chem.E.Symposium 1986.

3. J.A. Heck: First U.S. Polybed PSA unit proves its reliability. Oil & Gas Journal, February 1980.

4. D.J. Stookey, C.J. Patten & G.L. Malcolm: Membranes separate Gases selectively. Chemical Engineering Progress, November 1986

FIGURE 1

THE ENSOL PROCESS

-PRINCIPAL CHEMICAL REACTIONS-

1. REFORMING/GAS SEPARATION

CH_4 + Steam ⟶ 3 Streams (i) CO Rich
(ii) Pure H_2
(iii) Residue (CH_4, CO_2, CO, H_2)

2. METHANOL SYNTHESIS

$$CO + 2H_2 \longrightarrow CH_3OH$$
$$CO_2 + 3H_2 \longrightarrow CH_3OH + H_2O$$

3. ACETIC ACID SYNTHESIS

$$CO + CH_3OH \xrightarrow{\star} CH_3COOH \text{ (>99\% Yield)}$$

(Byproducts - traces of H_2, CO_2, H_2O plus Ethanol and Propionic Acid)

★Catalyst: Rhodium Iodide + Methyl Iodide

4. ETHANOL SYNTHESIS

$$CH_3OOH + 2H_2 \xrightarrow{\star} C_2H_5OH + H_2O \text{ (>99\% Yield)}$$

(By-products - Ethyl Acetate and CH_4)

★Cobalt catalyst

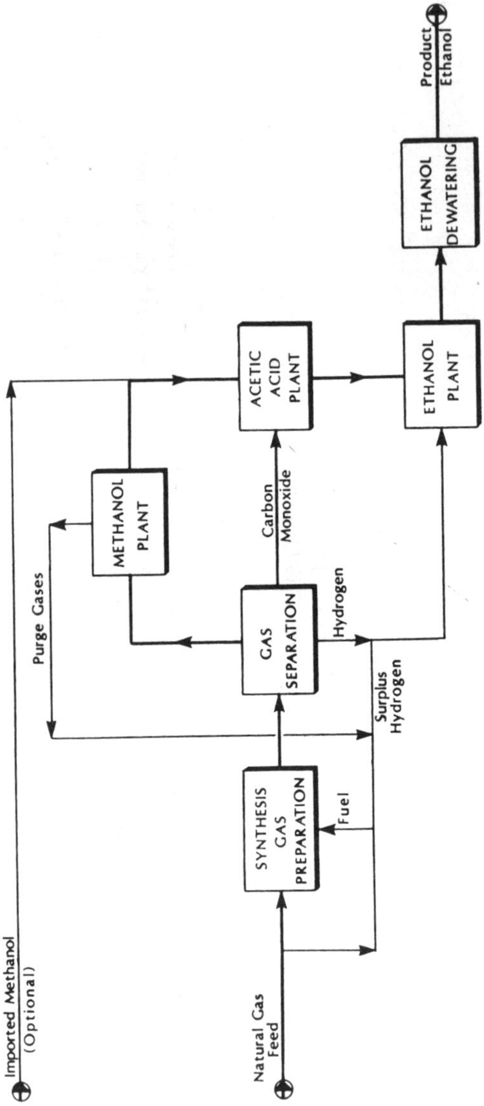

FIGURE 2

ETHANOL PLANT BLOCK FLOWSHEET

BASIC SCHEME

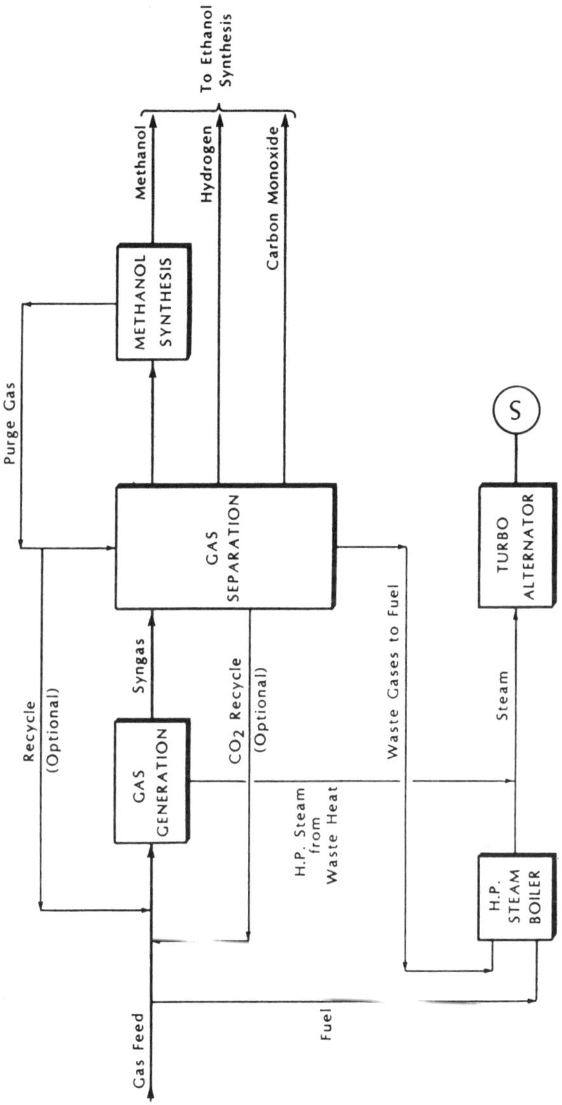

FIGURE 3

ETHANOL FRONT END
-SIMPLIFIED SCHEME-

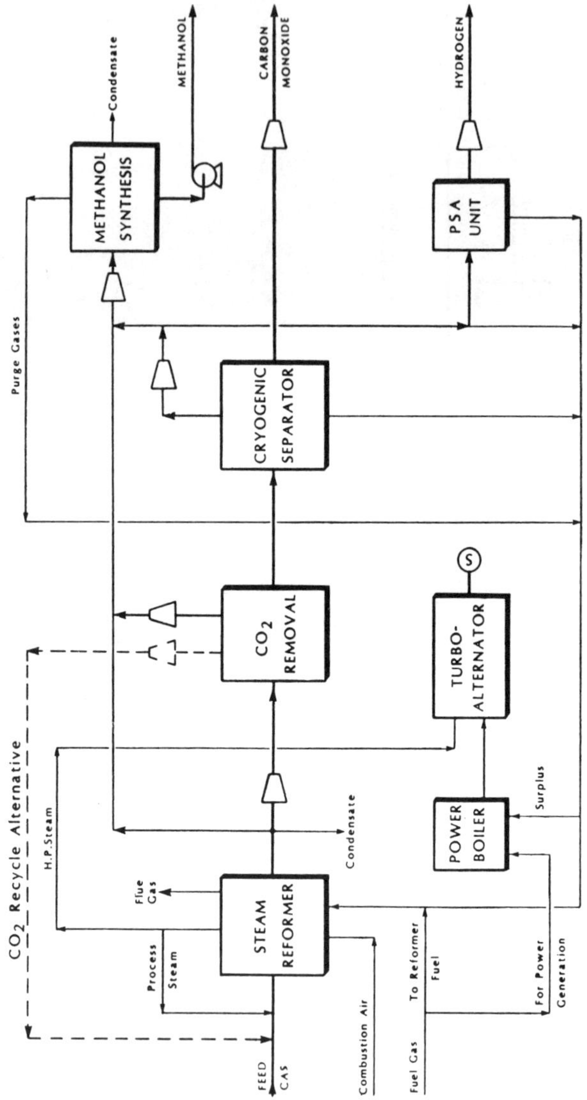

FIGURE 4

ENSOL FEED GASES

STEAM REFORMING FLOWSCHEMES

(CO_2 RECYCLE ALTERNATIVE)

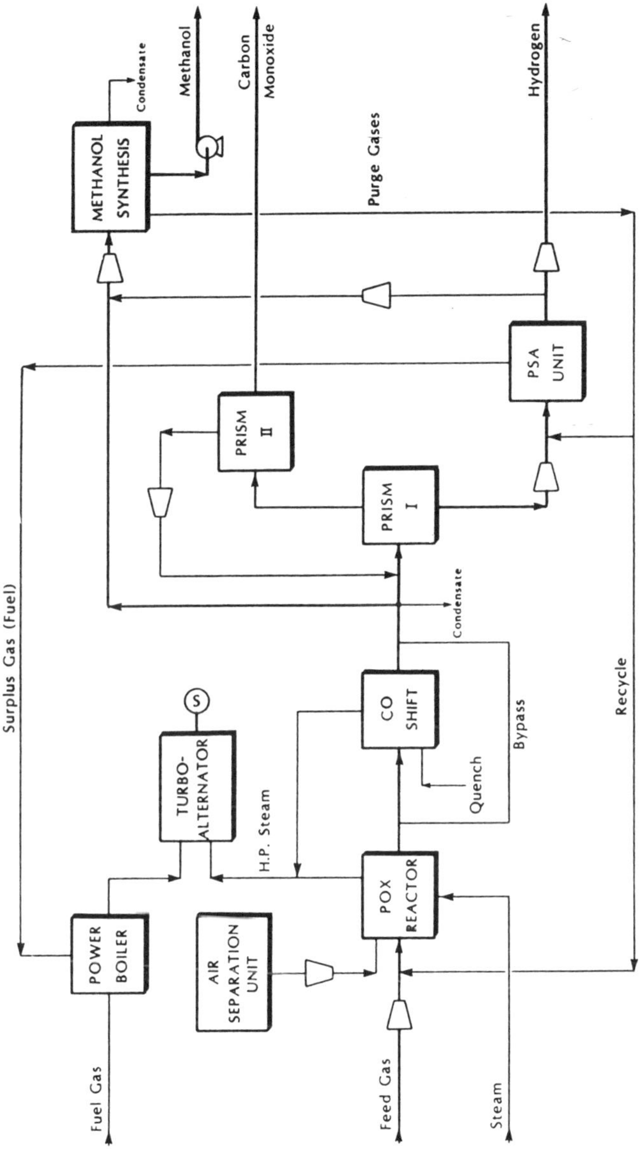

FIGURE 5

ENSOL FEED GASES

PARTIAL OXIDATION FLOWSCHEME

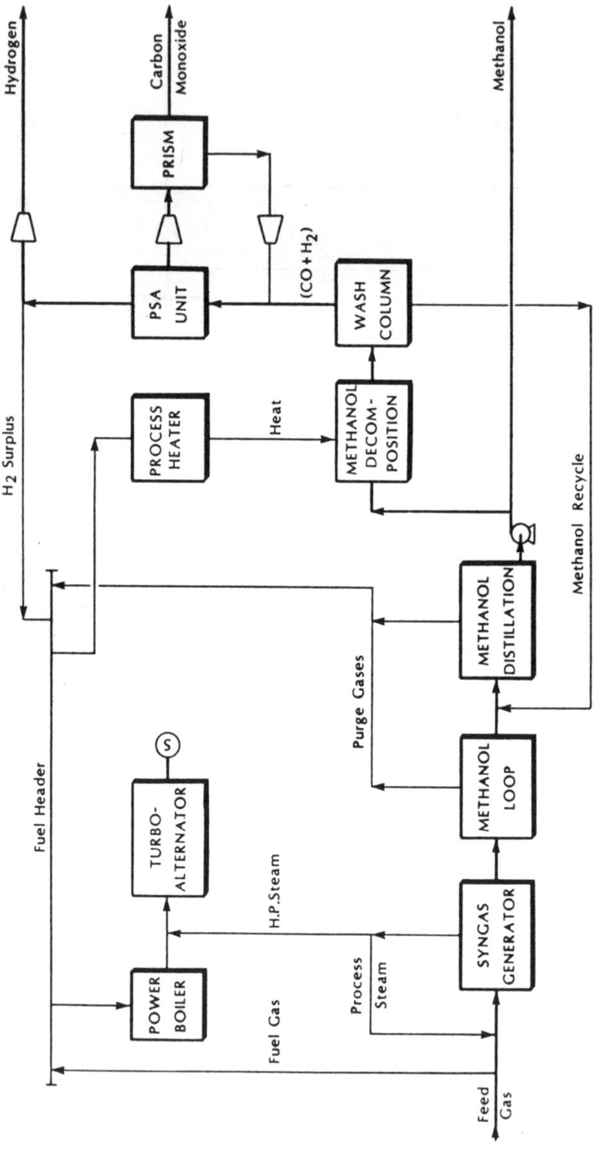

FIGURE 6

ENSOL FEED GASES

METHANOL DECOMPOSITION FLOWSCHEME

FIGURE 7

TYPICAL MODULE SEPARATION FACTORS

(mole fraction feed in each product stream)

Cryoseparation

Component	Hydrogen	CO	Waste
CH_4	0.15	0.05	0.80
H_2	0.90	0.005	0.095
CO	0.03	0.95	0.02
N_2	0.15	0.80	0.05
Argon	0.30	0.50	0.20

Prism Separator

Fraction of gas permeating

Feed Type	H2-rich	CO-rich
CH_4	0.05	0.24
H_2	0.79	0.99
CO	0.10	0.40
CO_2	0.54	0.95
N_2	0.05	0.23
Argon	0.23	0.33
H_2O	0.97	1.00

FIGURE 8

CAPITAL COST UNIT RATES

	Rate	Unit	Base
Reformer (complete with Heat recovery)	£150000	per MM kcal/h	3000
Gasifier + WHB	£ 1635	per dry kg mol/h+	12500
CO Shift + Heat Recovery	£ 800	per wet kg mol/h*	15000
Air Separation Unit	£ 8000	per MTD (99% O_2)	2000
Methanol	£ 11000	per MTD	1500
CO_2 Removal	£ 800	per kg mol/h*	11000
Cryo Unit	£ 2000	per kg mol/h*	6000
PSA	£ 1500	per kg mol/h*	9500
Prism	£ 1200-1800	per kg mol/h*	9000/5000
Decomposition	£ 12500	per MTD	2000
Compressor	£ 250	per kW	-
Turbine Drives	£ 150	per kW	-
Turbo-Alternator	£ 375	per kW	-
HP Boiler	£ 90000 ($C^{0.8}$ + $C^{0.5}$) (C=MTH)		50-500

\+ of output from unit.

* of feed to unit.

These quantities are derived from total installed capital cost estimates for plants of the base capacity indicated. For instance the methanol plant base estimate is £16.5 million at the 1500 MTD scale.

PRODUCT AND PROCESS SELECTION: ASPECTS OF EXPERT SYSTEMS IN DESIGN

S. A. Gregory

Optimality in the choice of product or process is a company decision. For top management today each product requires to be linked with a suitable competitive strategy. Engineers need to adjust their design thinking, whether in synthesis or evaluation, to match the new imperatives. This holds for design, whether manual or provided with computer-based designer support. We are at the beginning of a thrust to develop expert systems for design. It is important that value structures and other features needed to support competitive status should relate effectively to practical experience and be capable of embodiment within expert system protocols. This provides substantial challenges.

INTRODUCTION

This contribution sets out to reflect speculatively upon those considerations which would predispose towards optimality in design. Although alternative plans are possible under present-day conditions a successful manufacturing company requires to have:

* a technically adequate product supplied by a technically adequate process
* a product which provides an economic basis for entry into the market
* a competitive strategy which relates to the product, the market, and to potential competitors

These aspects provide primary elements of value in the absence of which action is only effective by chance. Within chemical engineering the technical needs are usually accepted without demur. The acknowledgment of a competent economic basis tends to be agreed but is often limited in scope. What is new for design people, although not for top managers, is the overt statement of strategic dimensions from a competitive standpoint.

To help with an appreciation it is useful to talk of the two faces of design. The first of these is the "3P" approach which deals with product, process, and plant. This covers the underlying technical function aspects of design. The second is the "2M" approach which brings together marketing and management and is of particular importance strategically. Technoeconomic evaluation helps to bridge the P-M gap. The requirement of designers is to embody the sum of wishes which lead to effective company performance. (Fig 1).

Although I shall assume that technical comment upon design can be avoided it is worth seeing that Ehrenspiel (1) has noted the shift which has become necessary within much of German engineering. In postwar Europe the Germans made their

* 22 Crescent Road, Stafford ST17 9AL

mark by emphasising technical excellence. Because of world competition they are now having to put emphasis upon excellence, cost, and reliability. This comes through not only from his book on cost-effective design but also from the V.D.I. guideline drawn up by a committee under his chairmanship (2).

In looking at some of the advanced countries on the world-scale it is clear that changes are being introduced which go in other directions. In the U.S.A., for example, arguments are being advanced for a strategy of rapid entry into a field in order to cope with competition. This comes strongly from Reinertsen (3) who provides plenty of information about design and development which can be traced back to useful earlier work on project time control. Emphasis is placed on aspects making designers more productive. In the U.K. concern with speed-up has been particularly focussed on the construction of buildings in the London financial area. Implications for design have been reported by Rippeteau (4). Gregory (5) has indicated that other kinds of strategy are possible, e g a strategy which aims to go up-market instead of aiming to offer lower price as a means of satisfying a market more effectively.

Within the chemical industry various moves are occurring which need to be accommodated. For some time the shift into specialisation has been taking place. Recently we have been impressed by the re-orientation within I.C.I. which brings together heavy products at one end of the business, and sets the growth products and specialist areas into a different part. Even at a smaller scale we have seen the launch on the public of the M.T.M. company with various aims which avoid some of the traditional approaches, where cheap plant is not the only rule but emphasis is placed upon flexibility. With a market-led company speed in conversion or build-up becomes important, as in the case of the electronics industry proposals from Reinertsen already quoted.

Gonzalez (6) has recently spoken of decisions within a company in relation to the use of continuous methods in the processing of pharmaceuticals. It is possible that what is being reflected is strategic rather than cultural as he claims, or culture derived from strategic awareness.

What is important is that we need to make a comparable shift in our engineering understanding and to integrate the conventional 3P approach with the revealed modern 2M approach of top management. The aim is to try to identify what new features have to be developed in synthesis and, even more so, the new kinds of evaluative considerations. Instead of seeking new technical solutions we may require to seek strategic solutions in which corporate competitive evaluations play a central role.

STRATEGY AND ITS DYNAMICS

Ohmae (7) has recently emphasised that the characteristics of a company may be construed from its stress upon:

* competition
* commitment
* creativity

Commitment is what the company attempts to provide through long-term attention to the key facets of the business. Creativity is supplied, other things being equal, to the fulfilment of the objectives of the business by the competent or outstanding professional enthusiasm of the members of the company. Much of what has been said in modern terms about competitive strategy comes from Porter (8). The dynamics which are generated depend partly upon the structure of the area of industry, whether fragmented or with major operators, and partly from the nature and performance of the competitors in the field. How your competitors behave and respond is a question which demands answers. According to Porter there are alternative strategic roles depending upon business perceptions of competitive capability, and given the capacity to make strategic changes as needed. If we take the case of polyethylene manufacture today, for example, it is only possible to stay in business if production costs can be kept as low as those of rivals and sufficient money can be put back into the business to keep the technology up to date. Operation has of necessity to be on a large scale. There are important alternatives to working in this manner. Business can be put into the development of new products, or it can be deployed into special product areas. By choosing conditions carefully small firms can outwit larger companies by going for special niches. Upon this range of possibilities other structures may be developed through salesmanship, technical service, speed of supply, corporate image, and the like. A typical diagram from Porter is given (Fig 2). Empirical data regarding the process industries are not well presented by him and some value would come from a suitable collection from current events.

In order to advance further it is useful to apply this kind of thinking to the analysis of product lifecycle patterns. From this it becomes clearer what might be the kind or kinds of effective action at different stages in the life of a product. In introducing lifecycle patterns it is important to emphasise that these are constructs or models which aid analysis and may not fit every case. In the attached diagram (Fig 3) products, processes, and plants go through stages of development. There is not necessarily a tight link between them, supply of the new input being based upon perceptions of various providers, from within the innovating firm or from outside.

We assume that the primary innovation is a product emergence. This is not always the complete case because, as in the development of the dyestuffs industry, a considerable pressure in part of the work was based upon finding some process of manufacture which displaced the original natural product, as in the case of alizarin which replaced the equivalent natural red dye, and, later, the similar stages in the development of synthetic indigo. The breakthrough was the working process in each case. Other dyestuffs were entirely novel, constituting material not previously worked, as in the case of the initiator - Perkin's mauve.

Furthermore, in such an introductory scheme invention may be at different levels ranging from:

* the rather novel product
* the product with major new functions but a prior background
* the product with a minimal change
* the product which is largely re-packaging

What a company does under this range of conditions will depend, in part, upon its professional level in the field. If the search for new products is unusual the launching procedures will tend to be different from the practices of the company which is continually developing new products.

At this stage, also, there is a need to look at the surrounding pressures. Who else is in the business? What kind of patenting or comparable protection can be deployed? Is there a heavy capital investment demanded in launching? Choices have to be made at a time when market uncertainty is highest. Do you go for:

* a slow and accurate launch, particularly if there are novel aspects
* a fast launch based on borrowed plant, etc.
* a fast launch using plant with plenty of flexibility?

Where possible the tendency will be to avoid anything which is unusual in manufacture in order to achieve the product launch, unless that is what makes the product right.

With respect to the next stage - here called product expansion - the primary aim will be to bring more of the product, should it be attractive, to the market. The effort is expansion rather than cheapening. But attention will have to be directed at who else is in the business and what kind of entry might be made by them. At this stage a first expectation will be the continuation of the initial manufacturing process but so arranged as to produce economies of scale. A result of the launch stage was to gain more information about the size and response of the market. Now design for product expansion has to deal with fewer uncertainties. Can any of your competitors leap in at this point, choosing to build a plant of about your capacity and, by some means, compete on costs? This will depend upon patent protection, special skills, process secrecy. At this stage possible answers include improvement in product quality. If there is more expansion envisaged measures will have to be incorporated which deal with the responses and potential responses indicated at this stage. Such will include going cheaper, finding new customers, using alternative feedstocks, by-products.

At the product cheapening stage, with a business which has a developing pattern, it becomes possible to make use of various economic analyses, e g the -4/10 rule, to suggest what extra new customers might be gained. At the product cheapening stage attempts will be likely to make radical adjustments to the process. Unless the process has few thermal inputs there will be thought about the conventional modes of thermal integration.

At the market saturation stage growth will be no faster than that of the rise of the national product, if any. Negotiations will be going on regarding joint operation on the large scale, or of possible management buy-out. There will be concern about overcapacity, the extent of other people's technology, specific location advantages, whether new products can be evolved in the family, and what kind of investments can be justified in order to upgrade the existing plants.

At the persist or dispatch stage decisions have to be made whether to stay in or get out if they have not previously come up. This may come about through the opening of other plants, possibly elsewhere in the world, or because of change in the conditions of supply of raw materials, or for other reasons. We may stay in if our technology is better than our competitors. Expansion of trade may be bad for us because of the locations in which the up-growth occurs. At this stage the number of competitors may be decreasing but the competition has become world wide.

What is emerging from this kind of analysis of the dynamics of trade and its strategic interconnections is that key objectives are changing and that major items require special treatment. For example we have to pay attention to the way in which the specific class of customer changes. First launch may be based on technical merit but later thought has to be given to the specialist needs of customers. Authors such as Engel and Blackwell (9) have suggested ways in which household consumers adopt choice patterns depending upon whether purchase needs high involvement or low involvement. The purchase of granulated sugar is a type case. The purchase of a car requires high involvement. Factors which attract industries and companies purchases have been studied by Rothwell and colleagues (10). Furthermore, these operations have to be set against a background which imposes social aspects of several kinds, dealing with safety, environmental pollution, preferred working conditions, and so on. According to the kind of business important extra criteria become added.

A METHOD OF REPRESENTATION

Up to the present methods of review of the technical products and processes have been available. Various methods have been advanced for the technoeconomic status of proposals. Rose (11), for example, has shown how to deal with some aspects of dealing with uncertainty in decision with respect to investment or cash flow. There is, however, remarkably little information available on ways of setting out or evaluating the status of strategic considerations. One of the few references which comes to hand is a chapter by Spitz in a book edited by Landau (12).

Although we can generalise the kinds of influence at work in bringing out cost estimates at the stage of concept consideration or at any other phase in the design or construction process Gregory (13) has attempted to summarise factors which are at work in any class of engineering (Fig 4). The intention here was to show the underlying common basis whether dealing with building and similar construction, mechanical engineering, and process engineering. Suitable inputs have to be made out for the case of batch or continuous manufacture in the mechanical industries, bringing together the recent work on cost estimating from Germany. The diagram aims to suggest the kind of approach which may be used depending upon the starting point chosen from which the project begins. It also attempts to show the kind of method which may be deployed depending upon the nature of the project, whether at the conceptual stage or beyond, emphasis being placed upon the difficulties of dealing with unusual schemes, whether novel or not.

In order to deal with the innovative side some emphasis has to be placed upon the extent of innovative content. Reference has already been made earlier to some kinds of procedure for the evaluation of new products or processes. A typical example comes from Habberjam (14). With such an approach it is possible to set out the kind of uncertainty which prevails at each stage of a development and to consider the extent of pilot work or otherwise which is needed to bring the project to an acceptable level of credibility. A simple approach is offered for commencement (Fig 5).

Finally we have to deal with strategic levels of work. We are looking for some kind of evaluative procedure which enables us to relate company posture with the state of competition envisaged. No one has attempted to do this, to the best of my knowledge, but a start is not without promise. In the diagram (Fig 6) special strategies are listed against which a measurement is suggested in terms of a suitable utility function or other proxy. As far as this paper is concerned this is speculation number one. In it some note is made of VDI Richtlinie 2221 (15) which is to be used as a reference basis for the development of a project. This aims to cover most areas of engineering but it is weak on modern features such as strategic alternatives. It has a strong but incomplete list of synthesis aids.

What we have done so far is to sketch out some of the features which have a possibility of helping us to deal with synthesis and, particularly with evaluing product and process schemes. This gives a basis for proceeding to examine the potential scope of overall design making use of expert systems.

INTRODUCING EXPERT SYSTEMS FOR OVERALL DESIGN

For those of you who are not directly at home with expert systems I offer an introductory quotation from Feigenbaum and McCorduck (16):

* "An expert system is a high-level intellectual support for the human expert"

This presupposes a suitable computer with adequate programs and suitably derived intellectual data of the various kinds needed to constitute the significant base. This introduces questions about:

* harnessing experts
* in suitable ways
* for specific purposes

With respect to engineering design the application of expert systems is seen as a step towards an improved designer support environment. An introduction to the scope of such systems in engineering design has been offered by Gregory (17) with a particular emphasis upon designer tasks rather than artificial intelligence requirements. The rise in availability of expert system shells has been of particular value in stimulating initial interest in design.

With respect to chemical engineering design various efforts were undertaken, from the mid-1960s onwards, to study process design. From this early period we can pick up, for example, the work of Gregory and Bridgwater (18), King (19), and Rudd and co-workers (20). Gregory and Bridgwater tackled a particular class of process in principle, making the important heuristic assumption of a common and definable process pattern or paradigm which contained possibilities of variation by a computational device. The American workers selected paradigm schemes from across a range of process layouts and listed recommended process constitution heuristics. This work was later much extended by Wells (21). In this sector of work various assumptions were made about technical or economic validity. Work of this kind continues today. Because of the way they worked Gregory and Bridgwater used an economic selection process which it was possible to reduce to a simple form. This led to step-counting and various later developments, now available in modern cost-estimation.

Within the field of chemistry the collection of methods of process operation began also in the 1960s, with particular emphasis upon Corey (22).

The recent spread of interest about expert systems in engineering design in various practical domains has opened since the early 1980s. Much of the early enthusiasm has been linked with electronics design. In a practical paper McDermott (23) has given details of a way in which computer systems are "configured" - linked into a customer installation. This is a practice which has been taken up by computer manufacturers around the world. This does not give a best solution but offers the first one out. It is a one-level design scheme.

More abstractly, Lenat and Sutherland (24) showed how to come up with a new semi-conductor system at the conceptual level. This approach began initially with attempts by Lenat to find new mathematical schemes and reached the point that although mathematical proposals were discoverable in principle no novel schemes were revealed. Although very expensive computer operations were used the old-style manual mathematicians had covered their ground thoroughly. This kind of system worked by generating heuristics. In comparison with the schemes evolved in the 1960s the advance is substantial but requires human intervention through the use of evaluation schemes with terms such as "interesting". What Lenat and Sutherland did was to transfer their efforts to the design of potential micro-circuit elements which were constructed around processes defined by word structures which defined functions.

On the mechanical engineering side there have been at least two pieces of publication in some depth. The first comes from Dixon and Simmons (25) and the second from Mistree and colleagues (26). The first of these is linked with practical design, aiming to provide component-level products which aid manufacture. They work on the principle that a prior design exists and that design is concerned with variation. Mistree's work is based on the notion that design is only selection and that suitable versions of mathematical programming can be used to bring out the best alternative.

Several reviews of engineering design with expert systems have been offered. Of these the offering by Rychener (27) is of interest since it comes from the large group at Carnegie Mellon University which ia tackling design themes.

Within the period 1984-1985 it was becoming apparent that certain simple design efforts were possible within chemical engineering by making use of available expert system shells. In a conference in 1985 we had some exploratory schemes which dealt with:

* materials of construction
* elementary plant selection

On the side dealing with materials of construction it was possible to cope with technical solutions and with certain kinds of economic solution. In a general way it was then possible to see that selection schemes might be applied within limited frameworks at all kinds of level in the design process. The method of linking them together remained as yet unclear although experts from the field of artificial intelligence were talking of "blackboard" schemes. The problem was not one of technical frames but of hierarchical evaluation.

It is clear from other contributions at this conference that people are going ahead with the development of narrow-front single-level expert system schemes. The big question is whether it will be possible to integrate them into overall schemes which have adequate evaluative credibility.

DO EXPERTS NEED SUPPORTING MODELS?

In the 1985 conference an emphasis was put upon the need for some kind of aid to knowledge engineers in trying to identify useful aspects of knowledge in a highly technical field. In order to supply experts in suitable ways it seems almost impossible to avoid the use of applicable models. Among these we may expect to find high-level meta-models about the nature of design activity. A first suggestion here would be to try to make use of the work undertaken in Germany, particularly in V.D.I Guideline 2221 (15), already cited, in which an attempt is made to express a model suitable for most types of engineering design but this may well fail on several grounds:

* it is not specific enough for some engineering domains
* it makes use of features which some designers find objectionable such as a linear flow-plan, without parallellism
* it makes no use of heuristics
* no account is taken of alternative business strategies and their accompanying design variations
* there is no treatment of product lifecycle needs

The German specialists are not unaware of some of these needs. Ehrlenspiel's work in V.D.I. Guideline 2235 (2) has provided excellent research back-up on the question of heuristics in design for manufacturing.

The second major speculation is that overall design using expert systems will take many years to come to fruition and that, in this time:

* better and more applicable design activity models will emerge
* people working on narrow-front design will become aware of a need to link up
* useful ways will come forward to deal with short-cut modes of evaluation which do not negate the requirements of integration in an overall design
* in the period of development design requirements will shift to need more sophisticated modes of evaluation

ASSEMBLING KNOWLEDGE ON VALUE STRUCTURES

A central feature of the present contribution is that, although knowledge on methods of synthesis will tend to go forward because of the technical thrust which characterises engineers, there will be a tendency to lag on the side of value structures. For this reason interest requires to be stirred up here with hints of possible short-term industrial benefit.

In the 1960s when Gregory and Bridgwater (18) were preparing their rapid cost-estimating approach a number of subsidiary research studies were set up to collect and analyse relevant data. The events referred to had all taken place in the past. Is it possible that value structures identified as important in the first part of this paper, particularly those which deal with competitive status, might comparably be analysed from historical data?

Bearing in mind the difference in scope from cost-estimating data it should be possible to attempt to re-trace the history of the process industries. Although some of the earliest entrepreneurs said remarkably little there are still good hints to be picked up. Roebuck, for example, relied on secrecy, as did Young. Muspratt said little about keeping his factories going but he went for low cost by economies of scale, and aimed to elude any plant additions to deal with environmental problems. Can we use such a method to put together layers of industrial history covering such features as:

* the rise and fall of the Leblanc soda industry
* the synthetic dye industry
* penicillin and the other antibiotics
* the extensive story of petrochemicals

In addition to written knowledge it is possible to disengage what would appear otherwise as tacit knowledge. The rise of process knowledge in an organised way seems to have been based much upon the efforts of people like Gossage, Deacon, Solvay, and Castner, although many of the successful entrepreneurs had little to say. Davis made sense of this in a roundabout way by constructing a book about how much of, say, the Leblanc industry worked, at the end of mature life. What we received from him was a first mode of attack upon the technical tasks required for the solution of any process system problem. Davis had little to offer on the matter of process economics which has developed largely since the 1940s. He and his colleagues were broadly aware of the competitive pressures at work but did not have a competent mode of discussion.

What has developed considerably over the last 25 years has been a system view of process problems with an emphasis upon the technical system content. The construction of heuristic synthesis modes around this technical base is now widely accepted. What has been deficient has been the comparable development of technoeconomic heuristics. Although what has been stated earlier has to deal with the study of suitable competitive schemes, with some justification through historical and case studies. It will be necessary to try to build up or invent suitable heuristics which aid the incorporation of the new knowledge. We need to be able to deal with the three areas already denoted.

RELATIONSHIP WITH GENERAL COMPUTER ADVANCES ETC

Although this paper has been concerned with the needs of design it is useful to bear in mind that computer specialists and others may make useful inputs. There is already interest in the use of parallel processing, particularly for dealing with shapes. With object orientated environments we may be able to deal with whole technical systems. Fuzzy system methods will help deal with uncertainty. The use of repertory grid measurements (personal construct theory) may help to deal with the analysis of experts or to construct potentially suitable outputs.

CONCLUSIONS

Our challenges centre upon competitive behaviour. There is a growing interest here shown by Gregory (28) and particularly by Pettigrew (29) who studied I.C.I.

REFERENCES

(1) Ehrlenspiel, K. "Kostengünstig Konstruieren" Springer, Berlin, 1985.
(2) V.D.I. Richtlinie 2235, Verein Deutscher Ingenieure, Düsseldorf, 1982.
(3) Reinertsen, D.G. "Whodunit? The search for the new-product killers" Electronic Business, 1983, July, 62-66.
do. "Blitzkrieg product development: cut development times in half" Electronic Business, 1985, January, 102-109.
(4) Rippeteau, J. "Fast track to big savings and quick earnings potential" Financial Times, 1986, October 2.
(5) Gregory, S.A. "Price, choice, and design" Financial Times, 1986, August 26.
(6) Gonzalez, V. "Company culture and the choice of technology" Chemistry and Industry, 1985, May 6, 301-303.
(7) Ohmae, K. "The Mind of the Strategist" Penguin, Harmondsworth, 1983.
do. "Triad Power" (No U.K. publication yet).
(8) Porter, M.E. "Competitive Strategy" Free Press, New York, 1980.
(9) Engel, J.F. and Blackwell, R.D. "Consumer Behavior" Holt-Saunders, Hinsdale, 4th edition, 1982.
(10) Gardiner, P. and Rothwell, R. "Tough customers: good designers" Design Studies, 1985, 6, (1) 7-17.
(11) Rose, L.M. "Engineering Investment Decisions" Elsevier, Amsterdam, 1976.
(12) Landau, R. and Colan, A.S. eds "The Chemical Plant", chapter by Spitz, Reinhold, New York, 1966.
(13) Langdon, R. and Gregory, S.A. eds "Evaluation", chapter by Gregory, Design Council, London, 1984.
(14) see Malpas, R. "Chemical technology: scaling greater heights" Chemistry and Industry, 1977, Feb 5.
(15) V.D.I. Richtlinie 2221, Verein Deutscher Ingenieure, Düsseldorf, 1985.
(16) Feigenbaum, E.A. and McCorduck, P. "The Fifth Generation" Pan Books, new and updated, 1984.
(17) Gregory, S.A. "Designer support environment: expert systems" Design Studies, 1987, in press.
(18) Gregory, S.A. and Bridgwater, A. "Some approaches to new process design" in "Design, Decision, and the Computer" I.Chem.E Symposium Series No. 35, 1972.
(19) King, C.J. "Understanding and conceiving chemical processes" A.I.Chem.E Monograph Series 8, 70, 1974.
(20) Rudd, D.F., Powers, G.J. and Siirola, J.J. "Process Synthesis" Prentice-Hall, Englewood Cliffs, 1973.
(21) Wells, G.L. "A study of some process heuristics" in "Current Design Thinking", I. Chem.E. Midlands Branch, 1979.
(22) Corey, E.J. and Wipke, W.T. "Computer-assisted design of complex organic syntheses" Science, 1969, 166, October 10, 178-192.
(23) McDermott, J. "Domain knowledge and the design process" Design Studies, 1982, 3, (1) 31-36.
(24) Lenat, D.B., Sutherland, W.R. and Gibbons, J. "Heuristic search for new microcircuit structures" A.I. Magazine, 1982, 3, 17-33.
(25) Dixon, J.R. and Simmons, M.K. "Expert systems for mechanical design: a program of research" A.S.M.E., 1985.
(26) Mistree, F., Karandikar, M. and Kamal, S. "Knowledge-based mathematical programming: a hybrid approach for decision-making in design" IFIP-WG, Design Theory for CAD, Tokyo, 1985.
(27) Rychener, M.D. "Expert systems for engineering design" Expert Systems, 1985, 2, (1), 30-44.
(28) Gregory, S.A. "Corporate policy and design jobs" DESIGN 82, EFCE Symposium Series No.22, 1982, 53-63.
(29) Pettigrew, A.M. "The Awakening Giant" Blackwell, Oxford, 1985.

Fig 1 DESIGN INFLUENCES: 3P input; 2M input

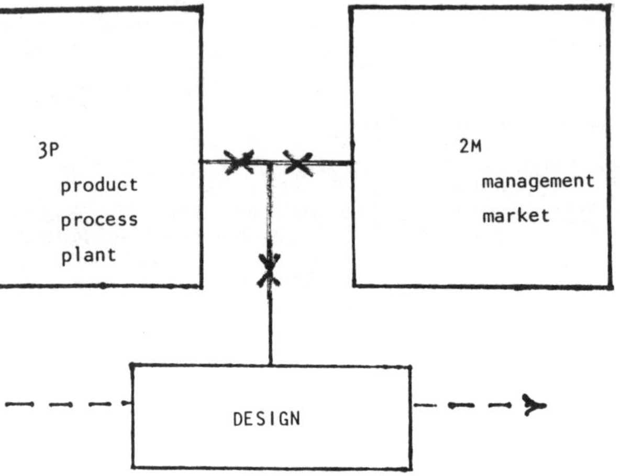

Fig 2 STRATEGIC DISCRIMINATION

(ex Porter)

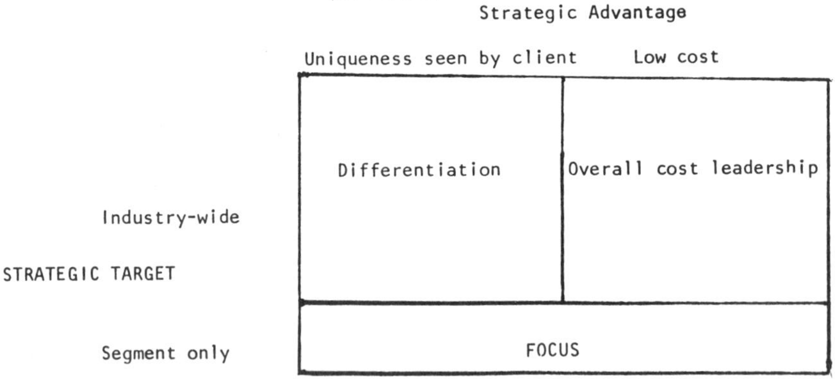

Fig 3 LIFECYCLE PATTERNS AND PROCEDURES (take as contingent)

1. Product introduction (may be at successive stages)
(may relate to process or plant)
- radical
- analogy
- concept scheme development
- design for development path ('stretch')

2. Product expansion
- increased level of operation

3a. Market extension
- changed approach to use by cost level
- personalisation of use, etc

3b. Cheaper product
- attribute analysis
- functional analysis
- full-range functional analysis
- mfg process improvement: yield/energy/labour
- mfg process change: input/capital cost
- buying in instead of making

3c. Differentiation
- added service content
- better product e g cheaper in use
  - reliability
  - technical performance
  - easier to handle
  - user quality
  - psychological attractiveness
- system provision
- product image design
- adjustment to lifestyle

(Bracketed alongside items 2 to 3c:)
- speed to market
- flexible plant
- more derivatives
- market location
- market linkage
- safer in use
- environmental checks
- packaging

4. Product maturity
- emphasis upon marginal adjustment
- labour reduction
- joint operation
- reduction of technological input

5. Product disposal
- dispose or overwhelm

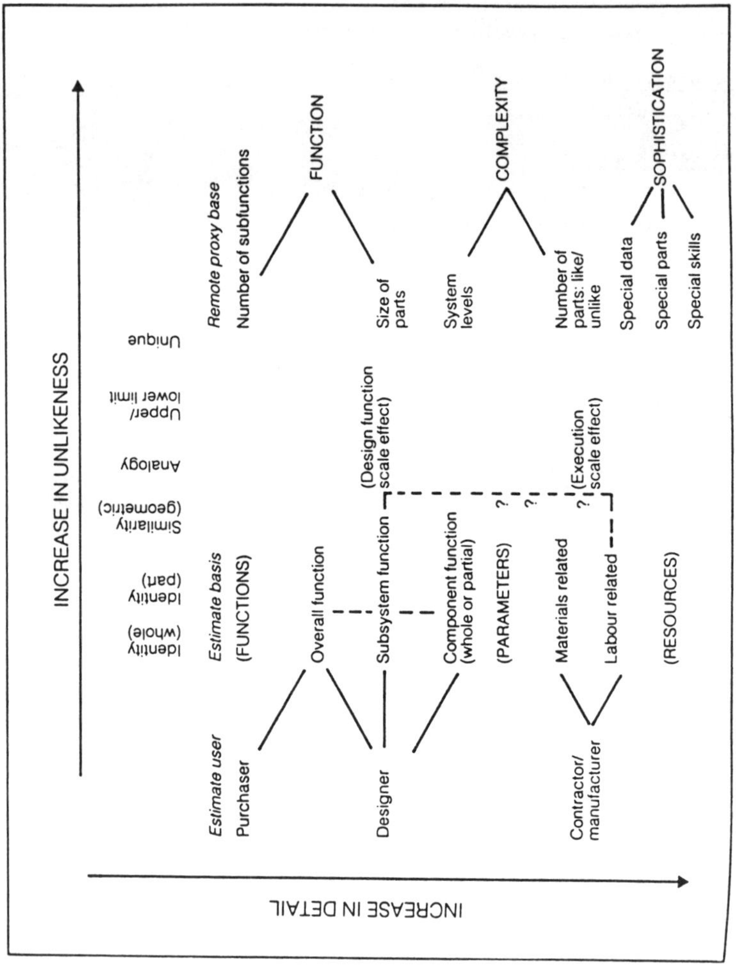

Figure 4
Suggested interrelationships: estimate user / estimate basis / extent of unlikeness.

Fig 5 TECHNOLOGICAL TASKS VS OTHER FACTORS

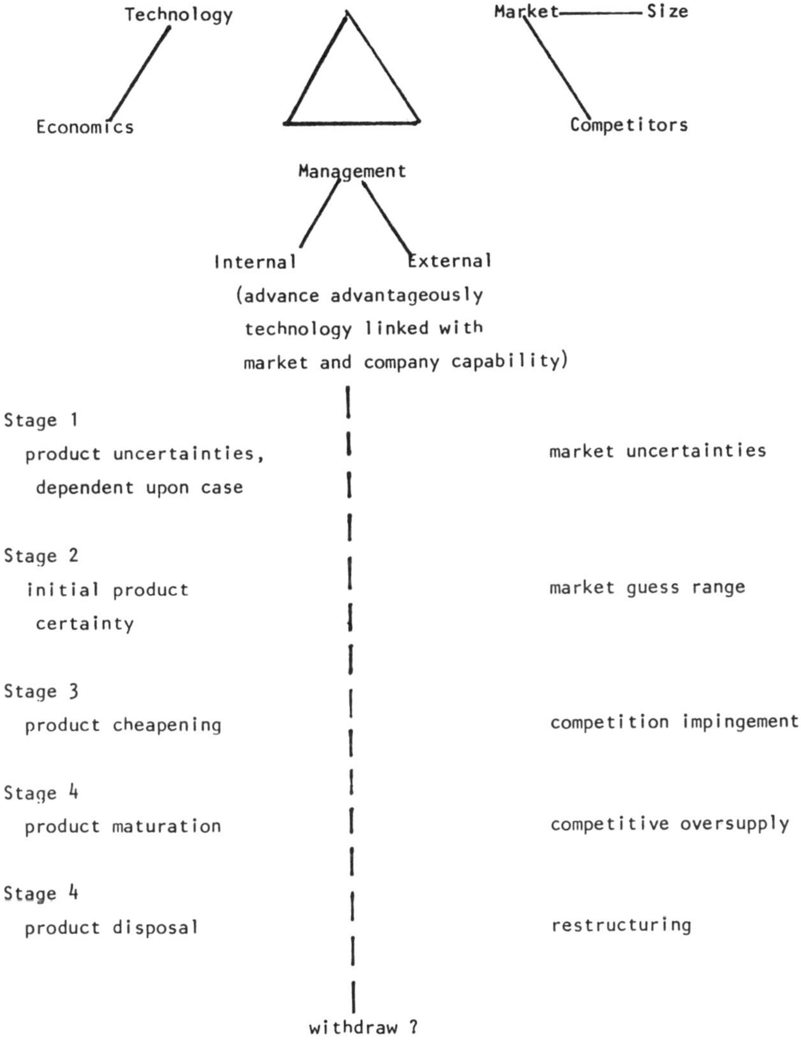

Fig 6 DESIGN STRATEGY ELEMENTS CHOICE related to MANAGEMENT M A U FUNCTION

(perhaps 20+ elements and sub-elements)

	M A U F or proxies
a. VDI Guideline 2221 i. new product - successive stages	
b. VDI Guideline 2221 ii. new process - successive stages	
c. VDI Guideline 2221 iii. new plant - successive stages	
d. Scale up	
e. VDI Guideline 2235 i. design for product cost/manufacture	
f. Lowest capital cost	
g. Lowest operating cost	
h. Flexible design, e g for starting, etc	
i. Speedy design, for various stages as a competitive or other feature: speed vs cost/overall time	
j. Value creation: safety in use ease in use superior quality reliability in use	
k. Safety in operation	
l. Environmental design	
m. Design integration: chosen aspects	

FORMULATION OF THE LARGE-SCALE NLP PROBLEM FOR OPTIMISATION OF REACTOR-SEPARATOR PROCESSES

P. FLOQUET*, L. PIBOULEAU* AND S. DOMENECH*

The optimal synthesis of reactor-separator processes can be expressed as a large-scale mixed-integer programming problem (MIP). This MIP problem is transformed into a large-scale NLP problem, whose formulation is rather difficult, by means of structural parameters. A lagrangian-based procedure is used to solve the NLP problem. The synthesis of benzene chlorination processes involving one or two reactors, illustrates this continuous approach.

INTRODUCTION

The optimum Computer-Aided Design (CAD) of reactor-separator processes can generally be stated as a large-scale mixed-integer programming (MIP) problem (NISHIDA et al., 1981), where the integer variables represent the process topology and the continuous variables refer to the design parameters of unit operations.

This optimisation problem can be defined as : given a set of goal products (composition, flow rate, pressure and temperature), systematically synthesize a process leading to this production at a minimum total annual cost (sum of investment and operating costs), from known reactants and reaction paths. In order to reduce the combinatorial aspect of the problem, we consider only well stirred tank reactors and simple distillation columns (a simple separator splits a single feed into two product streams) not thermally coupled.

By using a structural parameter approach, the initial MIP problem is transformed into a large-scale NLP problem. In this continuous approach, the distillation columns do not perform complete separation, and all the possible structures (recycles and direct connections) are imbedded into a superstructure defined by means of structural parameters. The structure and the operating conditions are simultaneously optimised by using a lagrangian-based algorithm, where the linearized subproblem is solved with a large-scale projected reduced gradient procedure. Although, in this approach the process as a whole is optimised, an important formulation effort is required to obtain a continuous problem from the initial MIP by using an approximation strategy of all the discontinuities (corrective factors in cost evaluation and specification or purity constraints).

The methodology is illustrated by the synthesis of benzene chlorination processes, involving one or two reactors.

* E.N.S.I.G.C. - UA CNRS 192
Chemin de la Loge
31078 TOULOUSE Cedex - FRANCE

PROBLEM FORMULATION

The general flowsheet of plant under consideration is shown in Figure 1. Each continuous reaction unit of the process is an ideal isothermal well stirred tank reactor (CSTR) and after the reaction step the output components are separated in a separation sequence and then the products are either removed or recycled to the reactors. The simple separators of the sequence are plate distillation columns, because distillation is the unit operation that is most commonly used for separating mixtures. The computation of the total annual cost of the process first requires the design of each unit operation. The reactors are designed by means of the model of ideal CSTR given by LEVENSPIEL (1962), and the well known method of Fenske-Underwood-Gilliland (PIBOULEAU, 1985) has been retained for computing the main technical characteristics of distillation columns (minimum reflux-R_{min}, number of theoretical plates and feed plate location). By using a thermodynamic model appropriate to the nature of the process components, the condenser, reboiler and reactor duties can be evaluated. The process may also involve equipment for changing temperatures and pressures (flashes and compressors). Insofar as the cost of flashes is negligible compared to reactor and separator costs it is not considered in the costing procedure. The compressors are designed according to the total adiabatic power as described by CHAUVEL et al. (1976).

The technico-economic criterion associated with the process is the total annual cost C, representing the sum of investment costs C_{in} and operating costs C_{op}. The plant investment calculation is a rather difficult task because it depends on technical design of unit operations, but also on several external factors, for example, the process location and various taxes.

The cost C_{in} is computed by using the FOB (Free on Board) costs of all the equipment (reactors, agitators, distillation columns, heat exchangers, compressors, platforms, ladders and internals). The operating cost C_{op} is the sum of the energy and utilities cost, the labor-related cost and the raw material cost. All the correlations employed for calculating the various costs are given by FLOQUET (1986). Thus the synthesis problem can formally be expressed as follows :

$$\min_{I, x_i} C(I, x_i) = \sum_{i \in I} \left(C_{in\,i}(x_i) + C_{op\,i}(x_i) \right) \qquad (1)$$

$$x_i \in D$$

$$I \in S$$

where i represents a feasible unit operation (reactor or separator) ; x_i, the design variables (temperature, pressure, conversion rate...) of unit i ; D, the domain where the design variables are defined (temperature and pressure positive, conversion rate bounded, ...) ; I, a possible unit operation configuration chosen in the set S of all the configurations that can produce the goal products.

The above problem is a large-scale mixed-integer programming problem (MIP), where the discrete decisions are made over all the subsets I, and the continuous decisions are made over the range of values of the design variables.

CONTINUOUS FORMULATION

Structural parameter apprcach

For fixed numbers r of reactors and d of distillation columns, the set S of all the feasible flowsheets can be represented by a superstructure (ICHIKAWA et al., 1969) ; an example is given in Figure 2. Within this superstructure, each element I (i.e. feasible flowsheet) corresponds to given values of the components of the structural parameter vector s which are continuous variables used to represent all the possible connections between unit operations in the superstructure. For reactor-separators systems, the structural parameters are the recovery rates of intermediate components into the distillate of a column and the split fraction of components between unit operations. So the initial MIP problem is transformed into a large-scale NLP problem, formulated as follows :

$$\text{Min}\, C(x_j, x_k, s) = \sum_{j=1}^{r} \left(C_{in\,j}(x_{j,s}) + C_{op\,j}(x_{j,s})\right) + \sum_{k=1}^{d} \left(C_{in\,k}(x_{k,s}) + C_{op\,k}(x_{k,s})\right) \qquad (2)$$

$$x_j, x_k \in D$$

where s is the vector of structural parameters whose components are bounded by 0 and 1 (ICHIKAWA et al., 1969).

General assumptions

The superstructure is defined by using the following assumptions.

(a_1) The flowsheets under consideration only involve reactors (CSTR), distillation columns and equipment for changing pressure and temperature (flashes, compressors and heat exchangers). The distillation columns are assumed to be simple (one input and two outputs) and the reactors (involving one input and one output) are connected in series.

(a_2) In order to obtain a finite number of distillation columns, it is assumed that only the lightest and heaviest components of a column feed are completely separated. So it is shown (PIBOULEAU, 1985) that the number d of distillation columns is given by :

$$d = 2^{(N-1)} - 1 \qquad (3)$$

where N is the total number of components to be separated.

(a_3) The order of component volatilities is invariant ; so the components can be stored in a ranked list.

Optimisation variables

The optimisation variables are as follows.

a/ Reaction level

the operating pressure and temperature of each reactor

the partial molar flows of the output stream of each reactor

the partial molar flows of the input feed stream of the first reactor

the structural parameter defining the split fraction of the output stream of each reactor.

So, for a given number r of reactors in series and a number N of components, the number n_r of optimisation variables for the reaction level is (FLOQUET, 1986) :

$$n_r = r(N+3) + N - 1 \qquad (4)$$

b/ Separation level

The optimisation variables associated with a distillation column are the following :

the operating pressure and reflux ratio

the structural parameters defining the recovery rates of intermediate components into the distillate

the structural parameters representing the split fractions of the top and bottom components which are recycled to the reactor sequence.

The number n_S of optimisation variables for the separation level is equal to (PIBOULEAU, 1985) :

$$n_S = \sum_{i=1}^{(N-1)} 2^{(i-1)} (2r + N + 1 - i) \qquad (5)$$

Constraints definition

The range of values of all the variables above mentioned is bounded : all the structural parameters lie between 0 and 1, and all the design variables are physically bounded.

The set of constraints can be partitioned into three subsets (FLOQUET, 1986) :

- inequality (linear) constraints on structural parameters ; the number m_{li} of these constraints is given by :

$$m_{li} = 2^N - 2 + \sum_{i=1}^{(N-3)} 2^{(i-1)} (N - i - 2) \quad \text{if } r > 1 \qquad (6a)$$

$$m_{li} = \sum_{i=1}^{(N-3)} 2^{(i-1)} (N - i - 2) \quad \text{if } r \leq 1 \qquad (6b)$$

- nonlinear equality constraints giving, for each reactor, the output stream composition as a function of the input stream and the design variables ; the number of these constraints is :

$$m_{nle} = r \, . \, N \qquad (7)$$

- for each component a production constraint (nonlinear equality) may be imposed ; so the number m_{nle} becomes :

$$m_{nle} = r \, . \, N + N \qquad (8)$$

Criterion discontinuities

The implicit nonlinear criterion (Eqn 2) contains some discontinuities essentially due to :

corrective factors in cost evaluation (type of equipment, inflation, ...)

specifications : for example when a component in the feed stream of a column reaches a given purity specification, this separator and all its successors are removed from the superstructure and some costs are set to zero. Furthermore when compressors are introduced into the process, the criterion discontinuously increases.

These discontinuities can be expressed as :

$$Y = f_1 (x) \qquad \text{if } x < x_0 \qquad (9a)$$

$$Y = f_2 (x) \qquad \text{if } x \geq x_0 \qquad (9b)$$

and are approximated by the following continuous function :

$$Y = f_1(x) . (1 - \text{Jump}(x)) + f_2(x) . \text{Jump}(x) \quad (10a)$$

where

$$\text{Jump}(x) = 1/2 + 1/\Pi \text{ Arctan}(\theta (x - x_0)) \quad (10b)$$

θ being a large positive scalar, determined by successive trials to get an approximation error less than 3%. In order to avoid numerical difficulties when $x = x_0$, the smallest value of θ giving this arbitrary approximation error is chosen. In the following examples, the value 100 is assigned to θ.

Problem formulation

In the following, the index r_i corresponds to the reaction step (r_i = 1 to r) and the index s_i, to the separation step (s_i = 1 to N-1). The number 1 is assigned to the first column of the plant, and the two successors of the separator p are numbered 2p and 2p+1.

The feed of the separator 2p is the top product of the pth separator and its bottom product feeds separator 2p+1. The separation of a feed containing N components into two product streams with N-1 components forms the first separation step s_i = 1. In general, the separation step number s_i ($1 \le s_i \le N-1$) contains the set of the separations of feeds with $N+1-s_i$ components into $N-s_i$ component products (see Figure 2). So the NLP problem can be stated as follows (FLOQUET, 1986).

$$\text{Min } C(P_r^{ri}, T_r^{ri}, X^{ri}, \gamma_{ri,ri+1}, P_p, t_p, \beta_p, \gamma^0{}_{p,ri}, \gamma^1{}_{p,ri})$$

$$\text{with } C = \sum_{r_i=1}^{r} (C'_{ri} + g_{ri}^{comp} C_{comp,ri}) + (C'_1 + g_1^{comp} C_{comp,1})$$

$$+ \sum_{s_i=2}^{N-1} \left(\sum_{p=2^{(si-1)}}^{2^{si}-1} \prod_{i=1}^{N-s_i+1} g_{i,p} (C'_p + g_p^{comp} C_{comp,p}) \right)$$

under the constraints:

$$\left.\begin{array}{l} \sum_{r_i=1}^{r} \gamma^0{}_{p,ri} \le 1 \\ \qquad\qquad p = 1 \text{ to } 2^{(N-1)} - 1 \\ \sum_{r_i=1}^{r} \gamma^1{}_{p,ri} \le 1 \end{array}\right) \text{ (linear constraints)}$$

$\beta_p(i) \geq \beta_p(i-1)$ $\quad$ $i=1$ to $N-s_i-2, p=2^{(si-1)}$ to $2^{si}-1$ and $s_i=1$ to $N-3$ $\quad$) (linear) constraints)

$X(i)^{ri} = g_i\,(Q_{in,ri}, T_r^{ri}, P_r^{ri})$ $\quad$ $i=1$ to N and $r_i=1$ to r $\quad$) (nonlinear constraints)

and the following bounds :

$X(i)^{ri} \geq 0$ $\quad$ $i=1$ to N et $r_i=1$ to r

$0 \leq \gamma_{ri,ri+1} \leq 1$ $\quad$ $r_i=1$ to $r-1$

$T_{rmin} \leq T_r^{ri} \leq T_{rmax}$

$\quad$ $r_i=1$ to r

$P_{rmin} \leq P_r^{ri} \leq P_{rmax}$

$0 \leq \beta_p(j) \leq 1$ $\quad$ $j=1$ to $N-s_i-1$, $p=2^{(si-1)}$ to $2^{si}-1$ and $s_i=1$ to $N-2$

$0 \leq \gamma^0{}_{p,ri} \leq 1$ $\quad$ $r_i=1$ to r, $p=2^{(si-1)}$ to $2^{si}-1$ and $s_i=1$ to $N-1$

$0 \leq \gamma^1{}_{p,ri} \leq 1$

$P_{min} \leq P_p \leq P_{max}$

$\quad$ $p=2^{(si-1)}$ to $2^{si}-1$ and $s_i=1$ to $N-1$

$t_{min} \leq t_p \leq t_{max}$

The previous terms are defined as follows.
- P_r^{ri} (respectively P_p) : operating pressure of reactor r_i (respectively of column p)
- T_r^{ri} : temperature of isothermal reactor r_i
- X^{ri} : partial molar flow of the output stream of reactor r_i
- $\gamma_{ri,ri+1}$: split fraction of the output stream of reactor r_i going to reactor r_{i+1}
- t_p : reflux ratio of column p
- β_p : vector of recovery rates of intermediate components into the distillate of column p
- $\gamma^0{}_{p,ri}$ (respectively $\gamma^1{}_{p,ri}$) : split fraction of the top (respectively bottom) component which is recycled from column p to reactor r_i
- C' : continuous approximation of the sum of investment and operating costs for a reactor or a column
- g_k^{comp} : continuous approximation of the 0-1 function associated to the compressor (of cost $C_{comp,k}$) of the kth unit operation
- $g_{i,p}$: continuous approximation of a 0-1 function that activates or deactivates the cost term C'_p when some given specifications are reached or not by the feed components of the column p
- $Q_{in,ri}$: feed stream of reactor r_i (expressed in terms of decision variables)
- g_i : nonlinear function giving the output stream composition of reactor r_i, as a function of

the input stream $Q_{in,ri}$ and the design variables T_r^{ri}, P_r^{ri}

- T_{rmin}, P_{rmin}, P_{min}, t_{min} (t_{rmax}, P_{rmax}, P_{max}, t_{max}) : lower bounds (upper bounds) on T_r^{ri}, P_r^{ri}, P_p and t_p.

PROBLEM SOLUTION

The problem is solved by using the lagrangian-based procedure described by PIBOULEAU et al. (1985). The solution consists of a sequence of major iterations, each one involving a linearization of the nonlinear constraints and the minimization of an augmented lagrangian (MURTAGH and SAUNDERS, 1980). For each new major iteration, by choosing the initial lagrange multipliers to be the lagrange multipliers corresponding to the linearized constraints at the solution of the previous major iteration, it can be proved (GILL and MURRAY, 1974) the sequence converges superlinearly to an optimal solution of the initial problem (x^*, λ^*), provided that the initial estimates (x_0, λ_0) are sufficiently close to that solution. A positive value of the penalty term ρ is also necessary when the problem is known to be highly nonlinear and/or nonconvex. The importance of the choice of the initial values of the penalty term ρ_0 and of the lagrange multipliers λ_0 is shown in the following section.

EXAMPLE

The benzene chlorination process under consideration (FLOQUET, 1986) is shown in Figure 3. After the reaction, the product (monochlorobenzene-MCB) is separated from the unreacted benzene-B recycled to the reactors, and from the by-products (di and trichlorobenzene-DCB and TCB). Hydrochloric acid is eliminated at the reaction output by a stripping operation, whose cost is not taken into account in the criterion evaluation. The problem consists of determining the optimal process (structure and operating conditions) for a total production of monochlorobenzene of 30.10^6 kg/year.

Process involving one reactor

The initial flowsheet can be determined by using a breadth-first algorithm (FLOQUET, 1986) (in that case, all the operating conditions are fixed at nominal given values), of cost 32.90 10^6 \$/year and it is given in Figure 4. The components are completely separated, so the flowsheet contains 3 columns.

The problem involves 40 variables and 6 constraints (1 linear and 5 nonlinear ones). As it is shown in Table 1, the problem can be solved by using four strategies : minimisation of the criterion C alone, of the lagrangian function $C - \lambda^T(q-\bar{q})$, of the penalty function $C+\rho/2(q-\bar{q})^T(q-\bar{q})$ and of the augmented lagrangian $C-\lambda^T(q-\bar{q})+\rho/2(q-\bar{q})^T(q-\bar{q})$, where q is the vector of nonlinear constrains and $\bar{q}$ is the linear approximation of q in each major iteration. When the criterion alone, or the lagrangian function is minimised, the problem does not converge, so a penalty term is needed. Too small or too great a value of the penalty coefficient ρ_0 causes divergence or very slow convergence of the search. Furthermore, the choice of the initial values of the lagrange multipliers is preponderant on the convergence rate. A procedure of choice may consist in making one or several major iterations with the criterion alone, and then in initializing the lagrange multipliers with the values obtained at the end of these major iterations.

The optimal value of the cost C, reached when minimising the penalty function, is 29.7 10^6 \$/year ; the corresponding optimal plant is shown in Figure 5-a. Since the values of recovery rates of components into the distillate of columns are either 0 or 1, we find again complete separations. However the partial molar flow of TCB being very small (< 10^{-3}) in the output stream of the first column, it not separated from the DCB.

Process involving two reactors

The problem is initialized by using the same procedure as in the previous case, and it involves 75 variables and 24 mixed linear and nonlinear constraints. The optimal value of the cost C, obtained by minimising the augmented lagrangian function, is 29.8 10^6 $/year ; the optimal flowsheet is reported in Figure 5-b.

CONCLUSIONS

Like many problems of CAD of chemical engineering processes, the synthesis of reactor-separator systems can be expressed as a large-scale MIP problem. By using a structural parameter approach, the initial MIP problem can be transformed into a large-scale NLP problem, whose formulation is rather difficult. The NLP problem is then solved with a lagrangian-based procedure, where the importance of the choice of the initial values of the penalty term and of the lagrange multipliers is shown with the CAD of a benzene chlorination process. Processes involving one or two reactors have been considered. In the two cases, the optimal processes have the same structure, and the values of the optimal costs are very close. Nevertheless, for simplicity reasons, it seems that the plant involving only one reactor must be chosen.

REFERENCES

CHAUVEL, A., P. LEPRINCE, Y. BARTHEL, C. RAIMBAULT and J.P. FAURE
Manuel d'Evaluation Economique des Procédés, Technip, Paris (1976)

FLOQUET, P.
Procedures discrète et continue d'optimisation et de CAO en génie chimique. Etude de cas. Thèse de Doctorat de l'Institut National Polytechnique de Toulouse, France (1986)

GILL, P.E. and W. MURRAY
Numerical methods for constrained optimization, Academic Press, NY (1974)

ICHIKAWA, A., N. NISHIDA and T. UMEDA
An approach to the optimal synthesis problems, 34th Annual Meeting Soc. Chem. Engineers, Japan (1969)

LEVENSPIEL, O.
Chemical reaction engineering : an introduction to the design of chemical reactors, John Wiley, NY (1962)

MURTAGH, B.A. and M.A. SAUNDERS
The implementation of a Lagrangian-based algorithm for sparse nonlinear constraints, Report SOL 80-1, Stanford Univ., CA (1980)

NISHIDA, N., G. STEPHANOPOULOS and A.W. WESTERBERG
A review of process synthesis, AIChE J., 27, 321 (1981)

PIBOULEAU, L.
Mise en oeuvre d'une procédure arborescente et d'une méthode de gradient réduit pour la CAO de cascades de colonnes de rectification, Thèse de Doctorat d'Etat, Institut National Polytechnique de Toulouse, France (1985)

PIBOULEAU, L., P. FLOQUET et S. DOMENECH
Optimisation de procédés chimiques par une méthode de gradient réduit
Partie I : Présentation de l'algorithme, R.A.I.R.O., Recherche opérationnelle, 19, 247 (1985)

PIBOULEAU, L., P. FLOQUET et S. DOMENECH
Optimisation de procédés chimiques par une méthode de gradient réduit
Partie II : Exemples d'illustration. Comparaison avec d'autres méthodes
R.A.I.R.O., Recherche opérationnelle, 19, 321 (1985)

Criterion	ρ_0	λ_0	Number of major iterations	Number of minor iterations	Optimal value of the criterion (10^6 \$/year)	Optimal value of the cost C (10^6 \$/year)	Maximal value of the constraints	Number of criterion calls
C			NON CONVERGENCE					
$C - \lambda^T(q-\bar{q})$		$0 \in R^6$	NON CONVERGENCE					
$C+\frac{\rho}{2}(q-\bar{q})^T(q-\bar{q})$	1		NON CONVERGENCE					
$C+\frac{\rho}{2}(q-\bar{q})^T(q-\bar{q})$	10		5	45	32.4	32.4	10^{-3}	4775
$C+\frac{\rho}{2}(q-\bar{q})^T(q-\bar{q})$	100		17	132	29.7	29.7	10^{-4}	14009
$C+\frac{\rho}{2}(q-\bar{q})^T(q-\bar{q})$	1000		NON CONVERGENCE					
$C-\lambda^T(q-\bar{q})+\frac{\rho}{2}(q-\bar{q})^T(q-\bar{q})$	10	$0 \in R^6$	23	225	31.7	30.7	10^{-4}	23850
$C-\lambda^T(q-\bar{q})+\frac{\rho}{2}(q-\bar{q})^T(q-\bar{q})$	100	$0 \in R^6$	12	115	31.1	31.7	10^{-4}	12190
$C-\lambda^T(q-\bar{q})+\frac{\rho}{2}(q-\bar{q})^T(q-\bar{q})$	1000	$0 \in R^6$	NON CONVERGENCE					
$C-\lambda^T(q-\bar{q})+\frac{\rho}{2}(q-\bar{q})^T(q-\bar{q})$	10	λ_1	9	89	29.3	30.7	10^{-4}	9434
$C-\lambda^T(q-\bar{q})+\frac{\rho}{2}(q-\bar{q})^T(q-\bar{q})$	100	λ_1	10	92	30.7	31.2	10^{-4}	9762

Table 1 - Optimisation of the process involving one reactor.
λ_1 is the vector of lagrange multipliers at the solution of the first major iteration corresponding to the first problem (criterion C)

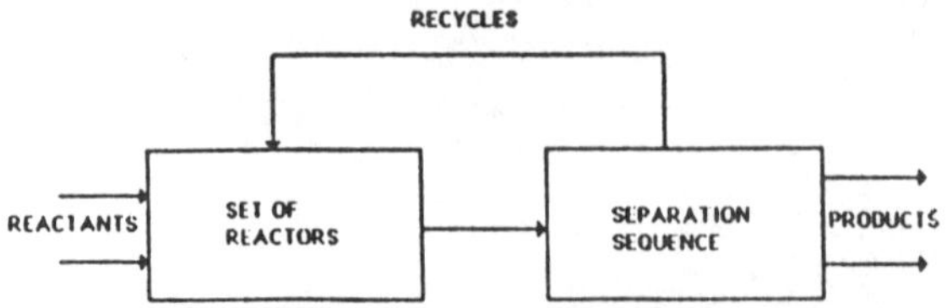

Figure 1 General flowsheet for reactor-separator processes

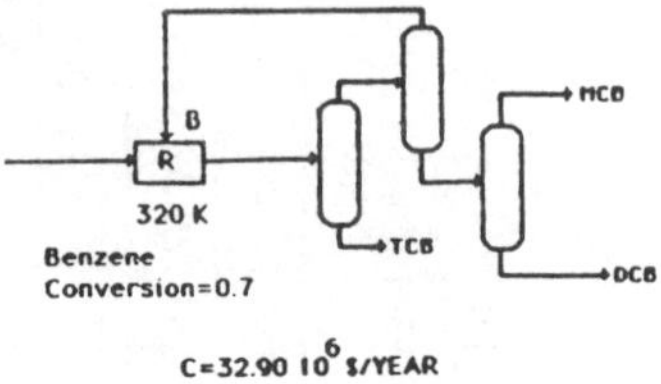

Figure 4 Initial flowsheet for the process involving one reactor.

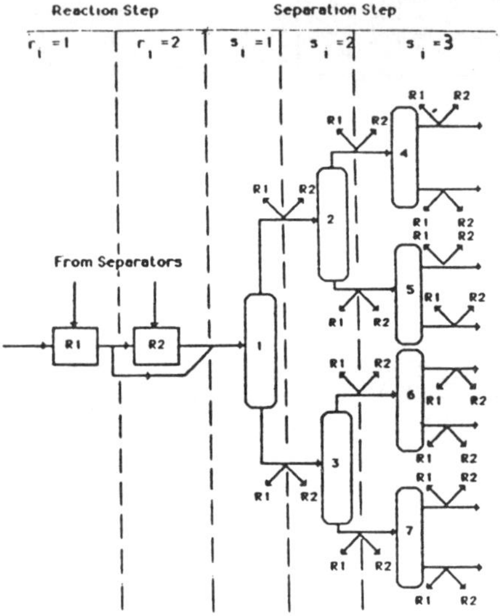

Figure 2 Superstructure for a process involving 2 reactors and 4 components to be separated

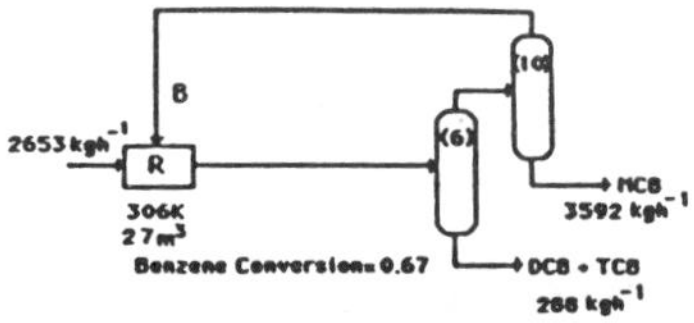

Figure 5-a: Optimal process for one reactor
Optimal value of cost = 29.7 10^6 \$/year.

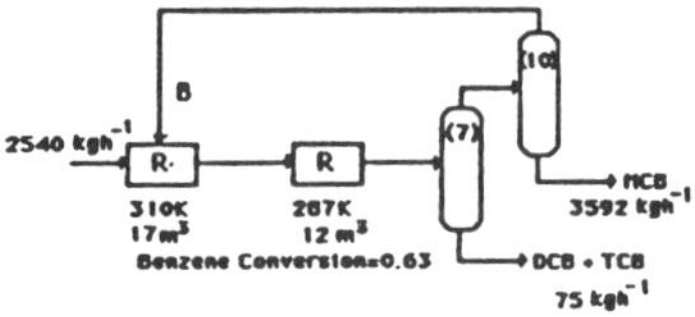

Figure 5-b. Optimal process for two reactors
Optimal value of cost = 29.8 10^6\$/year

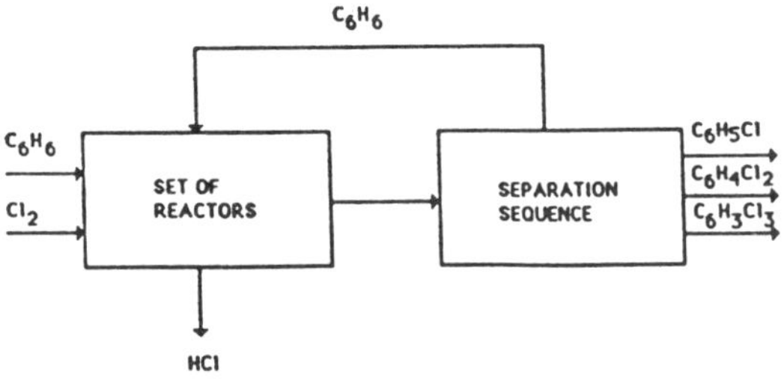

Figure 3: General flowsheet for the Benzene Chlorination process.

PROCESS INTENSIFICATION : FLUID BED HEAT TRANSFER

A Murphy, R Sibbet, C Ramshaw

An exploratory study assessed the feasibility of using vapour compression with a fluid bed drier. The attraction of the scheme was critically dependent upon the intensity of heat transfer in the bed, since large temperature differentials implied unacceptable pressure ratios across the compressor. A heat transfer matrix was required which interfered with the bed circulation as little as possible, while giving the desired heat transfer intensity. The arrangement adopted consisted of thin vertical slats approximately 1 cm high spaced at roughly 1 cm intervals. Some experimental results are reported which confirm that this arrangement can provide the necessary heat transfer while allowing adequate bed mixing.

INTRODUCTION

A standard approach to the problem of drying granular materials is to use a fluid bed. Wet feed is continuously supplied and the heat necessary to dry the material is usually provided by a tubular steam condenser immersed in the fluidised zone. An obvious improvement to this scheme would be to compress the water vapour generated in the bed and then to condense it within the heat exchangers immersed in the fluid bed (Fig. 1). In effect this amounts to running the process with an open cycle vapour compression heat pump using water as the working fluid.

In order to achieve an acceptable bed heat transfer intensity with conventional technology, very high temperature differentials (>50°C) are needed. These in turn require an expensive compressor and high compression costs. On the other hand if temperature differences of only 20°C were needed, then there could be substantial capital and running cost savings. The speculative work described below explored heat transfer matrix designs which were capable of transferring ≈1 MW/m^3 for ΔT = 20°C, which held the prospect of rendering the vapour compression approach fully economic, without inhibiting particle circulation to an unacceptable extent.

BACKGROUND

Heat transfer from fluid beds to immersed surfaces has been extensively researched, though understandably most of the work has involved tubes and tube arrays.

Botterill et al (1) have measured the local heat transfer coefficient around the circumference of a 3.5 cm tube immersed in a fluidised used bed of alumina or sand particles up to 3 mm diameter. Fig. 2 gives a polar plot of the heat transfer coefficient, which is least at the tube top and bottom, and reaches its maximum just above the centre line. These findings agree with the studies of Gelperin (2). Glass and Harrison (3) explored the flow patterns of the solid particles near a horizontal tube in a two dimensional fluidised bed. Three distinct flow regimes were evident. On the upstream side there existed a thin air pocket of variable thickness while on the downstream side a cap of defluidised particles was observed. At equatorial lateral zones of the cylinder, chains of bubbles were formed from the air in the upstream air pocket. These findings largely explain the observations, since the defluidised cap and the absence of particles in the immediate vicinity of the tube base would be expected to depress the local heat transfer coefficient.

Most workers who have measured overall heat transfer coefficients for immersed horizontal tubes or vertical surfaces, agree that short vertical dimensions improve the heat transfer performance. However Gelperin (4) states that above 10 mm diameter the dependence on tube diameter disappears. On the other hand Xavier and Davidson (5) give a correlation for the surface/particle heat transfer coefficient which show it to be inversely proportional to the square root of the vertical surface length. Since tubes are relatively ineffective at their top and bottom surfaces, it seems reasonable to conclude that an array of short vertical strips would be the best configuration for a fluid bed heat exchanger - provided the particles can circulate freely.

Although many workers have studied particle mixing in fluid beds, none appears to have explored the effect of immersed arrays of tubes. Intuitively it would be expected that the total plan area of the array should be minimised in order to reduce the obstruction effect and facilitate axial mixing. In addition, lateral particle transfer should be improved when the heat transfer elements have short vertical dimensions and are in well-spaced layers.

These considerations led to the heat transfer matrix consisting of an array of narrow vertical metal strips with spacing and width in the region of 1-2 cms.

EXPERIMENTAL

Experiments were performed in two phases. The first studies were aimed at establishing the influence of various arrays on the particle circulation time in the bed. Subsequently the heat transfer coefficient was measured with a view to achieving the target of 50 $kW/m^3 {}^\circ C$ with the high specific surface area which the envisaged matrix design would allow.

MIXING STUDIES

Bed mixing was characterised in terms of the circulation time of particles within a 17.8 cm square bed consisting of 2 mm urea prills which had a minimum fluidisation velocity of about 0.8 m/s. One of these particles was coated with a phosphorescent pigment, and a photomultiplier detected the particle when it reached the bed surface which was irradiated with light from a U/V source. An automatic data logging system was set up to count the number of appearances the particle made in a given run time.

A plot of the circulation time results is presented in Fig. 3A, 3B as a function of the gas fluidising velocity and the number of layers of strips in the bed. As expected, the circulation time becomes large for beds containing the matrix layers, as the fluidising velocity approaches the minimum value for the particles. On the other hand the circulation in an empty bed was much less sensitive to gas flow and, as expected, was better than that in the presence of immersed matrices at all flows. The influence of the number of layers of matrix on the circulation was not clear - the expected consistent increase of mixing times with extra layers did not materialise. However there was a clear rapid reduction in circulation time as the fluidising velocity increased beyond around 1.1 m/s. Hence it would be reasonable to expect the proposed heat exchanger design to give acceptable mixing rates at about 1.5 times the minimum fluidiising velocity.

HEAT TRANSFER STUDIES

These studies explored the heat transfer behaviour of a matrix of 1 cm wide strips in a 30 cm square fluid bed of either sand or Pruteen particles. Two principal matrix configurations were used as shown in Fig. 4. In the first (Matrix A) the strips were spaced at 2 cm intervals with layers being placed immediately above each other and offset by one half a strip spacing. The second arrangement (Matrix B) was similar, with a spacing of only 1 cm. Some very limited experiments were also performed with a single 2.5 cm diameter circular tube and with Matrix C which consisted of 2.5 cm strips at 2 cm spacing.

Up to 5 strips were coated with heater foils which allowed a known amount of electric power to be dissipated over the available strip area. The strip temperature was measured with a platinum resistance thermometer embedded in a groove immediately under the foil. Since the bed film heat transfer was limiting, this construction gave a good approximation to the surface temperature and allowed the heat transfer coefficient to be calculated.

The heat transfer coefficients presented in Tables 1, 2 are averages based on the number of strips equipped with heater foils. As expected for the modest gas velocities involved, the heat transfer coefficients with no particles in the bed were about 35 $w/m^2 °K$. Several repeat runs were performed to establish the reliability of the results. Bearing in mind that the matrix B had a specific surface area of 200 m^{-1}, the observed heat transfer coefficient of around 600 $w/m^2 °K$ with a sand bed, corresponds to a heat flux of over 2 MW/m^3 with ΔT = 20°K. Even matrix A, with half the specific surface area of B was able to achieve the target heat transfer intensity with sand in the bed. The limited results obtained with a bed of Pruteen particles show a very significant fall in performance, with coefficients for matrix B in the region of 150-200 $w/m^2 °K$. Thus with these particles the performance target would not quite be reached.

However the performance of the 2.5 cm tube was very poor. With a square tube pitch of 5 cm the bed specific area would be 31 m^{-1}. A 20°K temperature difference in this case gives a heat flux of 0.2 MW/m^3 with sand and 0.1 MW/m^3 with Pruteen. Further support for the strategy of exploiting short vertical strips is provided by the limited results with matrix C shown in Table 2. A very significant reduction in heat transfer is evident with both sand and Pruteen.

DISCUSSION

The work described above was an exploratory exercise to assess the feasibility of a proposed idea. Although the bed mixing characteristics with a 2.5 cm diameter tube matrix were not investigated, it is anticipated that they would be very poor compared with the acceptable mixing demonstrated with the strip matrices. Ideally, the mixing runs should also have been performed with the particles actually used in the heat transfer tests - sand and Pruteen. However there were sound experimental reasons for restricting the scintillation technique to relatively large particles. In order to bring the concept to fruition, the production/value engineering of the strip matrices must be carefully considered - perhaps borrowing some of the construction techniques used in refrigeration heat transfer.

Nevertheless, the main objective of the work has been achieved, since the target heat transfer intensity has been obtained with acceptable bed mixing.

Recommendations

1. More extensive mixing studies should be performed, particularly those including closely spaced arrays of tubes and perhaps horizontal strips.

2. The heat transfer behaviour of even finer arrays could usefully be explored in order to establish a practical upper limit to the heat transfer intensity which might be acheived.

3. The production engineering of the proposed matrices should be assessed with a view to their economic manufacture.

TABLE 1

HEAT TRANSFER RESULTS - SAND

UMF sand = 0.06 m/s
Mean particle size = 184 microns

Matrix Type	Superficial gas velocity (m/s)	Heat Transfer Coeff. $W/m^2 {}^oK$	Bed Material
	0.16	34	
A	0.33	37	None
	0.49	40	
B	0.16	35	
	0.33	39	None
	0.49	43	
A	0.16	35	
(repeat)	0.33	38	None
	0.49	42	
B	0.16	37	
(repeat)	0.33	39	None
	0.49	44	
	0.16	593	
B	0.33	606	Sand
	0.49	596	
	0.16	663	
A	0.33	676	Sand
	0.49	670	
A	0.16	600	
(repeat)	0.33	633	Sand
	0.49	638	
A	0.16	622	
(repeat)	0.33	615	Sand
	0.49	622	
B	0.16	563	
(repeat)	0.33	595	Sand
	0.49	591	
2.5 cm tube	0.16	336	
	0.33	355	Sand
	0.49	361	

TABLE 2

HEAT TRANSFER RESULTS - PRUTEEN

Pruteen mean particle size = 820 microns

Matrix Type	Superficial gas velocity (m/s)	Heat Transfer Coeff. W/m²°K	Bed Material
B	0.33	194	
			Pruteen
	0.33	213	
	0.49	209	
	0.65	200	
	0.98	163	
	1.60	149	
2.5 cm tube	0.16	18	
	0.33	27	None
	0.49	30	
	1.60	40	
	0.33	151	
	0.49	158	
	0.65	171	Pruteen
	0.98	165	
	1.60	164	
C	0.33	293	Sand
	0.16	117	)
	0.33	120	) Pruteen
	0.49	119	)

REFERENCES

1 J.S.M. Botterill, Y. Yeoman, K.R. Yuregir, Powder Technology 39 177 (1984)

2 N.I. Gelperin et al, I Eng. Phys. (Engl. Trans.) 10, 473 (1966)

3 D.H. Glass, D. Harrison, Chem. Eng. Sci. 19, 1001 (1964)

4 N.I. Gelperin, V.G. Einstein, pp 470 in 'Fluidisation' 1971 Academic Press (Eds. J.F. Davidson, D. Harrison)

5 A.M. Xavier, J.F. Davidson, 'Fluidisation' Camb. Univ. Press 1978

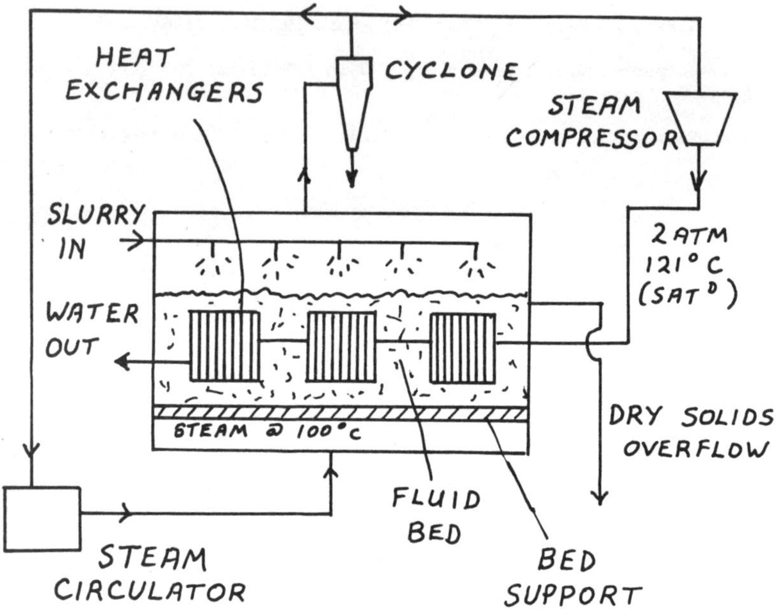

FIG 1 DRYING/GRANULATION WITH VAPOUR RECOMPRESSION

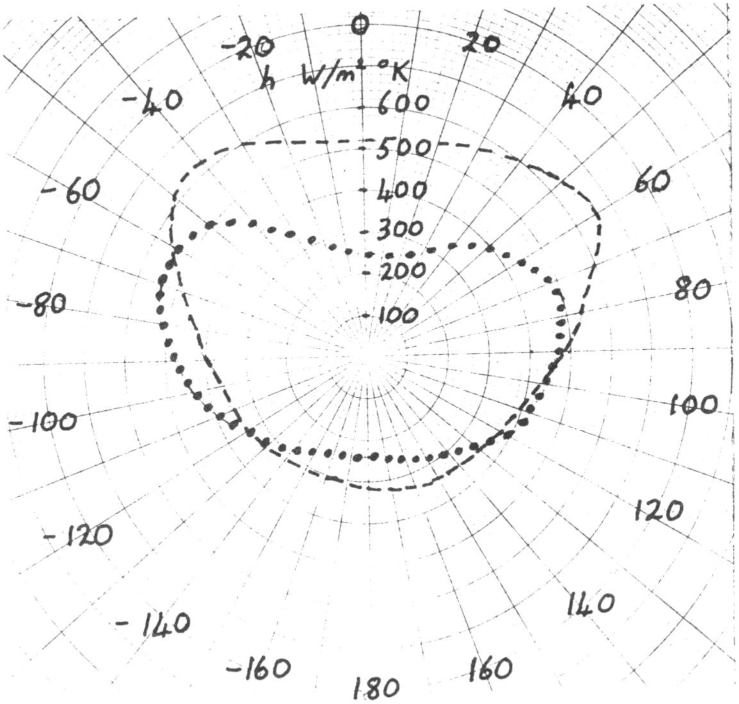

FIG 2 LOCAL HEAT TRANSFER AROUND A 3·5 CM TUBE IN A FLUID BED

U_{MF} = 0·11 M/S
- - - - u = 0·89 M/S
• • • • • u = 0·23 M/S

FIG 3A PARTICLE CIRCULATION IN A FLUID BED

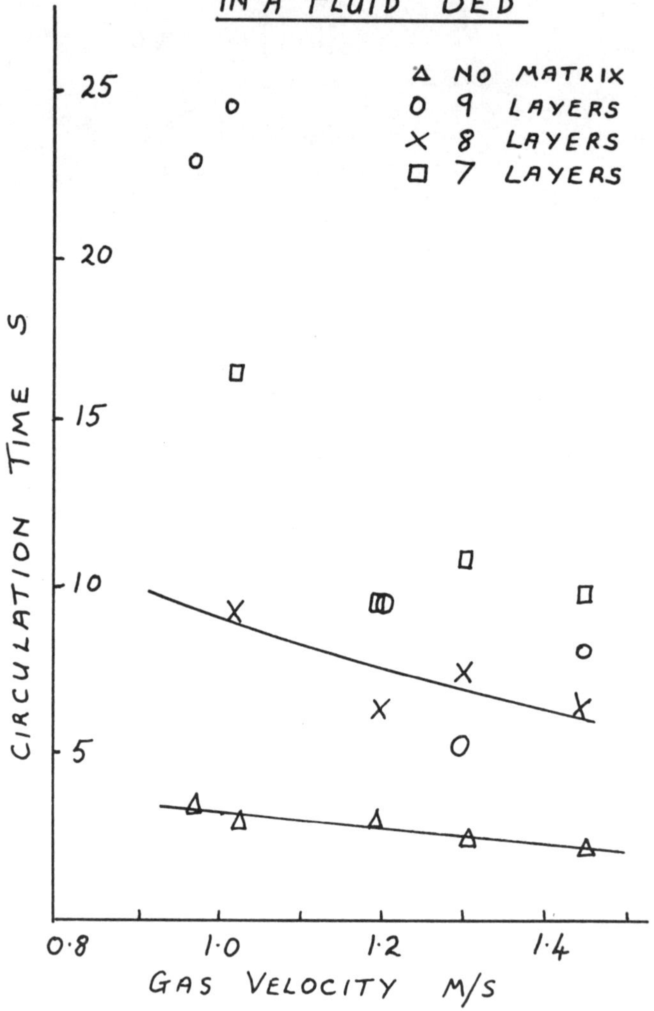

FIG 3B PARTICLE CIRCULATION IN A FLUID BED

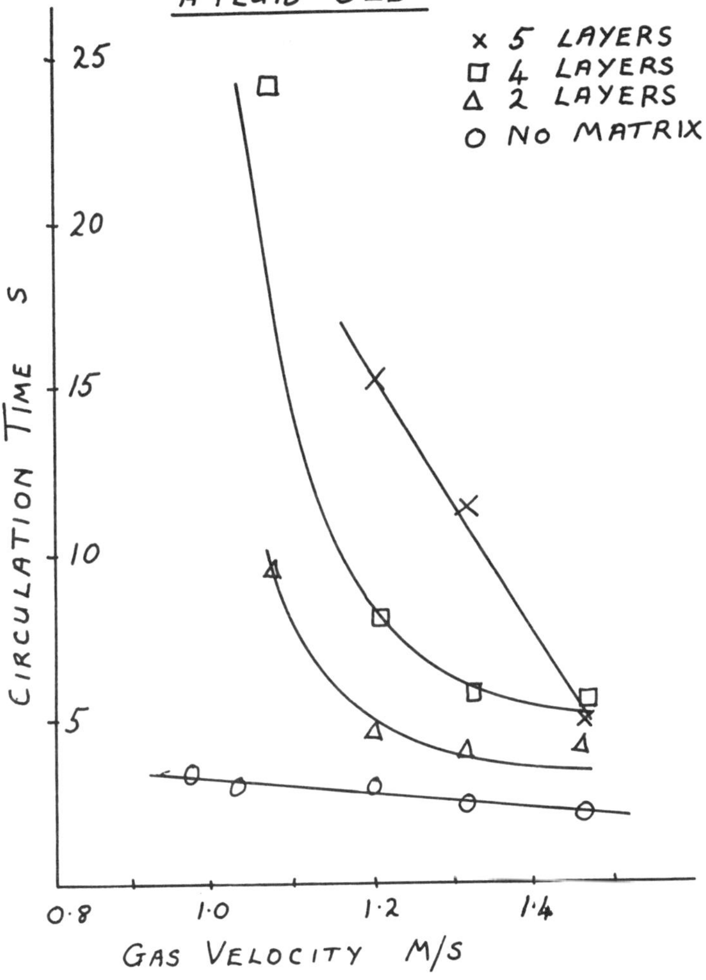

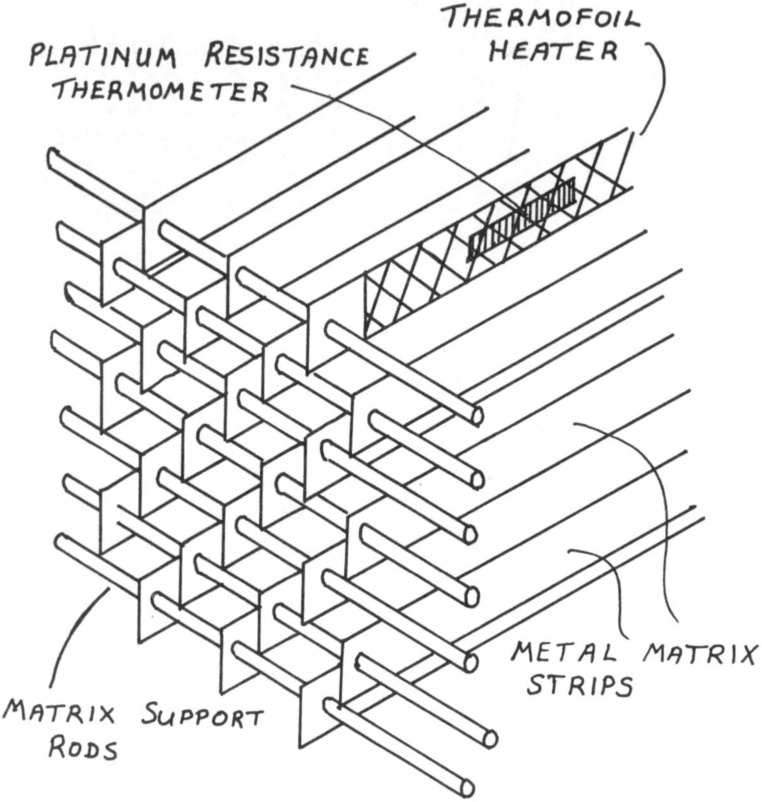

FIG 4 THE FLUID BED HEATER MATRIX

MODERN PROJECT INVESTMENT APPRAISAL : RETURN TO SIMPLICITY

C.L. Hicks *

For most chemical engineers the effectiveness of their technical work can only be ultimately assessed via an economic appraisal. If they lack the skills to do this themselves then all kinds of problems can subsequently arise. The principal purpose of this paper is to explain, justify and demonstrate a very simple method of project investment appraisal that any chemical engineer should be readily able to understand and apply. This method is also compared with the most frequently used current alternatives.

INTRODUCTION

However ingenious or effective a new technical idea may seem, it is not likely to get very far in any commercial organisation unless it is eventually shown to be economically attractive as well. Hence an engineer (chemical or otherwise) is always faced with the dilemma of either trying himself to evaluate the economic worth of any technical invention or improvement that he has in mind or leaving it to someone else, who may have little real interest in or understanding of his particular expertise and ideas. If an engineer in such a situation prefers to carry out his own economic evaluation, he is then faced with the task of deciding how to do it - that is, what data to obtain, how to use the data to achieve a correct result and how to interpret the result in a meaningful way.

In this day and age he may well be faced with a request to produce a profitability estimate that is an After-grant-and-tax-in-real-terms Discounted-Cash-Flow-Rate-of-Return for comparison with a stated average-long-term-real-terms-weighted-cost-of-capital. Although this is a valid enough requirement, the information needed for such an evaluation can be formidable, the necessary data handling techniques are somewhat complex (if the absolutely correct mathematical answers are to be obtained) and the interpretation of the result can sometimes necessitate the application of considerable knowledge and experience in relation to compound interest calculations. Hence the engineer's 'economic evaluation dilemma' can become very real and frustrating. The only easy solution would be a method of project investment appraisal that is simple to apply and understand, yet is reliable enough to give estimates that are comparable to those that would be obtained using the most up-to-date and sophisticated techniques. The basic purpose of this paper to to describe, justify, and demonstrate just such a method of project investment appraisal.

* BP Chemicals Limited, 76 Buckingham Palace Road, London SW1 OSU.

DATA AND DISCUSSION

The Range of Methods

Over the years many different methods for assessing profitability have been developed (1,2,3) to cope with the very wide range of factors that now affect investment projects (4). In essence all these methods can be sorted into one or other of 3 categories, i.e. Value-related (e.g. Cash Flow, NPV) or Time-related (e.g. variations on Payback Time) or Ratio-related (e.g. Ratio-on-Sales-Value, Ratio-on-Capital). However since most projects require some initial investment capital, which has to be either borrowed or diverted from an alternative use, then Ratio-on-Capital methods are the most fundamental primary measure of project attractiveness, since a project can normally only be worthwhile if it is going to earn more than the cost-of-capital. Currently there are 3 basically different Ratio-on-Capital-type methods of assessment available to any project assessor, i.e.

i. The official, legally-approved accounting technique used in company financial reports and known as the Historic Return-on-Capital-Employed (ROCE) method.

ii. The modern technique much used by economic analysts and based on discounting and known as the Discounted Cash Flow Rate of Return (DCFRR) (5,6) method.

iii. A long-established, but currently little used, technique that was once employed frequently in relation to technical projects and known as the Return-on-Investment (ROI) method.

Commenting briefly on the above, the historic-capital-based ROCE method has become increasingly controversial since the early 1970's, when high rates of inflation showed up its deficiencies all too clearly. The DCFRR method undoubtedly represents the most sound measure of profitability, in that it takes account of all cash movements in or out and their timing and integrates them on the basis of the rate of interest that could be paid on all capital outstanding at the end of each year of life of the project. It is thus the most fundamental measure of project profitability, but it does require all aspects of a project to be covered in detail and the answer obtained is of the 'black-box type', i.e. it is not always easy to see precisely why it is at the particular level shown. The ROI method is the simplest method of the three and was frequently used in the past by technologists who wanted quick estimates of the attractiveness of a potential project whilst still at the conceptual stage. It has however become less used in recent years due to the pressure to employ only 'modern' compound-interest-type methods of evaluation such as DCFRR and NPV.

In comparing these three Ratio-on-Capital methods, there can be little doubt that, for estimating the profitability of a past project where all the data are 'facts', a real terms DCFRR result gives by far the most definitive answer. However when dealing with the future, where many of the data are hypothetical and debatable, the use of a very detailed method such as DCFRR becomes more questionable. The real-life situation is that all three methods are used by different groups who are concerned in different ways with the vital task of selecting future projects. Therefore, in such circumstances, how does the effectiveness of these techniques for profitability estimation compare? Can one select one of them as being much better than the other two in relation to two of the most important criteria, i.e. simplicity of use and accuracy of result? In order to review this particular aspect, the way in which these methods work is examined more closely below.

General Comments on the Methods

A fully-detailed step-by-step illustration of the way in which the 3 Ratio-on-Capital methods are used is given in Reference 3 (N.B. Copies will be available at the Conference). The main features that are relevant in this context are that, even under the ideal conditions of a steady-state project with zero inflation, the ROCE method relates the Annual Profit to a declining Fixed Capital (i.e. reduced by the amount of the previous year's depreciation charge) thus giving an ever-rising year-by-year profitability indication, despite the fact that the Annual Profit sum is constant. The DCFRR method relates only to annual cash flows, completely avoids notional charges and gives a single, overall profitability result. The ROI method gives a steady year-by-year result, but with an average that is different to that produced by the DCFRR method. Although the DCFRR method is the most accurate method of calculating project profitability in principle, the lack of any year-by-year trend data is a drawback, especially when related to projects with a 10 to 20 year life. The relationship between the 3 methods is now explored more systematically.

Comparison of the Results of the Methods

Two of the most important economic criteria for an evaluation of any project are the level of Annual Net Cash Flow and the ratio of Fixed Capital to Working Capital. Hence a survey has been carried out into the comparative results obtained using these three profitability measurement methods over a range of Annual Net Cash Flows from 10-40% of capex (since below this range a project is very unlikely to be acceptable and above it the accuracy of the profitability measurement is no longer one of the most important issues) and over the full Fixed-Capital-to-Working-Capital-Ratio range of 0-100%. All the calculations are related to basically the same 10 year project.

Starting with the 0% Fixed Capital Case (i.e. 100% Working Capital), the comparative results for the three methods are shown in Figure 1. This shows that, rather remarkably, in this situation all 3 methods give absolutely identical results across the whole income range. The basic reason for this is because they all treat Working Capital in the same way and it is the treatment of Fixed Capital that is mainly responsible for the problems that arise with regard to the inter-relationship between the 3 methods; as a consequence in the 100% Working Capital Case the results are the same throughout (N.B. The ROCE method was originally devised for merchant traders, who of course work in a largely Working-Capital-type investment situation).

At the other end of the scale (i.e. 100% Fixed Capital) one can see, via Figure 2, the results are very different. Thus all of the methods start from zero % p.a. instead of 10% p.a., due to the need to repay the Fixed Capital over the 10 year project life. Then, due to the direct effect of the ROCE method's approach to Fixed Capital, the ROCE trend rises rapidly to 88% p.a. for a 40% capex p.a. Cash Flow. At that level the true profitability result, as represented by DCFRR, is 39% p.a. (N.B. Full details of the numerical calculations will also be available at the Conference). The relevant ROI estimate goes the opposite way compared to

ROCE and comes out relatively low at 30% p.a. Hence at the 40% capex p.a. Cash Flow level, ROCE gives a result that is over 49% points too high, whilst ROI gives one that is 9% points too low. The Average ROI figure is low compared to the DCFRR result because the sum deducted from the Net Cash Flow to cover the Plant Depreciation Charge required for the ROI method is rather too high. This is because the compound interest benefits that arise from investing this sum to provide replacement capital in the future are not allowed for in the ROI method as they are for the DCFRR method. As a result the residual Net Profit for ROI is lower than it should be in comparison to the DCFRR equivalent.

Real-life projects have a mixture of Fixed and Working Capital and in the process industries that ratio is typically around 80%:20% respectively. Also many plants take some years to build and some time to fully load the capacity. When this type of more realistic, mixed capital project is assessed by all 3 methods, the results are as shown in Figure 3. Here it can be seen that the Average ROI result is close to the true DCFRR result right across the range, whilst the Average ROCE result is still relatively high, e.g. at 56% p.a. for 40% capex p.a. Cash Flow, compared to the DCFRR result of 29% p.a. (i.e. still 27% points too high). In addition to the Average ROI line, the Ideal-Year ROI trend is also shown (i.e. based on an even simpler single-ideal-year calculation) and can also be seen to be relatively accurate (i.e. to within 3% points) until the profitability is rising well over the 30% p.a. level. The reason for the significant changes to the position shown in Figure 2 is that the DCFRR method is very sensitive to timing effects and hence the spread-out phasing of both capital expenditure and the gradual build-up of output leads to a marked reduction in the DCFRR result, but the other two results are not so severely affected because of the different way they are calculated (e.g. they ignore the phasing of capex and the lower output in the early years has a less significant effect). Hence over the critical measurement range (i.e. 10-40% capex p.a. Net Cash Flow levels) either Average ROI or Ideal-Year ROI are very close indicators of the true DCFRR result.

A number of other factors (e.g. level of inflation rate, length of project life) can also affect the accuracy with which ROCE and ROI predict profitability in relation to the absolute standard of the DCFRR method. A wide range of checks has been carried out and all have led to the same general result, i.e. both that DCFRR and Average/Ideal-Year ROI results relate fairly closely, especially under realistic phasing conditions, and Average ROCE results are always significantly in error and on the optimistic side, except only where a very high ratio of Working Capital is concerned. Average Historic ROCE figures for Fixed-Capital-intensive projects are of course even more in error in inflationary conditions due to the historic monetary capital cost figure that has to be used.

The Reasons for the Compatibility of DCFRR and ROI Results

This leads inevitably to the question as to why two such apparently different methods as DCFRR and ROI should be so compatible in relation to their assessments of profitability. (N.B. It is easy to see why ROCE gives relatively high figures - i.e. due to the annually reducing fixed capital element). This is best understood by appreciating that the cash flow element of a DCFRR calculation, although normally viewed as a single entity, can in fact be regarded as a two-part payment. This is illustrated in Table 1 where it can be seen that an Annual Cash Flow of 50 units/year, for an example 3-year project with a DCFRR profitability of 12% p.a. on a

Fixed Capex of 120 units, can be split into a Depreciation Charge of 36 units/year and a Profit Element (at the calculated 12% p.a. return rate) of 14 units/year. The Depreciation Charge may seem at first sight to be too small to repay a capital of 120 units over only 3 years but, due to the compound interest nature of DCFRR, it should be regarded as a Sinking-Fund-type payment which earns interest at the stated rate year-by-year. Hence, as demonstrated in Table 1, the charge of 36 units/year does indeed build up to a total value of 120 units by the end of the project life. Thus although the DCFRR calculation is outwardly very different from the ROI-type technique, in fact it is in some important ways very closely analogous to it.

The Sinking-Fund-type Depreciation Charge difference is a major cause of what gap there is between DCFRR and ROI results. (N.B. This leads to a DCFRR single-year technique known as the 'Annual Capital Charge' method, where an annual Depreciation Charge is calculated using discounting tables that take account of compound interest). At the lower profitability end of the scale the gap is inevitably small anyway and practical realities (e.g. capex phasing, output build up) tend further to reduce this gap under all conditions. It is this real-life effect that makes the Annual Capital Charge method less accurate in practice. Interestingly, although in theory a very prolonged construction period, together with a very slow build-up to full output, could in fact reduce the DCFRR result to a level below the ROI, even then the difference would not be great and of course normal management effort is always strongly oriented towards stopping such a damaging situation developing in any case. Hence the combination of inevitable real-life phasing and normal management policy tends automatically to keep DCFRR and ROI results fairly closely in line for most projects. Where the time-phasing is unusually short the ROI estimate is always conservative and suitable allowance can be made.

Since the results are often so similar, how does the time and effort required to obtain the two types of answer compare? Whereas the DCFRR method inevitably requires a significant amount of detail to obtain the result, the ROI method can (especially if estimated in the single-column 'Ideal-Year ROI' form) be carried out with a remarkable economy of data and effort. Thus, as illustrated in Table 2, the amount of information required for the ROI evaluation is only circa 10% of that required for DCFRR and can be considerably less in complex cases! Thus for typically one tenth or less of the effort, one can get a profitability result which is as accurate as is necessary in the early and uncertain stages of project speculation or development. This does not mean to say that the full DCFRR technique is not a valuable technique at the end of the day, but rather that it is certainly not necessary to apply it all the time and at every step, particularly in the early stages of project development. In fact the two methods can complement each other very effectively.

Overall Comparison and Discussion

Thus it has been shown that the ROI method can simply and quickly give future project profitability estimates that are acceptably close to those that would otherwise be obtained from the much more laborious and time-consuming DCFRR method and both of these methods are significantly superior in this respect to the ROCE method whenever fixed capital is involved. However they are not the only advantages for the ROI method. As already mentioned, it gives information on year-by-year profitability trends which the DCFRR method cannot achieve in the same sense (i.e.

because it is an all-embracing, cumulative-type calculation). The ROI method can also deal with a corporate-type mixture of plant projects, where some are starting, some are in full production and some are nearing the end of their lives, which the DCFRR method also cannot cope with. In particular, due to its brevity and simplicity, the results from using the ROI method can be much more easily analysed and interpreted than most and similarly the reasons for its results can be much more easily explained to interested or concerned non-financial-experts. Again, because of the simple structure of the ROI system, sensitivity calculations can also be easily carried out (N.B. These also tend to agree closely with those obtained from using the more cumbersome DCFRR approach). Even tax effects can be dealt with in most cases via a simple percentage reduction factor unless there are very complex timing aspects. A comparative summary of the performance of the 3 methods in relation to a number of general features is shown in Table 3, which illustrates very clearly how superior ROI is overall on these practical and useful issues, as compared to the DCFRR and ROCE methods. It is only in situations where the cash flow profile is unusually erratic that an ROI evaluation would be misleading.

Overall it can be seen that, in terms of time, effort, cost, accuracy, intelligibility and versatility, the ROI method is remarkably advantageous. Although it is a simplified method and some care therefore has to be taken not to apply it in blatantly inappropriate situations (e.g. where cash flows are very erratic or unusual), its overall advantages when dealing with embryonic situations where the current data may be approximate and the future data are even more uncertain, make it an especially effective technique for potential new projects or situations. Although it has for some years been overshadowed by the superficially more sophisticated compound-interest-based Discounted-Cash-Flow-type methods, its virtues have remained unchanged and, in an era where time tends to be short and the cost of skilled manpower is high, its speed, clarity and utility give it an attractiveness that is hard to match via any other currently available technique!

CONCLUSIONS

1. The official, legal Historic ROCE method is accurate for Working-Capital-intensive projects, but it is dangerously misleading when applied to Fixed-Capital-intensive projects, especially under inflationary conditions.

2. For sound reasons the very simple ROI method of profitability assessment normally gives results that are close to those for the fundamentally accurate but much more time-consuming and less-easy-to-interpret DCFRR method and any abnormal situation where the correlation does not hold can normally be readily predicted and allowed for.

3. The ROI method is an admirable and effective method for chemical engineers to use for carrying out their own evaluations of their technical ideas at early stages of development, where important decisions may need to be made quickly on data that will often only be approximate.

4. The ROI method has a wide range of advantages in relation to profitability studies compared to either DCFRR or ROCE.

REFERENCES

1. "The Finance and Analysis of Capital Projects" by A.J. Merrett and A. Sykes, published by Longman (Second Edition, 1974).

2. "A Guide to the Economic Evaluation of Projects" by D.H. Allen, published by the Institution of Chemical Engineers (Second Edition, 1980).

3. "An overview of modern project investment appraisal - Part II", article by C.L. Hicks and A.V. Bridgwater in the Process Economics International Journal, Volume Six, Numbers 3 & 4, 1986.

4. "An overview of modern project investment appraisal - Part I", article by C.L. Hicks and A.V. Bridgwater in the Process Economics International Journal, Volume Five, Number 1, 1984.

5. "Investment Appraisal" by the National Economic Development Office, published by Her Majesty's Stationery Office (Fourth Edition, 1971).

6. "A Practical Approach to the Appraisal of Capital Expenditures" by C.G. Edge and V.B. Irvine, published by the Society of Management Accountants of Canada (Second Edition, 1981).

APPENDIX

GLOSSARY

1. Cash Flow - this is the net difference between annual cash payments and cash income for a project.

2. Capex - an abbreviation for 'capital expenditure' on a project, normally referring only to fixed capital and not including working capital.

3. DCFRR - Discounted Cash Flow Rate of Return, which is the steady annual rate of interest that can be paid throughout the life of the project on the debt outstanding at the end of every year of the project.

4. Depreciation - the notional annual charge that is necessary in order to recover enough capital to replace the plant (or its original capital value) at the end of its working life.

5. Fixed Capital - the capital that is invested in the process plant and the facilities and hardware that make it operable.

6. Net Present Value - normally the current value of a project obtained by discounting to the present and totalling the series of net annual values the project is expected to produce; the discount rate used is normally the cost-of-capital.

7. ROCE - Return on Capital Employed, the classical method of profitability evaluation as used by qualified Accountants and applied to actual, historic money values; the result represents the only legally acceptable view of a company's profitability.

8. ROI - Return on Investment, a traditional method long used by technologists that is similar to ROCE except that the Fixed Capital element is kept constant instead of being reduced year-by-year as for ROCE.

9. Working Capital - the capital that is temporarily tied up within a project whilst it is still operating, e.g. in the sunk cost of producing stocks of product on site and on the way to customers or in their hands; it is recovered by ultimately obtaining appropriate cash payments when the project terminates.

MODERN PROJECT INVESTMENT APPRAISAL - RETURN TO SIMPLICITY

Figure 1

The Relationship between Average ROCE, DCFRR and Average ROI for a 100% Working Capital Project

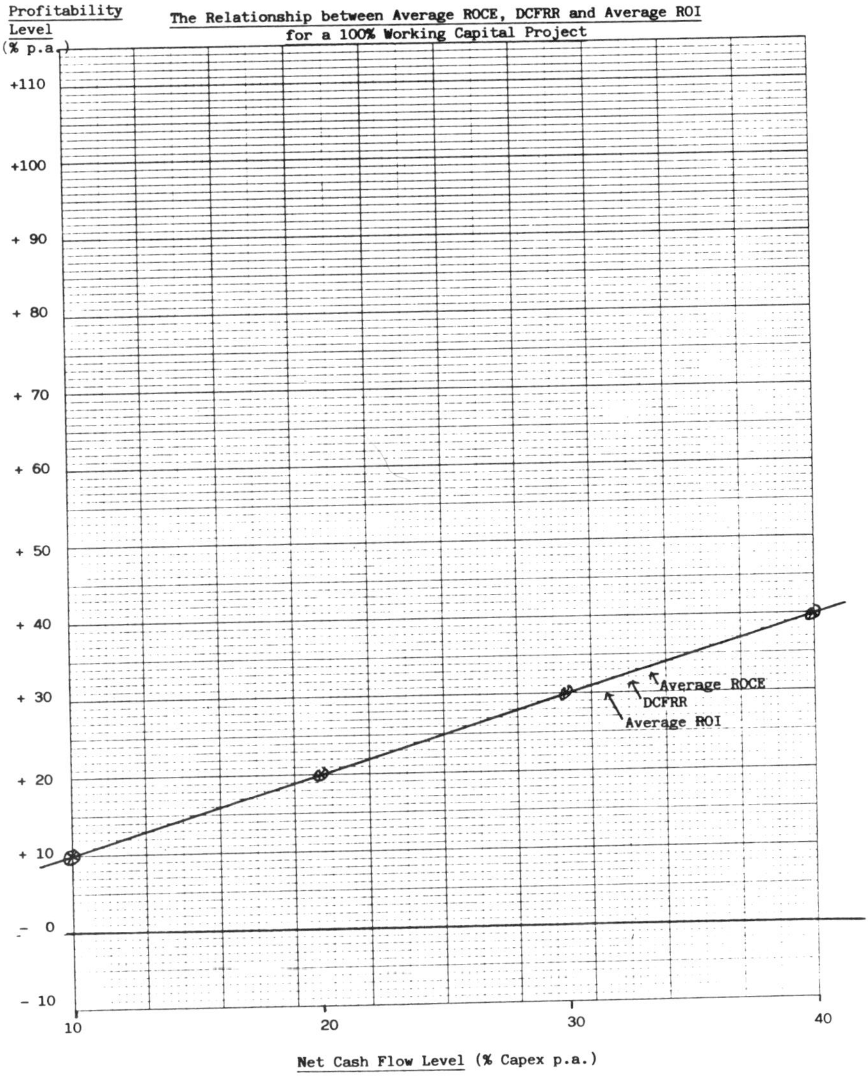

MODERN PROJECT INVESTMENT APPRAISAL - RETURN TO SIMPLICITY

Figure 2

The Relationship between Average ROCE, DCFRR and Average ROI for a 100% Fixed Capital Project

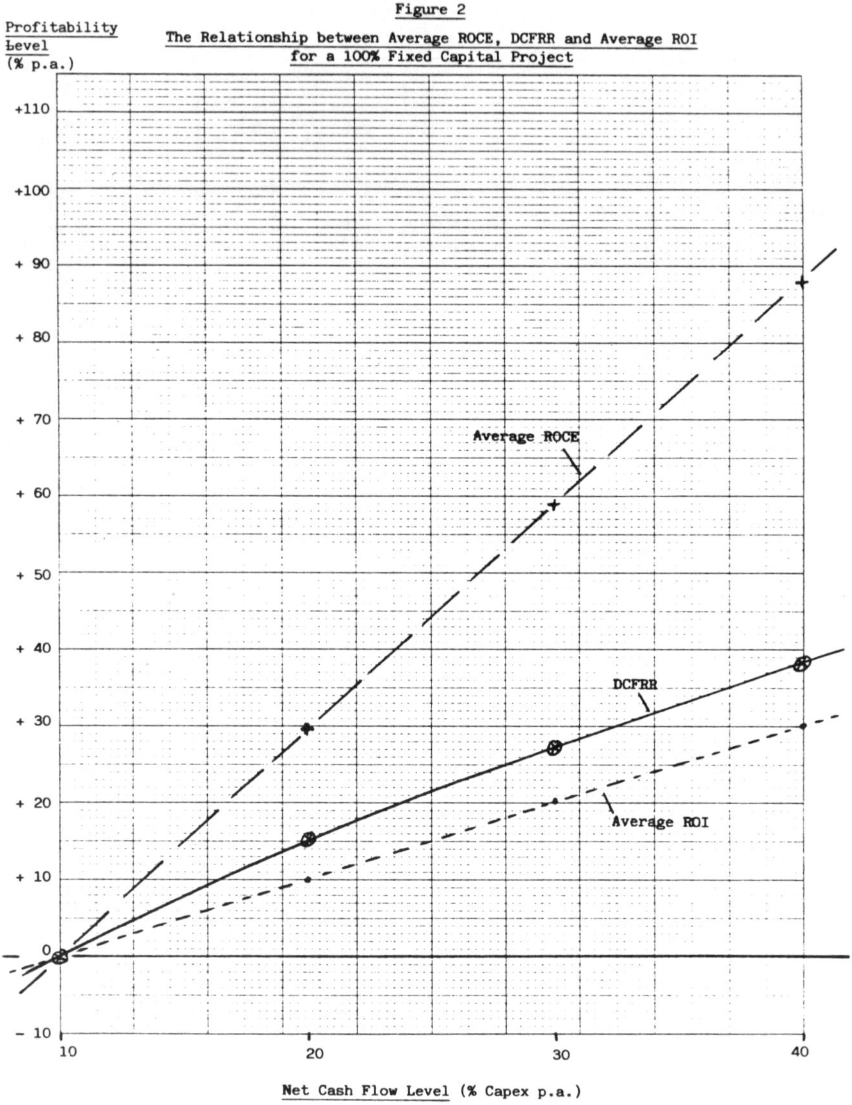

MODERN PROJECT INVESTMENT APPRAISAL - RETURN TO SIMPLICITY

Figure 3

The Relationship between Average ROCE, DCFRR and Average ROI for a Mixed* Capital Project

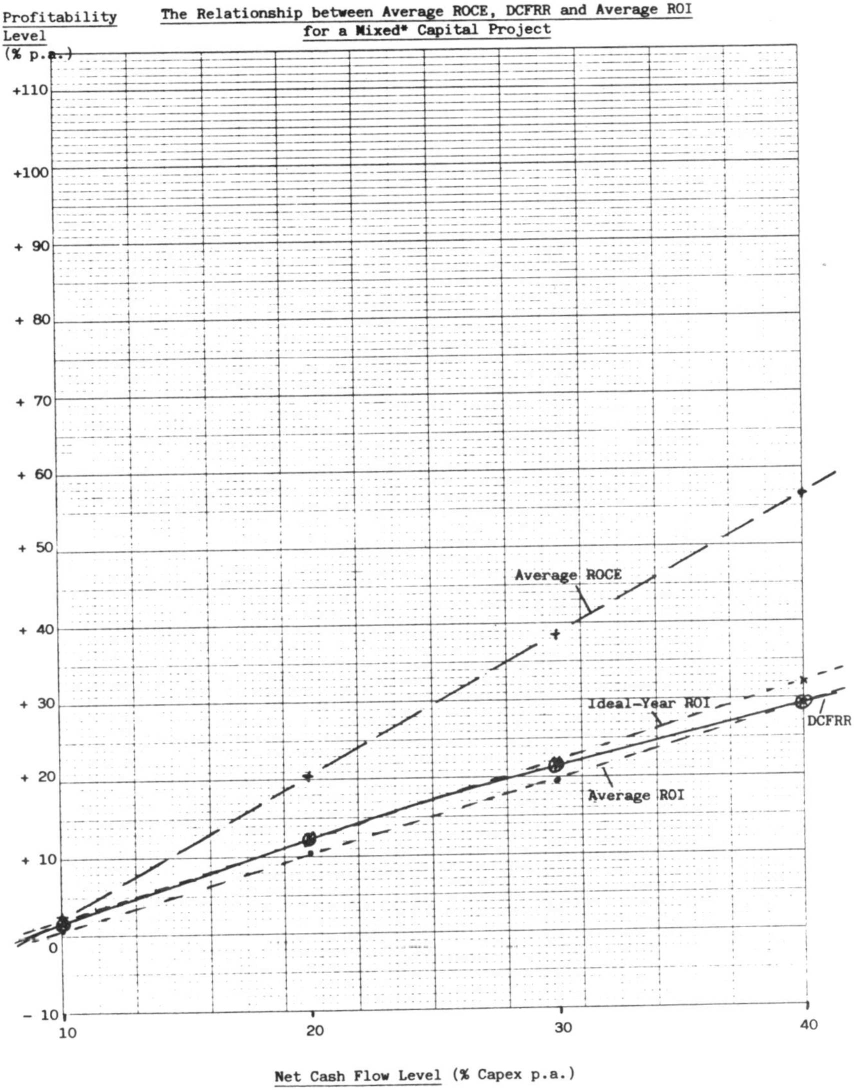

* 80% Fixed Capital/20% Working Capital.

MODERN PROJECT INVESTMENT APPRAISAL - RETURN TO SIMPLICITY

Table 1

An Alternative Interpretation of a DCFRR Result

Project Year	1	2	3	4	Totals
A. Net Cash Flow	(120)	50	50	50	-
B. Analysis of Net Income					
1. Depreciation Charge @ 12% p.a.[a]	-	36[b]	36	36	108
2. Profit Element @ 12% p.a.	-	14[c]	14	14	42
3. Total Net Income	-	50	50	50	150
C. Proof of Sinking Fund Result					
1. Start-of-Year Cash Position	-	-	36	76	-
2. End-of-Year Interest @ 12% p.a.	-	-	4	9	-
3. End-of-Year Cash Position I	-	-	40	85	-
4. End-of-Year Cash Payment	-	36	36	36	-
5. End-of-Year Cash Position II	-	36	76	120[d]	-

Notes
a. Based on 'Sinking Fund' system
b. ROIC Depreciation Charge = 40 units
c. ROIC Profit Element = 10 units
d. Rounded-down to 120 (due to cumulative rounding-off errors).

MODERN PROJECT INVESTMENT APPRAISAL – RETURN TO SIMPLICITY

Table 2

Comparison of the Minimum Data Requirements for the DCFRR and ROI Methods

Categories	DCFRR Method															Ideal-Year ROI Method
Project Year	1	2	3	4	5	6	7	8	9	10	11	12	13	14	Totals	1
Production Year	–	–	–	1	2	3	4	5	6	7	8	9	10	–		–
Cash Flows/Charges																
1. Fixed Capital	(120)	(480)	(480)	(120)	–	–	–	–	–	–	–	–	–	–	(1,200)	(1,200)
2. Working Capital	–	–	–	(180)	(60)	(60)	–	–	–	–	–	–	–	300	0	(300)
3. Variable Costs	–	–	–	(444)	(592)	(740)	(740)	(740)	(740)	(740)	(740)	(740)	(740)	–	(6,956)	(740)
4. Fixed Costs	–	–	–	(100)	(100)	(100)	(100)	(100)	(100)	(100)	(100)	(100)	(100)	–	(1,000)	(100)
5. Depreciation Charge	–	–	–	–	–	–	–	–	–	–	–	–	–	–	–	(120)
6. Sales Income	–	–	–	720	960	1,200	1,200	1,200	1,200	1,200	1,200	1,200	1,200	–	11,280	1,200
7. Net Cash Flow	(120)	(480)	(480)	(124)	208	300	360	360	360	360	360	360	360	300	2,124	–
8. Profit	–	–	–	–	–	–	–	–	–	–	–	–	–	–	–	240
9. Result	16.4% p.a.															16.0% p.a.[a]
10. Numerical Data	58 values															7 values

Note

a. ex $\frac{240}{1,200+300} = 16.0\%$

MODERN PROJECT INVESTMENT APPRAISAL - RETURN TO SIMPLICITY

Table 3

Comparison of the Versatility of the Principal Ratio-on-Capital Methods

No.	Feature	ROCE Method	DCFRR Method	ROI Method
1.	Simplicity	(√)	X	√
2.	Accurate Results	X	√	√[a]
3.	Speed of Results	(√)	X	√
4.	Sensitivity Checks	(√)	√	√[b]
5.	Fixed/Working Capital Ratio	X	√	√
6.	Year-by-Year Trend	(√)	X	√
7.	Competitor Comparisons	X	√	√
8.	Corporate Situations	(√)	X	√
9.	Ease of Interpretation	X	X	√
10.	Intelligibility to Non-Experts	X	X	√
-	Totals (reference √)	0	4	10

Notes

a. Not quite as accurate as the DCFRR method.

b. Much quicker than DCFRR.

√ = Very satisfactory.

(√) = Gives a result, but of dubious accuracy.

X = Not satisfactory.

TOTAL PROCESS PLANT ANALYSIS BY PROCESS INTEGRATION

Ian C Kemp

SUMMARY

This paper shows how process integration techniques can used to identify process changes in the separation and reaction systems which lead to reductions in the overall process energy usage, thus allowing total process plant analysis (TOPPA). The separation or reaction system is divided off from the background process and can be plotted on the graph of the background process grand composite curve (net heat flow profile). The best operating conditions to ensure maximum recovery of heat can then be determined. A general analysis of these concepts is supplemented by an in-depth analysis of flash systems including an actual case study.

INTRODUCTION

The basic ideas of process integration or "pinch technology" (Ref.1) were recognised as a major advance as soon as they appeared. The heat loads at given temperatures of all the hot streams and all the cold streams on a plant were combined to give the Composite Curves (Fig.1). Using these graphs or a simple calculation, one could for the first time obtain a practical yet scientifically-based target for the energy which the process requires, and also identify the critical point - the pinch - where heat exchange and network design are most difficult. The techniques were subsequently extended to heat and power networks (Ref.2) so that the level as well as the quantity of heat supplied could be determined.

These original techniques had one major limitation; they applied only to the process as it stood. However, our experience has shown that judicious changes to the process conditions often lead to further energy savings and an "improved target". A typical process can be divided into 4 sections - the chemical reaction system, the product separation systems, the heat exchanger network and the site utility/heat and power systems. Conventionally, plants have been designed section by section in this order, choosing the reaction system first and working through till the residual heating and cooling loads were mopped up by site utilities. However, it would also be useful to refer back from a partly developed flowsheet to see whether the sections previously designed can be further optimised to fit better with the process as a whole - recognizing that design should be an iterative process rather than a one-way street.

* Energy and Process Integration Service, Process Engineering Research Centre, B488T6, Harwell Laboratory, Oxon. OX11 ORA.

This paper provides a systematic approach to doing this by identifying the optimal placement of separation and reaction systems from an energy usage viewpoint and identifying possibilities for beneficial process change. The techniques are particularly valuable for retrofit of existing plants, where there is already a large amount of fixed equipment and hierarchical synthesis techniques are inapplicable; but they can just as easily be applied to new plants. They are a "first-stage" optimisation preceding detailed design but are at least as important as the unit operations analysis techniques commonly thought of as optimisation. To use an everyday example, it doesn't help to adjust the fine tuning on your TV if you're on the wrong channel!

The techniques are quantitative but mainly graphical. The emphasis is on giving the project engineer a clear understanding of how his process works and how the different parts relate to each other, rather than merely giving him a set of mathematical formulae or a "black-box" computer program.

MAIN DISCUSSION

The Grand Composite Curve

The key to process integration analysis is the net heat flux profile of the process. This is a plot of the heat transferred from hot sources (hot utility and hot streams) to cold sinks (cold utility and cold streams) against temperature. In a process which achieves the minimum energy requirement (the M.E.R. target) , the heat transfer across the pinch is zero and the graph meets the axis there. For a typical real process which is off target, the actual heat transferred through the process can be represented by shifting the graph to the right, when the hot utility and cold utility rise by the amount of heat transferred across the pinch.

The resulting graph is the Grand Composite Curve (Fig.2). This is most easily visualised as the heat which must be supplied to, or removed from, the process at any given temperature. It can also be seen as the horizontal distance between the Composite Curves (with a temperature correction added to allow for necessary driving forces).

The Grand Composite Curve can be expressed as a function G of interval temperature as follows;

$$G_T = G_0 + \sum_{COLD} (H_T - H_0) - \sum_{HOT} (H_T - H_0)$$

with interval temperature determined by

$$T = T_{HOT} - ½\Delta T_{MIN} \text{ or } T = T_{COLD} + ½\Delta T_{MIN}$$

and $G_T = 0$ at $T = T_P$, $G_T = Q_{H,MIN}$ at $T = T_{MAX}$, $G_T = Q_{C,MIN}$ at $T = T_{MIN}$

Note that all hot and cold streams in the intervals between T and T_0 must be included, not just those which exist at T.

For a temperature interval $(T - T_0)$ small enough for the heat capacity flow-rates CP of each stream to be roughly independent of temperature, this equation becomes:-

$$G_T = G_0 + \sum_{COLD} CP(T - T_0) - \sum_{HOT} CP(T - T_0)$$

The importance of the Grand Composite Curve was recognised by workers in ICI, not only for analysing heat recovery networks but also for determining utility supply levels (Refs.2,3,4) and for distillation column placement (Refs.3,4). EPI have now shown that it is also the vital tool in identifying the working temperature of any separation system (Ref.5). This analysis will be applied to reaction systems as well to complete the picture.

Moreover, as the Grand Composite Curve is best calculated for a flowsheet with most of the operating conditions known, it is ideally suited to retrofit studies, where the theories of process design analysis from scratch are less applicable because of the existing constraints.

Separation Systems

Any energy-based separation system works by degrading heat at a high temperature to a lower temperature to achieve the product separation. It thus incurs an exergy loss itself. All such systems are thermodynamically very inefficient; the exergy theoretically required to make the separation is usually less than 10% of that actually required. This means that thermal separation systems are major energy users, often in fact the dominant ones in a process. Clearly it is desirable to re-use as much of the degraded heat as possible.

To find how the separation system relates to the rest of the process (the background process) it is convenient to "split the Grand Composite". This was first done for distillation columns (Ref.4) where the latent heat loads of the reboiler and condenser were taken out and plotted as a separate rectangle on the background process Grand Composite. To be able to supply heat to or receive heat from the background process, the rectangle must fit above or below the grand composite, not to the side.

A similar analysis can be applied to evaporation (Ref.5) by considering the evaporative loads in the various effects; again these plot as rectangles. In fact, any separation system should be placed above or below the background process Grand Composite, so that the degraded heat is re-used (Ref.5). Otherwise the system is in fact transferring heat across the pinch - and thus of course wasting energy.

A general treatment of a simple process as in Fig.3 reveals that there are 3 cases to consider.

1. (a) $T_p > T_B$ and $Q_B < G_B$ - separation system fully below pinch

   (b) $T_p < T_C$ and $Q_C < G_C$ - separation system fully above pinch

   $Q_{xs} = 0$

All the heat passed through the separation system is used again in the process and the pinch temperature T_p is the same as for the background process.

2. (a) $T_p > T_B$ and $Q_B > G_B$ - separation system below pinch but not fully below Grand Composite

   (b) $T_p < T_C$ and $Q_C > G_C$ - separation system above pinch but not fully above Grand Composite

   $Q_{xs} = Q_B - G_B$ for (a) and $Q_C - G_C$ for (b).

The separation process uses more heat than can be handled by the background process. The remaining heat is transferred across the pinch and wasted. The pinch also moves, to T_B in (a) and T_C in (b).

3. $T_C < T_P < T_B$; separation system across pinch.

$Q_{xs} = Q_B = Q_C$.

None of the heat used to drive the separation system can be recovered. The overall pinch temperature is the same as for the background process.

Although this method provides a good first approximation in many cases, it is not a rigorous analysis because the reboiler and condenser loads of a distillation column are not usually equal. The top product may only be partially condensed, and there are product rundown streams and feed heating, which are the sensible heat demands and these also change if the column temperature is changed. We must find fixed points in relation to the rest of the process and calculate the effect of changing the entire separation system between these fixed points. Suitable bases are the feed stream's temperature as it emerges from the previous unit operation or storage tank, and the product temperatures as they pass to final storage or purification.

For a general separation system working between these 2 points with a feed being separated into 2 products, we can say:-

$$H_0 + \Delta H_{FEED} + Q_B - Q_C - \Delta H_1 - H_1 - \Delta H_2 - H_2 = 0$$

If we define $H_1 + H_2 - H_0$, the difference between the initial and final states, as ΔH_S, this becomes:

$$\Delta H_{FEED} + Q_B - Q_C - \Delta H_1 - \Delta H_2 = \Delta H_S$$

Where sensible heat effects are small compared with latent heats, this reduces to the simple rectangle representation with reboiler and condenser loads dominant. Otherwise, it plots as a polygon, as in Fig. 4. The fixed load ΔH_S is included in the background process grand composite, which will give a balanced set of heating and cooling loads for the separation system polygon.

The 3 cases identified on Page 3 still hold broadly; the pinch temperature is always the point at which the separation system touches the background Grand Composite Curve.

This rigorous treatment of separation systems allows us to determine, for example, the possibilities for load shifting between effects and sensible heat networks for inter-effect streams in evaporators; or the placement and sizing of pumparounds or intermediate reboilers and condensers on distillation columns. All these are identified by precise matching of heat loads and temperatures to the Grand Composite Curve.

To illustrate the in-depth analysis, we will use the simplest separation system of all - a flash. This is the basic building block from which analysis of more complex systems can be constructed.

Flash Systems

A flash vessel takes a liquid at high temperature T_0 and produces a vapour and a liquid at a lower temperature T_1; the vapour is then condensed and releases its heat at that temperature. The separation system thus plots as a horizontal line at T_1. Now the change corresponding to ΔH_S would be a linear release of heat over the complete temperature range between T_0 and T_1; this is added to both the background process and the separation system. The flash system therefore plots as a triangle, as in Fig. 5.

The simplicity of the flash system allows an exact analysis to be carried out, as follows:

$$h_o - h_1 = C_p (T_o - T_1) = \lambda_1 y = \Delta h_S$$

$$H_o - H_1 = w_o C_p (T_o - T_1) = w_1 \lambda_1 = y/w_o \lambda_1 = \Delta H_S$$

where y is the fraction separated. Thus the final composition is known for any flash temperature.

When the flash triangle just touches the Grand Composite, the heat released at T_1 from the flash can just - but only just - be used in the process. At this point:-

$$G_F = H_o - CP (T_o - T_F) = W_F \lambda_F$$

This equation can be solved numerically and the minimum flash temperature and maximum separation achieved at MER (maximum energy recovery or minimum energy requirement for the overall system) can be found.

What if the separation achieved is insufficient? In this case we repeat the process with another flash at T_2 and continue until we do achieve the desired separation. The analysis thus gives us the minimum number of flash stages required to achieve a given separation with maximum energy recovery.

Conversely, for a system with a given number of stages, we can estimate the extra energy which must be supplied above the background process MER to give a certain separation. The final composition tells us the lowest flash temperature T_N, and we find that Q_{XS} is the maximum value of the quantity $CP.(T_{i-1} - T_i) - G_i$. The various flash temperatures T_i are manipulated to minimise Q_{XS}, which is not a trivial task mathematically; however, a good approximation is obtained by looking at the Grand Composite and designing the system by eye.

A good first estimate of Q_{XS} can be obtained by taking $CP(T_{i-1} - T_i) - G_i$ equal for all the flashes. In this case $NQ_{XS} = CP(T_0 - T_N) - \sum G_i$, the intermediate flash temperatures being found by a simple hand calculation. This rough target can be bettered if there is a large "re-entrant" in the Grand Composite, where net heat flow is large; the flash can profitably be fitted in to that. It is desirable to have a flash close to a concave point in the Grand Composite Curve.

Case Study

A recent EPI study gives a good illustration of the techniques in action. This was on a batch process in which the major process steams were slurries - not on the face of it a promising situation for heat recovery. The system incorporated one flash which recovered atmospheric pressure steam for the hot slurry to heat process water.

Figs.1 and 2 are in fact the Composite and Grand Composite Curves for the original process. The target was close to the existing energy consumption (scaled here to 10 units) and the presence of slurries and solids made it impossible to find economic heat recovery projects to achieve the target. However, with the flash tank divided off it is immediately apparent that it is wasting a lot of energy (Fig.5). The energy bill would be more than halved if a heat exchanger could be used; but a batch exchanger turned out to be impossibly large and uneconomic, while changing to continuous flow was unacceptable for processing reasons. An improved flash system was clearly the most cost-effective heat recovery scheme and would also maintain product quality best.

The analysis showed that 6 flash stages would be needed to obtain maximum energy recovery and reduce Q_{XS} to 0, and this was uneconomic. By calculating Q_{XS} for different numbers of flash stages, EPI showed that a three-flash system gave the best rate of return, with one flash above and one below the existing atmosphere pressure flash, which was retained. The Grand Composite for this system is shown in Fig.6.

Two features deserve further comment. The pinch has moved and is now being caused by the vacuum flash; the other 2 flashes at first sight are not recovering the maximum amount of heat possible. This is because there were other restrictions on heat exchange in the plant; a constrained target was calculated which showed that some high-temperature heat would have to be used below the pinch, and the flash system was re-optimised accordingly. Secondly, the cooling load as well as the heating requirement has decreased, and since this limited the plant capacity in summer, the modifications have debottlenecked the process.

The final design saved nearly 40% of the energy previously used by the process and the payback was less than a year. Good operability was retained even for this batch process, by efficient sequencing of operations; throughput was increased and product quality maintained. So not only can a flash system be used for separation of volatiles, it is also an effective and often cheaper alternative to heat exchangers - if the system is optimised by energy and process integration principles. This is especially true for systems containing fouling or corrosive liquids or slurries.

Reaction Systems

A chemical reactor in energy terms is a device which releases or absorbs a large amount of heat energy in effecting a chemical transformation. So the heat loads on the hot streams (product rundown) and cold streams (feed heating) are not equal, as they were for the separation system after allowing or ΔH_S. The reaction system can again be plotted on the grand composite curve, but the unbalanced heat load means that it behaves more like a utility than a separation system.

For a general reaction system as in Fig. 7., a heat balance gives

$$\Delta\ (H_{FEED} - H_0) + \Delta H_R = (H_E - H_Z) + \Delta H_S$$

and usually the heat of reaction ΔH_R dominates the sensible heating terms $H_{FEED} - H_0$ and ΔH_S. This equation, and in particular the ΔH_S term, are equivalent to those formed earlier in obtaining an overall heat balance for separation system.

The internal behaviour of the reaction system is found to depend mainly on 3 quantities:-

ΔT_R - the temperature change through the reactor

ΔT_{MIN} - the minimum temperature difference chosen for heat exchange between cold feed streams and hot product streams

ΔT_S - the temperature difference between initial feed and final product streams.

The analysis assumes a single feed stream and a single product stream. By plotting $\Delta T_R - \Delta T_{MIN}$ against $\Delta T_S - \Delta T_{MIN}$ a truth table can be formed which gives the heating and cooling requirements at various temperatures. The feed and product streams will have similar CP's, so ΔT_R and ΔT_S will be almost proportionately related to ΔH_R and ΔH_S.

An exothermic reaction should ideally come above the pinch and the Grand Composite so that it can release all its heat to the process; conversely an endothermic reaction should lie below the Grand Composite. The ideal situation for an exothermic reaction is shown in Fig. 8, where the product rundown stream is released above the pinch, but the feed heating is below it so can be provided by waste process heat. For exothermic reactions, in fact, a reduction in reaction temperature (e.g. by using a more active catalyst) will not usually save energy and indeed may cost it because less useful heat is released to the process.

Obviously other factors such as reaction rate and yield will be of vital importance in designing the reactor and will also affect the separation system. However, the analysis above will show clearly how the heat of reaction ties in with the energy needs of the rest of the process and the design can then be refined by iteration. Alternative reaction routes can also be studied, sometimes with surprising results. For example, an apparently attractive biotechnology process which releases much useless low-grade heat and then requires a large separation system to concentrate the dilute product produced can be much less energy efficient than a conventional high-temperature reaction.

CONCLUSIONS

The development of a fully systematic analysis of separation processes and its extension to reaction systems are important conceptual advances. It is now possible to optimise the placement of all the sections of a process - the reaction and separation system, the heat exchanger network and the utility heat and power system - relative to each other by an interactive and iterative process. The energy and cost savings and the capital expenditure required for possible process changes can be calculated and used in an overall economic evaluation of the process. The optimum energy use for the overall process can now be found, and this Total Process Plant Analysis (TOPPA) is applicable to both new and existing plants.

ACKNOWLEDGEMENTS

The Author wishes to acknowledge the help received from Dr E.K. Macdonald, Dr D. Reay and other colleagues in EPI, in preparing this paper.

NOMENCLATURE

CP	Heat capacity flowrate	W/K	C_p	Specific heat capacity	J/kgK
G	Grand composite curve function	W	h	Specific enthalpy	J/kg
H	Enthalpy (flowrate)	W	w	Mass flowrate	kg/s
Q	Heat (flowrate)	W	y	Fraction separated as vapour	kg/kg

Subscripts

0	Initial conditions	1	Final or stage 1 conditions
2	Stage 2 conditions		
B	(Re)boiling conditions	C	Condensing conditions
C,MIN	Minimum cold utility	COLD	Cold streams
E	Reactor exit conditions	F	Flash conditions
FEED	Feed streams	H,MIN	Minimum hot utility
HOT	Hot streams	MAX	Maximum value of function
N	Number of flash stages	MIN	Minimum value of function
PRODS	Product stream	P	At pinch
R	(Due to) reaction	S	Across system (reactor or separator)
REACTS	Reactant stream		
T	Interval temperature	XS	Additional heat required
Z	Product conditions		

Greek Letters

ΔH	Enthalpy difference	W	ΔT	Temperature difference	K
λ	Latent heat of condensation	J/kg			

REFERENCES

1. A User Guide on Process Integration for the Efficient Use of Energy, 1982 (Institution of Chemical Engineers, Rugby).

2. Townsend D W, and Linnhoff B, 1983. Heat and Power Networks in Process Design (2 parts) AIChEJ, 29,5, 742/748.

3. Boland D and Hindmarsh E, 1983. Beyond HENs: A Total Thermodynamic Approach to High Energy Efficiencies. Paper presented at A.I.Chem.E. Diamond Jubilee Meeting, Washington, October 1983.

4. Hill J C, 1985. Process Integration - A Structured Approach to Operating Cost Reduction. Paper presented at Institution of Mechanical Engineers conference, "Energy and the Process Industries", June 1985.

5. Kemp I C, 1986. Process Integration applied to Separation Systems. Paper presented at Institution of Chemical Engineers Annual Research Meeting, Bradford, April 1986.

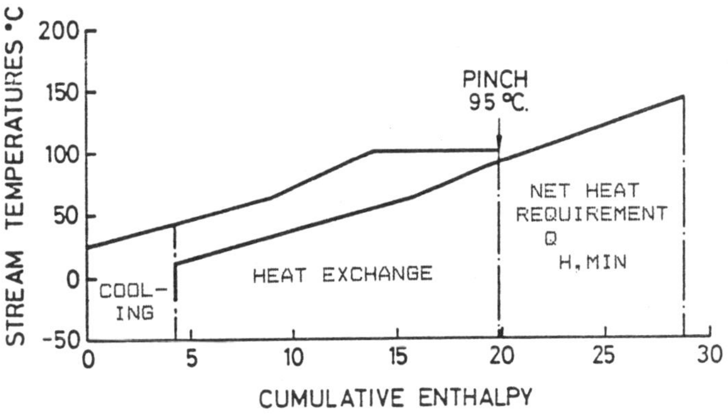

FIG. 1. COMPOSITE CURVES FOR PROCESS.

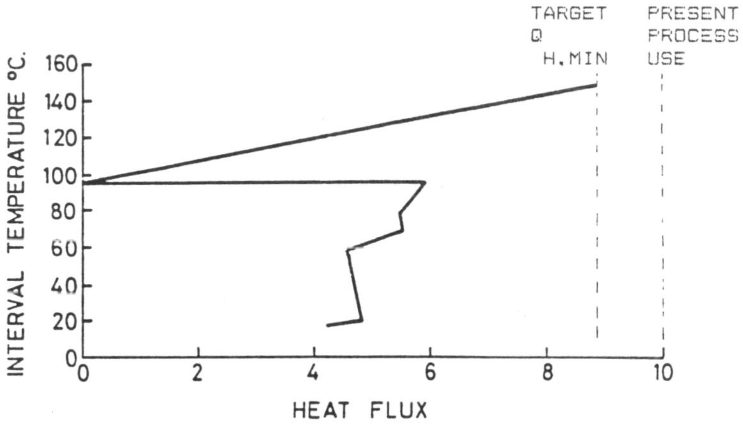

FIG. 2. GRAND COMPOSITE CURVE FOR PROCESS.

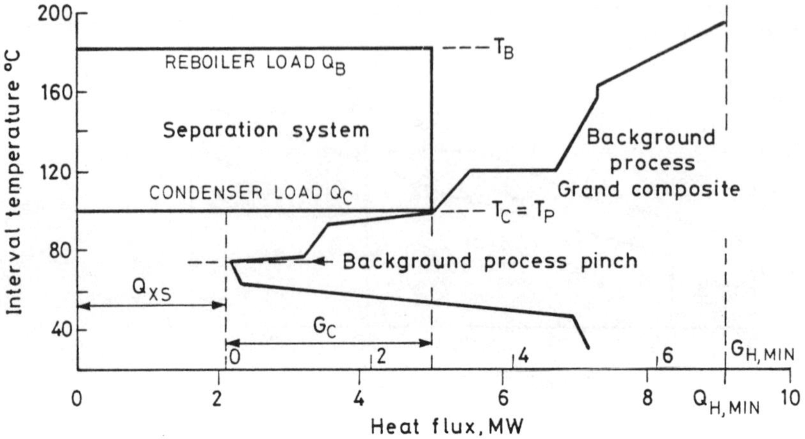

FIG.3. SPLITTING THE GRAND COMPOSITE - BASIC METHOD (SIMPLE DISTILLATION COLUMN)

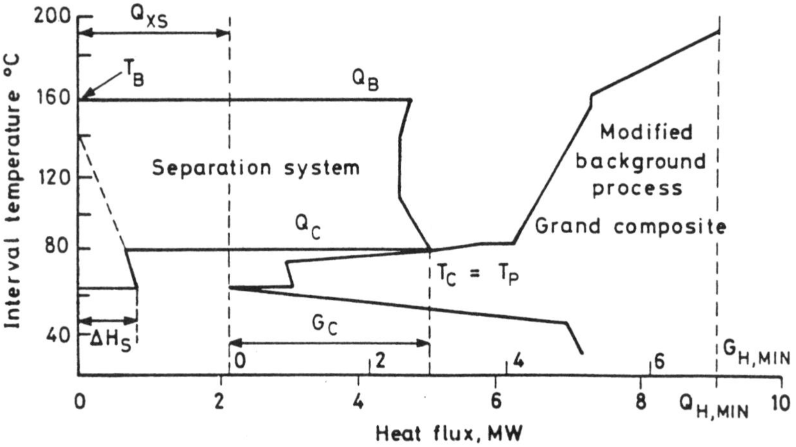

FIG.4. SPLITTING THE GRAND COMPOSITE - RIGOROUS TREATMENT (SENSIBLE HEATING; REBOILER AND CONDENSER LOADS UNEQUAL)

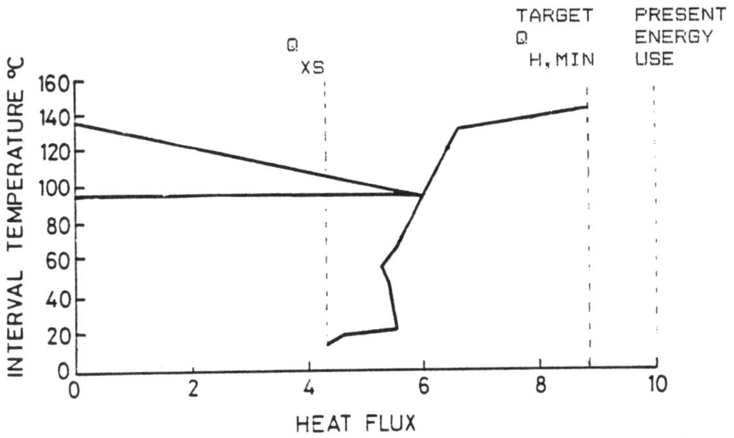

FIG. 5. SPLIT GRAND COMPOSITE - SINGLE FLASH

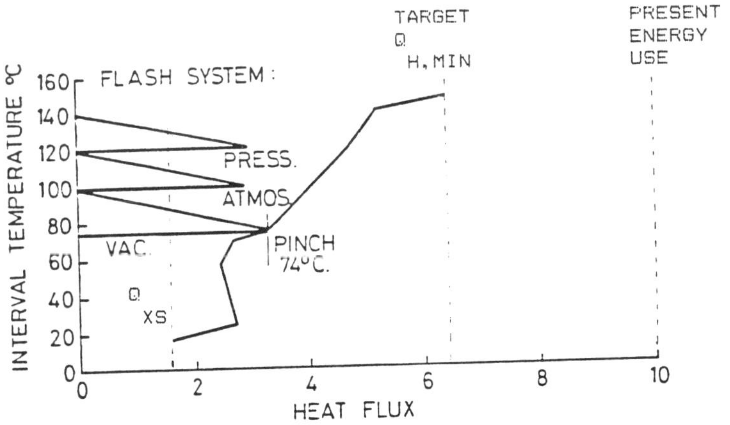

FIG. 6. SPLIT GRAND COMPOSITE - OPTIMISED TRIPLE FLASH SYSTEM

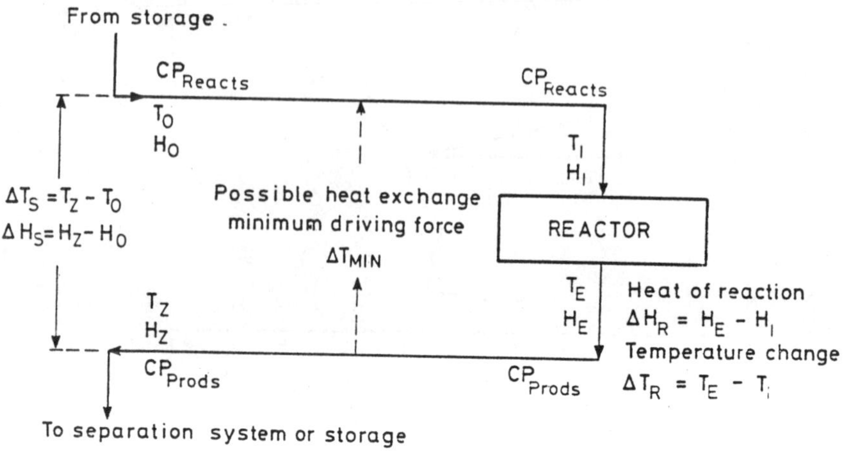

FIG. 7. GENERAL REACTION SYSTEM

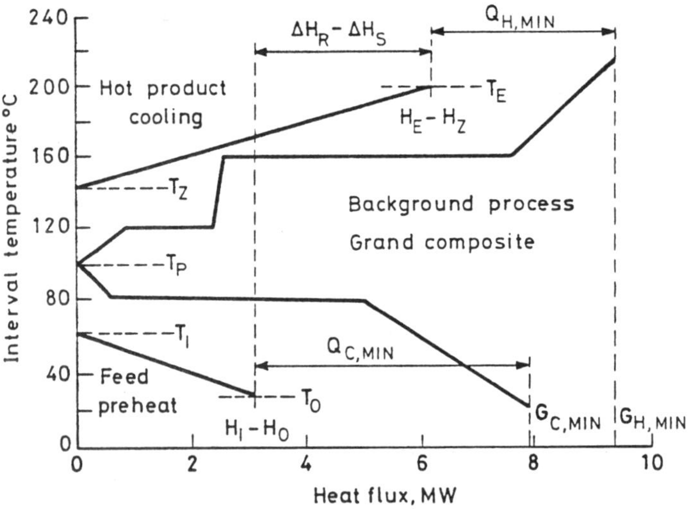

FIG. 8. EXOTHERMIC REACTOR IDEALLY FITTED TO PROCESS GRAND COMPOSITE CURVE

SYNTHESIS OF DISTILLATION SEQUENCES: A COMPARISON OF ENUMERATIVE AND AI METHODS

R.J. Best*, J.C. Conners**. N.S. Dhallu*, W.R. Johns* and V. Vadhwana**

A range of simple separation problems are synthesized by artificial intelligence and by optimization. The AI method employs established heuristics, based on sound physical principles, for each step in the synthesis. Quantitative expressions are developed for these heuristics that enable them to be "tuned" to give close agreement with the optimized results. The AI results prove to be sensitive to the choice and tuning of the decision rules. The ability of AI to predict promising new processes rather than merely interpolate current practice is thus suspect. We conclude, nevertheless, that there is scope for effective mixed procedures incorporating some optimization within an AI structure.

INTRODUCTION

A number of recent papers have proposed the application of Knowledge Based Systems (Artificial Intelligence) to the synthesis of chemical and bio-chemical processes, eg. Stephanopoulos and Townsend [1]. Other procedures have also been published in which heuristic rules play a key part, eg. Douglas [2] and Rossiter and Douglas [3]. These procedures are applied to synthesis problems in which the only yardstick against which they can be assessed is "engineering judgement" (ie the processes developed without the benefit of methodical synthesis tools). On this basis the procedures have been found effective, or show promise of being effective in matching or exceeding best "engineering judgement". To date, however, they have never been tested on problems which are at once sufficiently difficult to tax "engineering judgement" in reaching an optimum whilst at the same time sufficiently closely defined to permit a rigorous global optimum to be established. This work compares a well-formulated AI synthesis procedure with a mathematically rigorous optimal synthesis procedure. The comparison has been made over a sufficiently large range of synthesis problems for the conclusions to be meaningful.

For this comparative study, we have chosen the synthesis of processes for the separation of hydrocarbon mixtures by simple distillation. The process is well understood and reliable design and costing methods exist which provide a valid yardstick against which to evaluate AI methods.

The rigorous procedure with which the AI approach is compared achieves optimization by implicit enumeration and is described in References 4 and 5. The procedure is of general application and not restricted to distillation or other separation processes. It is, however, particularly well adapted to separation processes; the current version [6] includes well-tested routines for sizing and costing simple multi-component distillation columns.

* Chemical Engineering Department, South Bank Polytechnic
** Information Technology Research Unit, British Petroleum plc

We show that an AI program is capable of reproducing the results of a rigorous optimization sufficiently closely to form the basis for further, more detailed, engineering design and evaluation (eg parameter optimization). For the problems studied to date computer run times are similar for the two procedures. In principle, however, AI should have run times less sensitive to problem size. Against this advantage should be weighed the fact that AI decision rules contain many adjustable parameters; the predictive powers of any computational procedure that contains many parameters that may be adjusted to correlate known results must be suspect. We believe that the greatest advantage will be obtained from AI by incorporating known physical relationships and mathematical optimization procedures alongside the heuristic decision rules. In this way the number of arbitrary adjustable parameters will be kept to a minimum and the ability to synthesize completely novel processes reliably will be radically improved. We have approached the AI problem using LISP, PROLOG and POP11. Our experience to date indicates that POP11 will provide the most convenient tool for this integration of AI and optimization technology.

THE SYNTHESIS PROCEDURES

The processes synthesized have been simple hydrocarbon separation problems. For strict comparability between the two approaches, neither the integer optimization nor the AI technique consider heat integration between columns.

Implicit Enumeration

The implicit enumeration procedure for optimization centres around an efficient method for solving the following recursive equation [4]

$$Q(a) = \min_i \left[P_i(a) + Q(b_i) + Q(c_i)\right] \qquad \ldots 1$$

(See list of symbols)

For distillation "a" defines the feed stream to a column, $P_i(a)$ is the discounted cost of the i'th way of designing a column to process "a" and "b_i" and "c_i" are the corresponding top and bottom product streams. Q(a) is the most economic method of processing "a" to acceptable process outputs.

In order validly to reduce Equation (1) to an integer optimization, it is necessary to establish that the effect of parameter optimization of the precise recovery levels in each column is small. In Appendix 1 we show that providing splits are fairly sharp (>95 %), such parameter optimization has negligible effect on vapour and liquid rates or on reboiler and condenser duties. We further show that the effect on number of stages is sufficiently small to justify treating parameter optimization on a column-by-column basis. Thus a minimum P_i can be computed solely on the basis of the major components in "a" and Equation 1 can validly be used to derive the optimal separation train. The precision of the overall cost estimate should be good.

The cost correlations employed are based on the methods in References 7 and 8. Each option P_i is costed assuming a "sharp" split (99 per cent) between key components from which the minimum reflux ratio (R_{min}) is computed using Underwood's equation. the optimal reflux ratio is set to 1.1 R_{min} (6) from which vapour rates and heat loads are calculated. This information is sufficient to compute reboiler and condenser capital costs and estimate heating and cooling utility costs. It also enables column diameter to be calculated. Fenske's equation is employed to calculate the minimum number of separation stages which, with Gilliland's relationship enables actual stages to be obtained. The number of stages, together with the already computed diameter permits total column capital cost to be estimated. a target interest rate of 15 per cent per annum is employed to compute a Net Present Cost for each column. Such costs are simply additive over the whole process, as required by Equation (1).

Synthesis Using AI

The AI techniques used in the "Production System" - a particular type of Knowledge Base System. A production system is a means of applying a series of rules (in our case heuristics) in a sequence dictated by the problem being solved ("data-driven"). The complete system comprises:

(a) The problem database.

(b) A set of production rules.

(c) A program to interpret the rules.

For each of the languages used (LISP, PROLOG and POP11), we have used a pre-written rule interpreter program and have set up the database and production rules using the syntax allowed by the interpreter chosen.

The problem database initially consists of the raw facts known about the problem. For the separation system chosen, this comprised:

(a) The composition of the initial feed.

(b) The relative volatilities of the individual components.

(c) The flow rates of the individual components.

(d) Lists of groups of components that are acceptable as mixed products.

In the course of execution of the program, this database changes according to the inferences made.

Production rules are statements of the type:

IF (set of conditions) THEN (set of consequences).

The (set of conditions) is compared with the facts the database and the (set of consequences) consists of instructions to add new facts or remove facts from the database. The application of a rule takes place when a match is found between the (set of conditions) in the rule and a set of facts in the database. a record is kept of which rules and facts have been used and the system stops when all possible combinations have been tried or when some target has been reached (as specified by one of the production rules).

In the simple distillation synthesis problem, the target is to find the economic optimum split of a mixture into two fractions

$$b = B\,[a,\ i]$$
$$c = C\,[a,\ i]$$

where decision "i" is given by

$$i = I\,[a]$$

with a = [1,2,3...n] for an n-component mixture.

The primary heuristic rules which have been widely used in distillation train design are as follows:

(i) Favour "easy" separation first.

(ii) Favour 50/50 molar splits.

(iii) Favour taking one component from the top of the column.

(iv) Favour removing major components early in the separation train.

(v) Tend to reject splitting components that are acceptable as mixed products.

There is a sound theoretical basis for these rules, in that, under certain simplifying assumptions they tend to minimize total vapour rates in the system. Vapour rates determine both reboiler and condenser duties and also column diameters.

In order to implement these rules, they must be given quantitative expression. The decision "i" is, for this process, defined as follows. For the simple distillation of an "n" component mixture, (n-1) sharp splits are possible. thus stream

a = [1,2...n]

can be split into b = [1] c = [2,3...n]
b = [1,2] c = [3,4...n]

b = [1,2...(n-1)] c = [n]

These decisions are numbered as follows:

decision 1, take b = [1], from the bottom,
decision 2, take b = [1,2], from the bottom etc.

Using this convention, we now discuss, in turn, how we give quantitative expression to each rule.

(i) "Easy" Splits. The easy or difficulty of a split is conventionally defined by the relative volatility of the light key component with respect to the heavy key. Thus, the ease of split decision "i" is measured by $r_{i+1,i}$. This ratio can, in principle, take values from 1 (split impossible) to infinity (100 per cent separation in one stage without reflux).

Measured against the volatility of an arbitrary component, j, the split difficulty can be expressed as $(r_{i+1,j}/r_{ij})$. The easiest split will have the highest value of this ratio and the most difficult the lowest ratio.

We suspect that the scale of difficulty should not be measured linearly and have, therefore, decided to use the logarithms of the ratios, ie difficulty measure "i" (G[i]) is

$$G_1[i] = \ln\,[r_{i+1,j}/r_{ij}] = A_{i+1} - A_i \qquad \ldots 2$$

Unfortunately there is no intrinsic scaling that enables decision criteria such as $G_1[i]$ to be compared against other decision criteria. To facilitate comparison, we have decided to attempt to scale decisions such that a decision about which we feel neutral has weight 1.0. To scale Equation 2 we must measure $G_1[i]$ against the average difficulty of separation in any particular problem. The total range of values of "A" is $(A_n - A_1)$, from which the average difference is $(A_n - A_1)/(n-1)$. Any split for which $G_1[i]>(A_n - A_1)/(n-1)$ is easier than the average split and any less is more difficult. a scaled "ease of separation" factor has therefore been defined as

$$H_1[i] = (n-1)(A_{i+1}-A_i)/(A_n-A_1) \qquad \ldots 3$$

$H_1[i]$ has the property that values of $H_1[i] > 1$ imply that split "i" is favoured. Furthermore, if $H_1[p] > H_1[q]$, split "p" is easier than split "q" and hence "p" is more favoured than "q". The maximum range of values is H_1 is:

$$0 < H_1 < (n-1) \qquad \ldots 4$$

(ii) 50/50 Splits. We have chosen the following simple measure of a 50/50 split:

$$H_2[i] = \min\,(F[b]/F[c],\ F[c]/F[b]) \qquad \ldots 5$$

Equation 5 has the characteristic that:

$$0 < H_2[i] < 1$$

$H_2 = 1$ is favoured, $H_2 = 0$ is disfavoured. A 66:33 split has $H_2 = 0.5$. there is no intrinsic linearity to the scale, and we could equally put, for example

$$H'_2\,[i] = 2F[b]\,F[c]/F[b]^2 + F[c]^2) \qquad \ldots 6$$

H'_2 has the same limits: $H'_2 = 0$ is disfavoured, $H'_1 - 1$ favoured. H_2 does, however, not show such a strong preference for exact 50/50 splits, thus a 66:33 split has $H'_2 = 0.8$. We have no reason to employ an equation more complex than Equation 5.

(iii) Most Volatile from Top. If i = (n-1), this split is favoured. For other values of "i", it is not favoured. "Not favoured" is simply defined by

$$H_3[1] = 0, \qquad i = 1,2\ldots n-2 \qquad \ldots 7$$

and "favoured" by $H_3[n-1] = 1$

More subtle formulae taking account of the flow rate of component "n" have been tried but have given no benefit.

(iv) Major Components. The rule is clear enough if the "major" component is most or least volatile in "a"; it says remove the component immediately. If, however, there are many components the rule is less clear. We have interpreted it as follows, favour splits adjacent to major components. this rule encompasses splitting of least or most volatile components first. It does, however, also favour making splits adjacent to major intermediate components. such components will then be least or most volatile in one of the outlet streams and can then be split off at the next stage of distillation, ie early in the separation sequence. we have adopted the definition

$$H_4[i] = n \max [z_i, z_{i+1}] \qquad \ldots 8$$

If the largest component adjacent to cut (i) is of average concentration ($z=1/n$), $H_4[i]$ is then 1. Larger values are favoured, the largest being most favoured.

(v) Acceptable Mixtures. If (i,i+1) are acceptable as a mixed product, we simply set

$$H_5[i] = 0 \qquad \ldots 9$$

otherwise $H_5[i] = 1$. $H_5 = 1$, then means split favoured, $H_5 = 0$ means split not favoured.

Combining Decision Factors

The heuristics described so far can be used to establish new facts in the problem database relating to the desirability of a particular split. To achieve the target of finding the most desirable split, a set of rules for combining these facts must be given. The methods available range from the very sophisticated, employing mathematical techniques to manipulate confidence levels and probabilities, through to the very simple methods we shall describe here. In combining decision factors, the values obtained are frequently pre-treated. We discuss such possible pre-treatment before describing our chosen ranking procedure.

Pre-treatment

The decision factors can be left as real numbers or can be pre-treated to separate obviously good, obviously bad and indifferent values. This pre-treatment can take the form of targeting (on a pass/fail basis) or division into several bands. The mechanism for such targeting and banding is illustrated in Table 1.

The methods work as follows:

Using H_i (i), the splits are simply marked as H(2) better than H(4) better than H(1) better than H(3).

Using H^T(i), splits 1 and 3 would be rejected as "too difficult" whilst the other splits would pass.

Using H^B(i), split 3 would be rejected as "too difficult", splits 1 and 4 would be acceptable and split 2 would be favoured.

Having chosen one or other of thee measures, there are a variety of ways in which they could be used, the most usually considered are:

(a) Rank by weighted sum:

$$H[i] = w_1H_1[i]+w_2H_2[i]+w_3H_3[i]+ \dots$$

Develop synthesis tree further for any tied equal, such ties are likely with, for example, 0/1 target criteria.

TABLE 1

Component j	A_j	Split i	$H_1(i)$	$H_1^{(T)}(i)$*	$H_1^{(B)}(i)$**
1	0	-	-	-	-
-	-	1	1.0	0	0
2	0.5	-	-	-	-
-	-	2	1.4	1	+1
3	1.2	-	-	-	-
-	-	3	0.4	0	-1
4	1.4	-	-	-	-
-	-	4	1.2	1	0
5	2.0	-	-	-	-

Notes:
*Target value $H_i > 1$; acceptable, $H^T = 1$
$H_i < 1$; unacceptable, $H^T = 0$

**Reject value $H_i < 0.5$; $H^B = -1$
Neutral range $0.5 < H_i < 1.3$; $H^B = 0$
Favoured split $H_i > 1.3$; $H^B = +1$

(b) Rank in order of priority, eg.

Apply test H_1, if only one passes trace the split of the corresponding output streams. If none pass, stop the sythesis tree. If more than one pass, apply H_2 to all those that pass.

The same logic is applied to consequences of H_2 as was described for H_1, except that, if none pass test 2 develop the synthesis tree for all those that pass H_1, ie consider processing all those sets of (b,c).

Continue in this way until test H_5, when the synthesis tree is developed for all passing all 5 tests. This logic implies target weights H^T.

(c) Rank by weighted sum all those alternatives not rejected by any of the rejection criteria, eg use (H and H^T) or H^B.

Experiments that we have conducted indicate that, for our process synthesis problem a weighted sum is the most appropriate; all five decision rules should be given some weight in the synthesis. The only qualification to this conclusion is that very low values of H_1 should cancel any high values of other decision criteria. We can achieve this effect in one of two ways. Either set a pass/fail criterion applied before the weighted sum, or transform H_1 so that very low values give rise to large negative weights, eg set $H_1 = \ln (H[H_1]$; as H_1 approaches zero, then H_1' approaches minus infinity which then swamps all other decision rule values. We adopted this latter strategy.

RESULTS

A wide range of feed mixtures have been studied. Here we report results for mixture of the 10 components in Table 2. The flowsheets synthesized are given in Table 3 and the costs compared in Table 4. the distillation trains are described by a compact sequence of (n-1) digits, for an n-digit mixture.

TABLE 2

FLOW RATES OF COMPONENTS IN FEED MIXTURES USED FOR SYNTHESIS STUDIES

(Kmol/h)

Component	Mixture									
	1	2	3	4	5	6	7	8	9	10
1. n-nonane	1	1	1	1	1	1	1	1	10	1
2. n-octane	1	1	1	1	1	1	1	2	9	2
3. n-heptane	1	1	1	1	1	1	1	3	8	3
4. n-hexane	1	1	1	1	1	1	2	4	7	4
5. n-pentane	1	1	1	1	1	1	5	5	6	5
6. i-pentane	1	1	1	1	10	1	1	6	5	5
7. n-butane	1	1	1	10	1	1	1	7	4	4
8. i-butane	1	1	10	1	1	1	1	8	3	3
9. propane	1	10	1	1	1	1	1	9	2	2
10. ethane	1	1	1	1	1	1	1	10	1	1

The significance of these digits is as follows:

1) 1st digit, decision variable (i) for 1st separation
2) 2nd digit, decision variable (i) for separation of top stream from (1), until all tóp components separated, when

k) digit k, decision variable (i), for bottom stream from last column with a more than 1 component bottom stream.

Thus, for a 5 component mixture, the digit sequence (2, 3, 4, 1) defines the process.

1) Split inlet stream (12345) into (12) and (345).
2) Split stream (345) into (3) and (45).
3) Split stream (45) into (4) and (5).
4) Split stream (12) into (1) and (2).

TABLE 3

FLOWSHEETS SYNTHESIZED BY OPTIMIZATION AND BY AI

Separation Sequence	Optimization									
	1	2	3	4	5	6	7	8	9	10
1	9	9	9	9	9	9	9	9	6	9
2	8	8	8	7	6	8	8	8	9	7
3	5	5	7	8	8	6	5	5	8	8
4	7	7	5	6	7	7	7	7	7	5
5	6	6	6	4	5	4	–	6	4	6
6	4	4	4	5	5	–	4	4	5	4
7	3	3	3	3	3	3	3	3	3	3
8	2	2	2	2	2	2	2	2	2	2
9	1	1	1	1	1	1	1	1	1	1
	AI									
1	9	9	8	6	6	9	9	9	3	4
2	8	8	9	7	9	8	8	8	4	6
3	4	4	7	9	8	4	4	6	6	9
4	6	6	6	8	7	6	7	7	9	8
5	7	7	3	3	5	7	6	4	8	7
6	5	5	4	4	4	–	5	5	7	5
7	3	3	5	5	3	3	3	3	5	3
8	2	2	2	2	2	2	2	2	2	2
9	1	1	1	1	1	1	1	1	1	1

Two typical flowsheets are illustrated in Figure 1.

The AI procedure did not, in any case studied, generate the optimal sequence. In spite of this, the process costs were generally very similar (Table 4) with an average difference over all runs (including those not detailed here) of 5.7 per cent. Alteration of the parameters (eg cost coefficients or decision weightings) would make some difference in both methods of computation but would not alter the distinctive ability of the optimization program to look ahead to the consequences of its earlier decisions, compared with the discrete stepwise treatment in the AI method. This point is illustrated by reference to examples 1, 2, 3, 7 and 8 where the optimization method chooses early on to make the relatively difficult split between components 5 and 6: this amounts to a decision to incur a heavy cost at this point to be compensated subsequently by much reduced costs. By contrast, the AI procedure is conditioned by Heuristic 1 against making this choice. The detailed costing for example mixture 1 is shown in Table 5.

DISCUSSION

There are reasonable grounds for developing the heuristic rules of our AI program without recourse to empirical data correlation. Nevertheless, for the larger problems we found that considerable "tuning" was necessary before we were able to get the degree of agreement exhibited in the results. Consequently, we feel that AI methods are best applied in interpolating current practice rather than in developing radically new processes.

TABLE 4

COST DIFFERENTIALS

Mixture	Relative Cost Difference %
1	2.1
2	1.6
3	11.3
4	15.7
5	2.8
6	2.0
7	9.1
8	0.4
9	7.6
10	5.1

TABLE 5

STEP BY STEP COSTS FOR MIXTURE 1

Separation	Number of Cost Units	
	Optimal	AI
1	3663	3663
2	3004	3004
3	6769	4042
4	2421	2662
5	3418	5271
6	2000	3244
7	2526	2526
8	2681	2681
9	2424	2425
Total	28906	29517

It is difficult to quantify the number of variables available for tuning in an AI study of this sort. There are certainly five weighting parameters, but there are also decisions on the forms used to quantify the individual heuristic rules and qualitative decisions on the methods employed to combine the rules and hence rank decisions. The fact that we needed to experiment with the forms of the rules and tune the weightings places some doubt on the ability of more complex AI programs to come to good results without "tuning" against rigorous procedures. Once tuned, however, the AI program performance is good and we could now apply it to larger problems. It should be noted that an interactive facility is readily incorporated whereby the user can give guidance where the choice between two decisions is too close for confident automatic discrimination.

We attempted the synthesis problem using three different AI programming languages, LISP, PROLOG and POP11. All three worked satisfactorily. We found, however, that the superior general purpose programming facilities of POP11 made it more convenient for incorporating the relevant Process Engineering calculations.

We feel that the flexible data structures of POP11 make it particularly suitable for a symbiosis of AI and mathematical programming. The ability to handle recursion efficiently would, for example, enable us to modify Equation 1 such that the minimization over (i) was restricted to only those decisions meeting the heuristic criteria. All decisions meeting these criteria could be traced forward to generate a data structure incorporating all likely process variants; accumulating costs would then enable discrimination between them.

CONCLUSIONS

1. AI techniques can be used to synthesize process flowsheets. The processes synthesized are usually similar to these developed by mathematical programming. In all cases reported, however, the flowsheet developed by mathematical programming is more economic.

2. AI procedures do need tuning against reliable yardsticks before they can be employed with confidence. They should, at least in a Process Engineering context, be used with caution where such yardsticks are not available.

3. The language POP11 is convenient for combining data structures and decision handling facilities needed for AI with more traditional programming facilities needed to incorporate physical understanding of Process Engineering synthesis problems.

APPENDIX 1 - SENSITIVITY OF DISTILLATION COLUMN COSTS TO RECOVERY SPECIFIED

We make a simple analysis, based on Fenske-Underwood-Gilliland, for binary mixtures. Numerical experiments indicate that the conclusions are broadly applicable to multi-component systems.

1. Effect on Minimum Reflux Ratio

Consider a mixture of total feed rate F, mol fraction (z_1, z_2) and relative volatility r (ie r_{21}) from which is recovered Fz_2 (1-f) and Fz_1f from the top of the column. For bubble-point feed, Underwood's equations can be written

$$\frac{z_1}{1 - t} + \frac{z_2}{r - t} = 0 \qquad \text{...A1}$$

$$\frac{x_1}{1 - t} + \frac{rx_2}{1 - t} = R_m + 1 \qquad \text{...A2}$$

Equation (A1) gives

$$t = r / (z_1 + rz_2) \qquad \text{...A3}$$

Substituting in Equation (A2) gives

$$L(R_m+1)/F = \frac{(1-f)z_2(z_1 + rz_2)}{(r - 1)z_2} - \frac{fz_1(z_1 + rz_2)}{(r - 1)z_1} \qquad \text{...A4}$$

where

$$L = [fz_1 + (1-f)z_2]\ F$$

From which the minimum reflux rate is

$$LR_m = F\,[1 - (1+r)f]/(r - 1) \qquad \ldots A5$$

LR_m is the reflux rate. It is seen that, for an ideal binary mixture, it is:

i) Independent of the feed composition, and

ii) Only slightly dependent on the fractional recovery; thus for f in the range 0 - 5 per cent, LR_m would typically only change in the range 100 - 90 per cent.

The total vapour rate in the column is given by

$$V = LR_m S + L \qquad \ldots A6$$

where S is the multiple of minimum reflux ratio employed for practical distillation. It is not possible to derive such a simple expression for multi-component distillation but the same conclusions hold:

i) The minimum vapour rate varies very little with recovery in the column providing that the recovery is high.

ii) Similarly, if the feed composition only varies slightly as a function of recoveries in previous columns, the minimum total vapour rate in the column will vary proportionately less. Thus only parameters F and z_2 are affected which, in turn, have only a minor affect on Equation A6.

It may be noted that Dhallu [6] has shown that, with current energy prices, the optimal choice of S is to make it as close to 1 as will confidently leave a margin over R_{min} within the limits of precision of physical properties. A typical value would be S = 1.1 (in contrast to S = 1.2 to 1.5 quoted in the older literature). Optimizing S thus has negligible effect on total vapour flow.

The vapour flow in the column is the major factor that determines reboiler and condenser duties (and hence size, design and costs) and also column diameter, a major factor in column cost.

2. Number of Stages

The minimum number of stages for a 2 component distillation column is given by Fenske's equation

$$(N+1) = \ln\,[(x_2\,x_{1B})/(x_{2B}\,x_1)]\;/\;r$$

Typically x_2 and x_{1B} hardly change with recovery specified, these are the mole fractions corresponding to the bulk of the components, ie we can put, approximately, $x_2 = x_{1B} = 1$, when

$$N+1 = -\ln\,(x_{2B}\,x_1)/r$$

Values of x_1 and x_{2B} for "sharp" separations typically lie in the range 0.005 to 0.02, ie (N+1) typically lies in the range

$$- \ln (0.02^2)/r < N+1 < - \ln (0.005^2)/r$$

$$\text{ie} \quad 7.8/r < N+1 < 10.6/r$$

or (N+1) varies by only $\pm$ 10 per cent about some value (9/r). This small range of platage is maintained when N_{min} is converted to actual plates by applying Gilliland's relationship and realistic plate efficiencies. With column diameter fixed by vapour flow rate and, typically, column and plate (or packing) costs constituting about 1/3 of the total uninstalled cost of a distillation column, we see that optimizing product recovery levels is only likely to affect column costs by about 3 per cent.

3. Total Impact on Costs

As a consequence of consideration of vapour rate (1) and number of stages (height, 2), it is clear that for "sharp" separations, decisions made in one column have negligible effect in others so that each can be optimized separately as required by Equation (1).

APPENDIX 2 - COMPUTATIONAL EXPERIENCE

The enumerative optimization program (GENPROC II) is written in FORTRAN and has been developed at South Bank Polytechnic. AI programs have been written in LISP, PROLOG and POP11 and run at BP and at South Bank Polytechnic. Both the AI and GENPROC results reported here were obtained at the Polytechnic using a Digital Equipment Corporation VAX computer. Both programs included extensive diagnostic output which slowed execution significantly. Run times for both methods were neverthless acceptable, approximately 1 minute per run.

SYMBOLS

a	Vector defining feed stream to a distillation column of distillation train.
A_i	$\ln(r_{ij})$
b	Vector defining stream from bottom of distillation column.
B	Function that relates column operating (ie split) decision "i" to resulting bottom product stream, "b", for any given "a".
c	Vector defining stream from top of distillation column.
C	Corresponds to 'B' , for top stream.
F	Total molar flow rate of stream. F(a) flow rate of 'a' etc.
G_1	Measures 1st heuristic defined by Equation 2.
H_1	Weighted measure of 1st heuristic, see Equation 3.
H_2	Measure of 2nd heuristic, see Equation 5.
H_2'	Alternative measure of 2nd heuristic.
H_3, H_4, H_5	Measures of heuristics 3, 4 and 5 defined by Equations 7, 8 and 9 respectively.
i	Decision number, or component number.
j	Subscript.
I	Function defining optimal decision in relation to feed stream "a".
L	Reflux rate.
n	Total number of components.
N	Number of stages of distillation.

$P_i(a)$	Discounted cost of separating feed stream "a" using split decision "i" in an optimally designed single-stage distillation.
$Q(a)$	Minimal cost of processing feed stream "a" to acceptable product stream.
r_{ij}	Relative volatility of component "i" with respect to component "j".
R	Reflux ratio. R_m minimum reflux ratio.
S	R/R_m
V	Vapour rate in column.
x_i	mole fraction of component "i" in top product.
z_i	mole fraction of component "i" in feed.

REFERENCES

(1) Stephanopoulos, G. and Townsend, D.W., 1986, "Synthesis in Process Development" ChERD, 64, (No. 3) 160 - 174.

(2) Douglas, J.M., 1985, AIChEJ, 31, (No 3), 353.

(3) Rossiter, A.P. and Douglas, J.M., 1986, "Design and Optimisation of Solids Processes", ChERD, 64, (No 3), Pt 1 175, Pt 2 184.

(4) Johns, W.R. and Romero, D., 1979, "The Automated Generation and Evaluation of Process Flowsheets", Comp. and Chem. Eng., 3, 251.

(5) Evans, G.H. and Johns, W.R., 1982, "Automated Generation of Process Flowsheets including recycle and Un-predefined Unit Operations", I.Chem.E. Symp. Series, No 74.

(6) Dhallu, N.S., 1987, "Synthesis of Processes with Potential for Heat Integration", PhD Thesis, Polytechnic of the South Bank (CNAA) to be presented.

(7) Rathore, R.S. Wormer, K.A.V., and Powers, G.J., 1974, "Synthesis of Distillation Systems with Energy Integration", AIChEJ, 20, (No 5), 940.

(8) Haven, D.L., "Optimum Sequencing of Distillation Columns in Multi-Component Fractionation", 1969, MS Thesis, University of California , Berkeley.

ACKNOWLEDGEMENTS

The authors thank the British Petroleum Company plc for providing facilities for part of this work and for permission to refer to those results. We acknowledge the financial support of SERC, under grant nos 83306035 and 85304017, for work which contributed to this paper.

ADDRESSES

Correspondence should be addressed to Dr. W. R. Johns, Chemical Engineering Department, South Bank Polytechnic, Borough Road, London SE1 0AA.

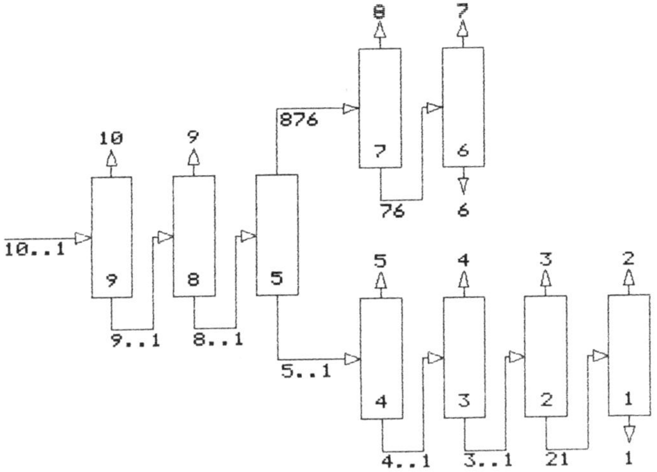

(a) Optimal Sequence for Mixture 1

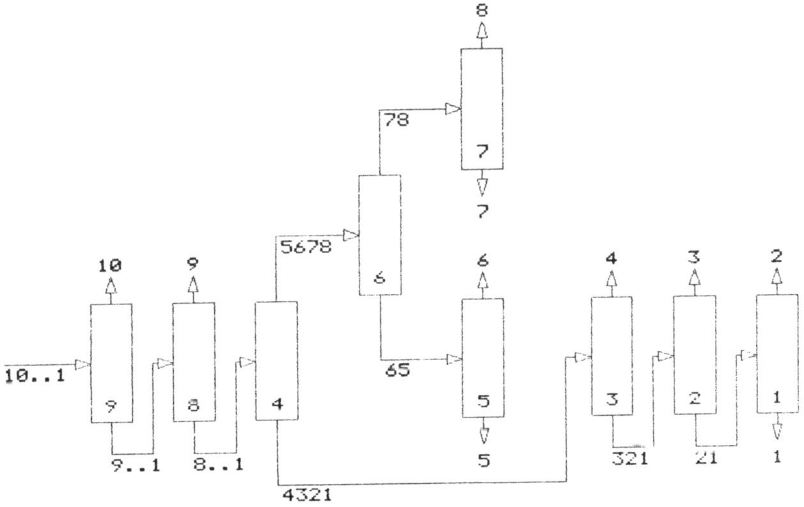

(b) AI Sequence for Mixture 1

Fig. 1 Typical Flowsheets, Illustrating Compact Numbering System

EXPERT SYSTEM FOR PRODUCTION SEPARATOR DESIGN

by D R Dick and M Zelouf
BP International Ltd, Engineering & Technical Support Department

ABSTRACT

BP has recently developed a large expert system, GASOIL, for the selection and design of hydrocarbon production separation equipment. This computer system draws together otherwise diverse knowledge and experience and gives advice in a manner resembling the human expert. The BP approach to expert systems; Knowledge Acquisition - getting the expert himself or herself to do the initial programming: System Construction and Delivery - using a modular and fully deterministic approach are described.

The system which is demonstrated is seen as an engineering tool to be used when needed which frees the expert from routine queries as well as giving him (or her) and other engineers greater insight into hydrocarbon production separation.

INTRODUCTION

This paper describes the development of an expert system for the selection, sizing, and design of hydrocarbon production separation equipment, i.e. the separation of gas/liquid hydrocarbon/water and solids at or near the wellhead. After a short introductory presentation, we will demonstrate the operation of this expert system.

WHAT DO WE MEAN BY AN "EXPERT SYSTEM"?

Our Expert System is an interactive computer program to assist, guide and inform the process engineer, who may not be skilled in production separator design, to select and design the most appropriate separator(s) for a particular duty. It embodies the knowledge and experience of experts in that field and can solve problems and give advice in a style resembling the human expert, in a consistent and auditable manner and can provide an explanation of how it reached its conclusion.

Expert System shells are available which generate a program from input rules or example data and provide the means to create a usable system in a relatively short time as we will demonstrate in our Expert System "GASOIL".

GASOIL is written in Fortran, contains 2800 rules and resides in 3 megabytes of memory. It incorporates not only the expert system for the separator selection and process design, but also calculation routines for vessel sizing, optimisation and weight calculation routines, text retrieval from BP Standards, and access to a computer aided graphics system to produce a two dimensional sketch of the vessel.

WHY AN EXPERT SYSTEM FOR SEPARATION EQUIPMENT DESIGN?

Separation vessels are often the largest and heaviest items of process equipment in a production system; particularly on an offshore production platform, minimising the size is of considerable value. A well designed expert system can assist in this task.

Production separator design is not well documented and what information is available is widely dispersed. The number of active practitioners in any given company, even major petroleum producing companies, is small.

The creation of an Expert System not only permits all the available expertise in this domain to be documented logically for subsequent re-use, but also encourages the engineer creating the system to organise his knowledge. This knowledge is then more widely and immediately available to his colleagues, even in his absence. The domain expert can then be freed from (a number of) routine enquiries and concentrate on the more demanding aspects of the job, such as development of improved designs, trouble shooting and extraordinary separation duties.

If the decision making process is traceable and consistent then any mistakes can be "learnt from", avoided in future and the system optimised.

A well designed Expert System is easier to use and easier to up-date than the alternative of bulky design manuals; it is also very simple to include feed-back of user comment and experience and amendments, as we will demonstrate in GASOIL.

HOW DID BP DEVELOP GASOIL?

By adopting a deterministic and modular approach BP developed GASOIL in about one man-year of effort.

A strictly logical and unique solution approach was used because engineering is generally deterministic in nature and because if a probabilistic approach were used a technical audit trail would be very difficult to establish, while the effect of subsequent changes would be very unpredictable.

A modular approach was used because this was required by the limitation of the initial shells used to develop the logic, to subdivide the work into reasonably manageable pieces and in order to facilitate subsequent editing and development into a larger system.

One of the keys to successful expert system development is to separate the knowledge acquisition phase from the system delivery. The initial knowledge acquisition was carried out by the domain expert himself, on a desk-top micro-computer, using the "Expert-Ease" shell from Intelligent Terminals Ltd., an example of which is shown on the screen. This is a very easy-to-use system; examples are input in a spreadsheet format and rules are induced from those examples giving an immediate feed-back to the programming domain expert at his own level. The initial stages of further developments in the expert system are still tried out using Expert-Ease.

The system was then transferred to a larger machine and GASOIL was developed from the pilot model by use of ITL's "EXTRAN-7" package, on a DEC VAX miniprocessor.

Calculation routines written in Fortran were interfaced with the expert system; some of these already existed in the technical calculations library but others were specially written for inclusion in GASOIL. The calculation procedures include independent routines for the gas/liquid separation and liquid sections and a preliminary reconciliation of conflicting requirements is automatically carried out; a more rigorous optimisation routine is available if required.

The enquiry text was added to the expert system and graded advice texts were also added as appropriate. A routine was added for text retrieval from the computer stored text of BP Codes of Practice and Standards, with a search target time of three seconds. An interface with the Design Office Graphics System - "DOGS" - by PAFEC was added to create a dimensioned sketch of the selected separator design.

GASOIL PROCESS CONTENT

The chemical engineering content of GASOIL was gleaned from many sources. Established theory and design methods were modified by practical experience within BP, augmented by maximum throughput field test runs over a range of operating conditions with a variety of different types and sizes of separator, plus experimental evidence from BP Research Centre, Sunbury. Technical literature and industrial/technical consortia publications and seminar presentations, again tempered where possible by experience, was also included. The technical base is not static and user comments are introduced progressively to refine the system and build-in the latest technology.

GASOIL considers the fluid properties (density, surface tension), the presence of solids such as sand and wax; it considers separation performance, space limitations turndown and pressure drop requirements and matches these to horizontal or vertical separator configurations, with or without a wide range of types of internals.

With the co-author of GASOIL, Mike Zelouf, at the keyboard we will now demonstrate the system.

A typical consultation might look as shown in the attachment.

CONCLUSION

We believe that Expert Systems such as GASOIL, by providing a logical, consistent and auditable decision making process, which is capable of pooling the knowledge of many engineers and of being "taught" to learn by mistakes, can assist in process design and optimisation of new facilities and in the revamp of existing facilities. We see GASOIL as a practical engineering tool to be used when needed. As well as freeing the expert from routine queries it gives him and other engineers greater insight into gas/oil separation, permits a more consistent approach to separator design and operational analysis, thereby optimising overall performance and effectiveness.

ACKNOWLEDGEMENT

Permission to publish this paper has been given by the British Petroleum Company PLC and is gratefully acknowledged.

DEMONSTRATION OF GASOIL FOR TYPICAL PRODUCTION SEPARATOR APPLICATION

All enquiries: D R Dick - Unit Operations CHE (x2434)
M Zelouf - Engineering Services (x3099)

This is an engineerng expert system for the selection and design of hydrocarbon production separation equipment (gas/oil separators) for offshore and onshore applications.

Assume current separator temperature and presssure for all questions. To get help at any stage, enter ? or ?advice for more expert advice on the curernt question.

Considering a production separator

Considering horizontal or vertical separator configuration

Question

There may be a space (weight/volume) limitation on the separator design, e.g. for an offshore installation.

Is there a significant space (weight/volume) limitation on the design?

Answer can be:

yes : there is a significant space limitation
no : there is not a significant space limitation
Choice ? y

Question

Are there any specific dimensional limitations on the separator design?

Answer can be:

height : The height requirement of the separator design should be limited
area : The plan area requirement of the separator design should be limited
none : There are no specific limitations on height or plan area
Choice ? ?advice

RULE PROCESS

This module is seeking to advise on the general form the separation equipment should take i.e. horizontal or vertical. If a compact design is being sought, details are required on the restriction i.e. general, in height or in plan area. The allocation is only preliminary and may be modified at a later stage when more information has been given.

The numeric information (flow rates, densities etc.) is used to calculate volumetric flow rates, gas/oil ratios for general type selection and is also stored for later use in the sizing calculation routines.

High volumetric flow rates i.e. >xx m3/hr suggest the use of horizontal vessels whereas low flow rates i.e. >xx m3/hr suggest the use of vertical separators. Very high and high gas/oil volumetric ratios i.e. >xx strongly favour the use of vertical vessels. Low gas/oil ratios i.e. <xx suggest horizontal separators. The separation quality requirements are the next input. High gas quality requirements slightly favour the vertical configuration while high liquid quality more strongly favours horizontal vessels.

The information on the separation quality requirements, together with the calculated gas/oil ratio and the total flow provide the basis for the preliminary horizontal/vertical type decision, and are stored for future use.

Question

Are there any specific dimensional limitations on the separator design?

Answer can be:

height : The height requirement of the separator design should be limited
area : The plan area requirement of the separator design should be limited
none : There are no specific limitations on height or plan area
Choice ? a

Question

What is the gas flow in kg/hr ? 2500
What is the gas density in kg/cubic meter? 14
What is the liquid (oil-phase) flow in kg/hr? 150000
What is the liquid (oil-phase) density in kg/cubic meter? 800

Question

Note, a liquid (aqueous-phase) flow of less than 0.1 kg/hr implies that no liquid/liquid separation is required - i.e. 2-phase separation only. What is the liquid (aqueous-phase) flow in kg/hr? 900

The total volumetric flow is between 0.1 and 3 m3/s

The gas/liquid volume ratio is between 5 and 25

Question

Liquid/liquid separation quality:

good	i.e. less than 1% water in 'oil' phase and less than 1500ppm 'oil' in water
rough	i.e. more than 1% water in 'oil' phase and more than 1500ppm 'oil' in water

What quality of liquid/liquid separation (if any) is required?

Answer can be:

good : Good quality liquid/liquid separation is required
rough : Rough quality liquid/liquid separation is required
Choice ? r

Rough quality liquid/liquid separation is required

Question

What quality of gas/liquid separation is required?

Answer can be:

best : Very high gas quality separation, i.e. < 1.0 mg liquid/kg gas
high : High quality gas/liquid separation, i.e. 1.0 mg liquid/kg gas
standard : Standard quality gas/liquid separation i.e. up to 10 mg liquid/kg gas
rough : Rough quality gas/liquid separation i.e. > 10 mg liquid/kg gas
Choice ? ?why

Since

The total volumetric flow is between 0.1 and 3 m3/s and the gas/liquid volume ratio is medium and Rough quality liquid/liquid separation is required the decision cannot be any of:

More than one vessel required for this separation
The strong preference is for a vertical separator
The performance is for a vertical separator

What quality of gas/liquid separation is required?

will help to:
establish the decision:

Liquid phase feed flow too high for filter/separator
or
The preference is for a horizontal serparator
or
The strong preference is for a horizontal separator

Question

What quality of gas/liquid separation is required?

Answer can be:

best : Very high gas quality separation, i.e. < 1.0 mg liquid/kg gas
high : High quality gas/liquid separation i.e. 1.0 mg liquid/kg gas
standard : Standard quality gas/liquid separation i.e. up to 10 mg liquid/kg gas
rough : Rough quality gas/liquid separation i.e. > 10 mg liquid/kg gas
Choice ? s

The preference is for a horizontal separator

Question

The process requirements suggest that a horizontal separator would be suitable, but you have said that there is limitation on the plan area

Would you like to consider a vertical or a horizontal separator?

Answer can be:

verti : consider a vertical separator, due to the area limitation
horiz : consider a horizontal separator, despite the area limitation
Choice ? ?advice

RULE: TYPE

This module checks that the preliminary vessel configuration obtained from the process information and any dimensional limitations are compatible. If the suggested form of separator "contradicts" the dimensional limitation the user is given a choice of which type of vessel to consider further. However, we suggest that for the first run through, for a given problem, the process advice is followed unless the dimensional restriction is very severe. The dimensional limitation should only be allowed to rule if the calculated dimensions at the end of the consultation are unacceptable.

If the separation requirements and/or the fluid flows given lead to a duty which is not practicable, the consultation ends with an appropriate warning.

To view the full rule type <?rule>

Question

The process requirements suggest that a horizontal separator would be suitable, but you have said that there is limitation on the plan area

Would you like to consider a vertical or a horizontal separator?

Answer can be:

verti : consider a vertical separator, due to the area limitation
horiz : consider a horizontal separator, despite the area limitation
Choice ? ? h

Examining properties of feed to investigate a horizontal design

Considering horizontal separators

Considering physical properties of feed

Question

Note, cutoff points for decisions based on the surface tension are:

low < 24 < medium < 28 < high

What is the surface tension of the ´oil´ phase in dynes/cm ? 25

Question

Note, cutoff points for decisions based on the viscosity are:

very low < 3.0 < low < 5.0 < medium < 10.0 < high < 100 < very high

What is the viscosity of the ´oil´ phase in centipoise (cP) ? 4

Question

Are waxy solids present in the feed stream?

Answer can be:

yes : Appreciable waxy solids are present in the feed stream
slight : Very little, finely divided wax is present in the feed stream
no : waxy solids are not present in the feed stream
Choice ? s

Formation of foam, probably stable, is possible

Question

Are inorganic solids (sand/scale) likely to be present in the feed stream?

Answer can be:

fine : fine inorganic solids (eg scale) are likely to be present
coarse : coarse inorganic solids (eg sand) are likely to be present
both : fine and coarse inorganic solids are likely to be present

none : no inorganic solids are likely to be present in the feed stream
Choice ? n

Consider more information in another module

Attempting to select a proprietary horizontal separator (HGG)

Defoaming and dewaxing internals will probably be required.

Due to the presence of wax all internals should be tolerant to wax and wire mesh and narrow slotted features should be avoided. The installation of heating or stream-out facilities to remove accumulated wax should be considered.

GH type gas/liquid separation, some liquid/liquid separation.

A separator with the following features is recommended.

A nominal liquid residence time of 210 seconds can be assumed.

Bulk separation (defoaming/degassing) and demisting type horizontal.
Bulk separation (defoaming/degassing) and demisting type horizontal separator.

Proprietary internals to enhance gas/liquid separation, including a demisting device on the gas outlet.

Due to the presence of wax all internals should be tolerant to wax.

The use of foamicide chemicals should be investigated as their use is probably essential. Defoaming internals and exit gas demister are required. Defoaming internals recommendations:

e.g. High surface area internals	BS & B	HSI
	Natco	Stockton Plates
	Robert Jenkins	HSI
	Howden	Labyrinth Pack

Exit gas demister e.g. vane, multicyclone type

Easy liquid/liquid separation. Consider the use of demulsifying chemicals and/or L/L internals, but only if space is at a premium.

Nominal residence time 210 secs.

<u>Question</u>

Is this separator recommendation acceptable?

Answer can be:

yes : The separator recommended above is acceptable
no : The separator recommended above is not acceptable
Choice ? y

The recommendation given is acceptable - vessel will now be sized

Calculation of vessel sizing will be based on the following parameters:-

Gas flow	=	25000.0 kg/hr
Gas density	=	14.0 kg/cubic meter
Liquid (oil-phase) flow	=	150000.0 kg/hr
Liquid (oil-phase) density	=	800.0 kg/cubic meter
Liquid (aqueous-phase) flow	=	900.0 kg/hr
Liquid (aqueous-phase) density	=	1000.0 kg/cubic meter
Surface-tension	=	25.0 dynes/cm
Viscosity	=	4.0 cP

No solids are present
Very little, finely divided wax is present

Formation of foam, probably stable, is possible due to the presence of wax all internals should be tolerant to wax and wire mesh and narrow slotted feature should be avoided

Design nominal residence time	=	210 seconds
Nominal residence time (to NLL)	=	240.00 s

Are these residence times satisfactory (y/n) ? y

Calculating...please press RETURN when ready for results
Enter separation constant (default - 3.5) :
Separator sizing results (to 2 d.p.)

Vessel diameter	=	2.20 m
Vessel length	=	7.16 m
Vessel volume	=	27.28 m3
Assumed effective gas length	=	5.73 m
Design gas height (above HLL)	=	1.10 m
Design liquid level (to NLL)	=	0.99 m
Design maximum gas velocity	=	8.79 m/s
Calculated gas velocity (to NLL)	=	0.26 m/s
Design nominal residence time	=	210.00 s
Calculated nominal residence time (to NLL)	=	275.56 s

<u>Vessel SHADOW/SKID sizes</u>

	Shadow size	Skid size
Length :	12.4 m	13.4 m
Width :	5.2 m	6.2 m
Height :	4.4 m	4.9 m

Is further optimisation required (y/n) ? y

OPTIMISING SEPARATOR GEOMETRY . . .

Optimised sizing results after 7 iterations:-

Vessel diameter	=	1.90 m
Vessel length	=	7.73 m
Vessel volume	=	21.98 m3

Assumed effective gas length	=	6.19 m
Design gas height (above HLL)	=	0.84 m
Design liquid level (to NLL)	=	0.95 m
Design maximum gas velocity	=	10.87 m/s
Calculated gas velocity (to NLL)	=	0.41 m/s
Design nominal residence time	=	210.00 s
Calculated nominal residence time (to NLL)	=	266.81 s

Vessel SHADOW/SKID sizes

	Shadow size	Skid size
Length :	12.6 m	13.6 m
Width :	4.9 m	5.9 m
Height :	3.8 m	4.3 m

Do these figures satisfy the PLAN AREA limitation specified earlier (y/n) ? y

VESSEL WEIGHT CALCULATION

Enter the design pressure (bar) or "help" : h

searching CP8 for design pressure
THE FOLLOWING OCCURENCES HAVE BEEN FOUND

2.1.4	2 OCCURRENCES
2.2.1	1 OCCURRENCES
4.1.1	2 OCCURRENCES
4.1.3	1 OCCURRENCES
5.2.5	1 OCCURRENCES

Finished 7 OCCURRENCES FOUND

Please enter a return to read paragraphs

2.1.4 A vessel may have a design pressure or calculating stresses too low to classify it within the scope of the preferred design code. To ensure a satisfactory standard of design and construction, the design pressure shall be raised to ensure the vessel falls within the scope of the code.

2.2.1 When vessels are to operate at varying conditions of temperature and pressure, the different pressure with their corresponding temperatures shall be examined to determine the controlling case for design purposes. In some interests it may be more economical to specify several design pressures and corresponding temperatures rather than the maximum pressure and maximum temperatuers.

4.1.1 In determining a vessel design pressure a margin shall be allowed above the operating pressure to accommodate normal presssure fluctuations and to allow for any pressure alarms and pressure-limiting instrumentation required. As a general guide the following shall apply:-

(a) For operating pressures up to and including 70 bar (1000 psig) this margin shall not be less than 10% of operating pressure with a minimum of 1 bar (14.5 psig).

(b) For operating pressures above 70 bar (1000 psig) the margin shall be approved by BP.

The major pressure vessel design codes require that the set pressure of any pressure relief valve shall not be higher than the design pressure.

4.1.3 When the relief valves of low pressure vessels vent into a common closed system such as a flare or blowdown drum into which relief valves from higher pressure vessels can also vent, the contractor shall evaluate whether raising the design pressure of the low pressure vessels will result in savings on the overall pressure relieving system that will more than offset any additional cost of the vessels.

5.2.5 Testing/Inspection

(a) Vessel butt welds to be covered by jackets shall first be 100% radiographed.

(b) Dimple plug/fillet welds shall be 10% dye penetrant checked after welding and if significant failures are found a 100% check shall be instigated.

(c) If a burst test is required to justify the design, the sample plate should contain approximately 20 dimples with the dimples positioned to within 1.5 mm (1/16 in) of the true pitch. Using the method of ASME VIII UG 101 the maximum allowable design pressure is set to:

$$\frac{\text{Burst Pressure}}{5} \times \text{joint efficiency} \times \text{UTS factor} \times \text{Temperature factor}$$

```
Press RETURN when page has been read

Enter the design pressure (bar) or "help" : 35
Enter the design temperature (deg C) or "help" : 455
Design temperature given is too high - please re-enter
Enter the design temperature (deg C) or "help" : 156

Actual (dry) weight     :  12910.7 kg
Hydraulic weight        :  37594.0 kg
Operating weight        :  25177.2 kg

Calculating vane demiser pad size...

Effective vane pad area              =       0.14 m2

A graphic representation of this vessel is availale.
Do you wish to see it (y/n) ? y
```

TYPICAL

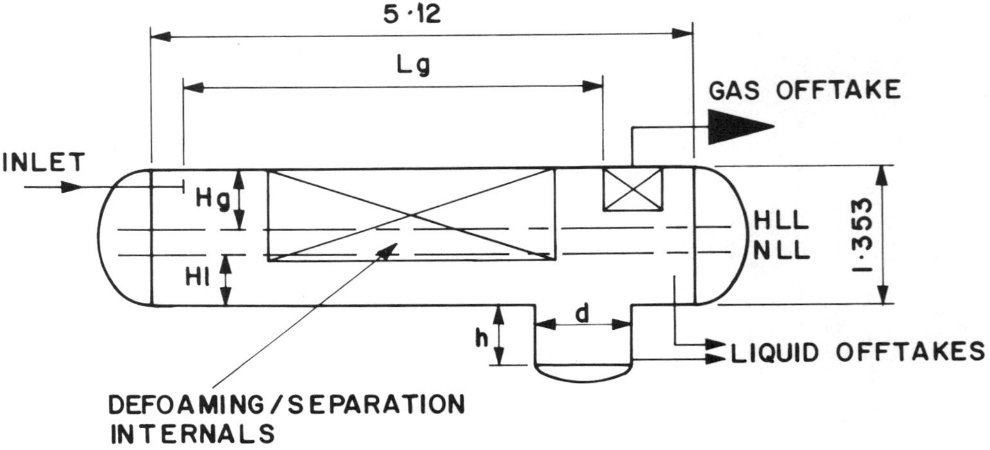

VESSEL TYPE		EXTVES2
Lg	(Effective gas length)	= 4·09
Hl	(Liquid level)	= 0·81
Hg	(Gas level)	= 0·35
h	(Boot height)	= 0·68
d	(Boot diameter)	= 0·68

OPTIMIZATION OF ACID GAS EMISSION CONTROL STRATEGIES IN THE UK POWER GENERATION INDUSTRY

W.S. Kyte[+] and J.R.P. Cooper*

In this paper some of the broader aspects of emission control are discussed and it is shown that there are various ways of achieving emission reductions. In the field of sulphur dioxide emission control the merits and optimization of coal cleaning, coal re-routing, flue gas desulphurization and advanced combustion technologies are discussed. It is shown that in all these areas a close integration and optimization of the total process from reagent supply to product disposal is required. Some of the problems of integrating these technologies into the existing chemical infrastructure are also discussed.

1. INTRODUCTION

One of the growth areas in chemical engineering is in the use of emission control technologies in the Electric Power Generation Industry. Because of the large scale of the electric power industry and the nature of the processes, these technologies have considerable economic implications and can have important implications for the chemical industry as well as possible environmental effects from the process wastes.

There is considerable scope not only for optimization of the individual process units but also for optimization with the power plant. The scale and breadth of the power generation industry offers further opportunity for optimization of emission control by the use of a combination of technologies rather than application of a single one.

In this paper some of the broader aspects of emission control will be discussed. The amount of optimisization possible will depend on whether emission reductions are required from existing power plants or only from new

*Central Electricity Generating Board
Generation Development and Construction Division
Barnwood, Glos

[+]Central Electricity Generating Board
Technology Planning and Research Division
Central Electricity Research Laboratories
Kelvin Avenue, Leatherhead, Surrey

stations. A further complication can arise from legislative requirements which may be considerably different for new and existing plant. It is because of all these interactions that emission reduction strategies are very different from country to country.

In the United Kingdom the Central Electricity Generating Board (CEGB) has recently announced that it will retrofit a sufficient number of generating units to maintain, taking one year with another, the decline in sulphur dioxide emissions which has occurred over the last decade. In addition all new power plants are likely to be required to be fitted with sulphur dioxide control technologies to the requirements of HM Industrial Air Pollution Inspectorate. The CEGB is also undertaking large scale experiments to establish optimum methods for reducing nitrogen oxides emissions for new and existing plant.

This paper will focus on emission control technologies for coal fired power stations.

2. GENERAL METHODS OF EMISSION REDUCTION

It is not generally appreciated that there are numerous methods of reducing emissions. Some of these methods are shown in Table 1 from which it can be seen that emission reductions can be achieved without the need for the treatment of flue gases. The considerable reductions that have occurred in UK sulphur dioxide emissions have mainly resulted from changes in fuel utilization and increased efficiency of combustion. Increasing use of nuclear power, renewable energy sources and energy conservation will have important implications for the future.

Table 1: Some Options for Reducing Acid Gas Emissions

- Energy conservation
- Increased combustion efficiency
- Alternative energy sources
- Alternative fuel sources
- Nuclear power
- Fuel cleaning
- Combustion modifications
- Flue gas treatment

3. REDUCTION OF NITROGEN OXIDES EMISSIONS

Nitrogen oxides (NO_x) are formed during the combustion of coal in the boiler. Some are formed in the furnace by gas phase reactions between nitrogen and oxygen but most come from oxidation, in the early stages of combustion, of nitrogen compounds contained in the fuel. Since this fuel nitrogen can be converted to either molecular nitrogen or to nitrogen oxides it is possible to reduce the formation of nitrogen oxides by modification of the combustion process. It is also possible to reduce nitrogen oxides after combustion by flue gas treatment processes.

3.1 Combustion Modifications

Nitrogen oxides formation can be reduced by ensuring that the early stages of combustion take place under fuel rich conditions. This ensures that some of the nitrogen compounds released with the volatiles in the coal are converted to molecular nitrogen instead of being oxidised to nitrogen oxides.

The combustion conditions can be changed by modifications to the furnace or to the burners alone. The full benefits of combustion modifications are obtained only when both furnace and burners are changed. Thus the scope for NO_x reduction is greater for new boilers than for retrofit applications. It is important to note that optimization for lowest NO_x formation may promote adverse factors such as enhanced boiler tube corrosion, loss of combustion efficiency and increased carbon content in the ash, with significant operational and economic implications. These must be considered in the overall optimization of boiler design and operation.

In order to quantify these effects and to optimize the degree of NO_x reduction the CEGB is undertaking a series of full scale boilers trials each of which will last for a period of three years.

3.2 Flue Gas Treatment

If high NO_x removal efficiencies are required then the nitrogen oxides will have to be removed from the flue gases after combustion as well as by combustion modifications. For modest degrees of removal, NO_x can be reduced by injection of ammonia or other reducing compounds into the furnace at the superheater level. For very high NO_x reductions selective catalytic reduction (SCR) with ammonia over a catalyst at 350-400°C must be used. This involves extensive modifications to the boiler and ductwork. This, together with the limited life of the catalysts, means that SCR is extremely expensive when compared with combustion modifications. For these reasons combustion modifications are seen as the most cost-effective method of reducing power station NO_x emissions.

4. REDUCTION OF SULPHUR DIOXIDE EMISSIONS

Sulphur dioxide emissions can be reduced by the removal of sulphur from the coal before combustion, by the removal of sulphur dioxide during the combustion process or by the removal of sulphur dioxide from the flue gases after combustion. The technologies for reducing sulphur dioxide have been described in detail elsewhere (1, 2, 3). They are generally complex physical and chemical processes and offer considerable scope for detailed optimization. However only the more general aspects of optimization will be discussed here.

4.1 Coal Cleaning

Sulphur occurs in coal in both organic and inorganic forms. The former is chemically bound to the organic coal matrix and is not accessible by physical separation methods. Inorganic sulphur is mainly present in the form of iron pyrite FeS_2, which is present in discrete particles throughout the coal mass. The size of these particles varies from sub-micron up to large lumps.

In principle it is possible to separate the pyrite from coal by making use of the differences in physical properties between the two materials (Table 2). In practice, however, the maximum amount of separation possible is limited by the degree of pyrite liberation which is determined by the crystal size distribution and the size to which the coal is ground.

Table 2: Coal Properties Pertinent to Coal Cleaning

Property	Units	Coal	Mineral	Pyrite	Extraction Option
Density	g cm^{-3}	1.15-1.5	2.4-3.9	4.8-5.0	Float/sink
Resistivity	ohm-cm	10^{8}-10^{13}	10^{4}-10^{5}	10^{2}-10^{4}	Electrostatic
Magnetic Susceptibility	cgs × 10^{6}	-0.5 to -0.6	-20 to +400	-40 to +65	HGMS
Wettability (contact angle)	(deg)	49-68	11-15	60-75	Froth flotation

A detailed investigation (4) has been carried out to estimate the degree of extra sulphur removal that would be obtained if current coal washing techniques were to be extended to the third of CEGB coal that is not treated at present. The results indicated that a maximum of 5% of the sulphur delivered to the CEGB could be removed with only about half of that being economic compared with flue gas desulphurization.

The coal in a modern power station is ground to a fine powder before combustion and hence a large proportion of the pyrite is liberated. The CEGB is investigating methods of exploiting this. One of the most promising uses the small difference in magnetic properties between coal and pyrite (5). A laboratory scale pilot rig is being operated to evaluate the potential of High Gradient Magnetic Separation of pyrite from coal. Preliminary laboratory results are encouraging and methods of integrating HGMS into the power station coal milling circuit are also being investigated.

Studies are being undertaken on the segregation of coal into high and low sulphur streams either by direct selection or by coal treatment at the pit or power station. The lower sulphur coals could then be burnt at an unmodified power station whilst the higher sulphur coals could be sent to a power station equipped with some form of flue gas sulphur dioxide control technology.

4.2 Sulphur Dioxide Removal During Combustion

In the dry injection method alkali, usually slaked lime or limestone, is injected into the hot boiler gases where it can react with the sulphur oxides and is then removed after reaction.

Under UK conditions it has been shown that such a process would be uncompetitive for UK power stations in comparison with wet flue gas desulphurization processes (6).

Two further processes, fluidised bed combustion and coal gasification, where the sulphur can be removed or converted during the combustion process, are being studied with a view to plant that could be built at the turn of the century and beyond. These technologies offer the possibility of both higher efficiency of coal conversion to electricity and much reduced emission of acidic gases (7,8).

4.3 Flue Gas Desulphurization (FGD)

There are well over one hundred different flue gas desulphurization processes available in various degrees of development, ranging from the laboratory scale to full scale commercial units. A detailed review of the more important processes available and papers discussing those processes applicable to UK power stations have been published (3, 9, 10). These processes are the limestone/gypsum, regenerative, spray-dry and sea-water processes. Detailed consideration of each of these technologies for a chosen situation involves optimization of all the units and sub-processes involved. In addition, the interaction between capital cost, plant lifetime and energy requirements must be carefully optimized.

Further factors emerge when considering an optimal sulphur dioxide emission reduction programme for a series of power stations. Several of the more important of these will be discussed here.

The CEGB has recently announced that it will retrofit three 2000 MW power stations with FGD (11, 12). The general geographical locations of these power stations have been chosen so that higher sulphur coals can be routed to these stations in order to boost the sulphur dioxide removal capacity of these stations. This rerouting must be carefully optimized subject to coal supply, coal quality and to transport and economic restraints. Consideration must also be given to the location of any new power stations which may be constructed and which would have to be integrated into the coal supply and FGD product disposal matrices.

A large modern 2000 MW power station burns some 4½-5 million tonnes of coal a year and thus the amounts of reagents required and products produced by FGD fitted to that station will be very large (Table 3). Supply of the raw materials and disposal of the products, by-products and wastes generated must be accomplished in an environmentally acceptable manner. For this reason it is the intention of the CEGB to market as much of the products of flue gas desulphurization as possible. However for marketable products, such as gypsum and sulphuric acid, there are existing markets for only limited amounts of FGD products and by-products (Table 4). For sulphuric acid it would be difficult to put the equivalent of one power station's FGD plant output into a single UK chemical works. Many sulphur burning sulphuric acid plants can only be turned down to approximately half capacity without upsetting the integrated operation of the whole chemical works. Thus complete replacement of the work's sulphuric acid demand with FGD product cannot be undertaken without considerable plant and operational modifications.

Two very important consequences arise from consideration of the imports and exports of these FGD processes. The first is that for any substantial FGD programme it is likely that more than one type of FGD process will have to be employed. The second consequence is that the location of a particular power station may strongly influence the process to be used due to the local marketability of the product and the acceptability of any wastes.

Table 3: Approximate Annual FGD Imports/Exports (tonnes) of a 2000 MW Power Station

(76% load factor, no S retention in ash, coal 2% S, 0.4% Cl)

Process		Wellman-Lord	Limestone/Gypsum	Lime/ Spray-dry
SO_2 Removal Efficiency		90%	90%	90%
IMPORTS				
Limestone	Solid	30,000	400,000	-
Lime	Solid	3,000	3,000	375,000
Caustic Soda	Liquor	40,000	-	-
Natural Gas	Gas	3×10^7 Nm^3	-	-
Sulphuric Acid	Liquid	-	30,000	-
EXPORTS				
Sulphur or Sulphuric Acid	Solid Liquid	75,000 280,000	- -	- -
Gypsum	Solid	-	560,000	-
Mixed Products (including fly-ash)	Solid	-	-	1.5×10^6
BY-PRODUCTS				
Calcium Chloride	Solid	25,000	25,000	-
Sodium Sulphate	Solid	30,000	-	-
WASTE WATER				
Treated Water		2×10^6	2×10^6	-
Waste-water Sludge	Solid	15,000	40,000	-

Table 4: FGD Products as a Percentage of Existing UK Markets

(2000 MW, 2% S, 0.4% Cl coal)

H_2SO_4	10-12%
S	10-12%
Gypsum	15-20%

The optimization of an overall FGD programme must be based on the requirements of the total CEGB system. The optimiziation of a specific process on a given site must not only be seen within the local supply and disposal market pattern but also within an overall cost optimization within the CEGB system.

5. CONCLUSIONS

A brief survey has been given of some of the many factors that will require consideration for the optimization of acid gas emission control strategies in the UK power generation industry.

6. ACKNOWLEDGEMENT

This paper is published by permission of the Central Electricity Generating Board.

7. REFERENCES

1. Gray, J.S., 1985, 'Static Combustion of Fossil Fuel - Present Technology', Conference 'Control of Acid Emissions in the UK', November 5, 1985, IChemE, London

2. Greenwood, C.W., Taylor, J.K., Woodruff, D.B., 1985, 'Developments in the Reduction of Acid Emissions from Coal Fuelled Power Generators', Conference 'Control of Acid Emissions in the UK', November 5, 1985, IChemE, London

3. Kyte, W.S., 1982, 'Transactions Review Paper - Some Chemical and Chemical Engineering Aspects of Flue Gas Desulphurization', Trans, I. Chem. E., 59, 219, 1982

4. Jenkinson, D.E., 1985, 'Coal Preparation into the 1990's', Colliery Guardian, 301, July 1985

5. Male, S.E., 1984, 'Magnetic Susceptibility and Separation of Inorganic Materials form UK Coals', Jl. Phys. D., App. Phys., 17, 155, 1984

6. Burdett, N.A., Cooper, J.R.P., Dearnley, S., Kyte, W.S., Tunnicliffe, M.F., 1985, 'The Application of Direct Limestone Injection to UK Power Stations, Jl. Inst. Energy, LVIII, 435, 64, 1985

7. Davidson, B.J., Holmes, J., Merrick, D., 1984, 'A Comparison of Combined Power Generation Cycles Using Coal: An Overview', Conference 'Energy Options', Inst. of Elec. Engrs., April 1984

8. Lawton, J., Dainton, A.D., Dawes, S.G., 'The CEGB/NCB Programme on PFBC', 8th Int. Conf. on Fluidized Bed Combustion, EPRI, Houston, March 1985

9. Kyte, W.S., 1986, 'Some Aspects of Possible Emission Control Technologies for Coal Fired Power Stations in the United Kingdom', Mire and Quary, 15, 10, 26, 1986

10. Kyte, W.S., 1985, 'Corrosion in Flue Gas Desulphurization Plant' in 'Dewpoint Corrosion', Ed. D. Holmes, Ellis Harwood, 1985

11. CEGB Press Announcement, 11 September 1986

12. Department of Environment News Release, 11 September 1986

OPTIMAL COST DESIGN OF A CYCLOHEXANOL PLANT

R. Scott *

Designs for processes with recycles may offer choices between low capital/high operating cost and high capital/low operating cost. A procedure for optimising these is described using the design of a cyclohexanol plant as an example. Costs are estimated for a base case design and then scaled using standard procedures to give costs for other cases within likely upper and lower limits. A more expensive design is justified when operating savings recover the extra capital within an acceptable pay-back period. At optimum design conditions, an inclined pay-back line is tangential to the curve relating capital and operating costs.

INTRODUCTION

Different process designs may achieve the same technical objective with equal satisfaction yet have different cost structures. Low capital may be associated with high operating cost and vice versa, presenting the designer with problems of reconciliation. An evaluation procedure is required which avoids replicate designs for the range of possible conditions and enables a justifiable decision to be readily made. This paper describes such a procedure which is generaly applicable and is exemplified by the design of a cyclohexanol plant.

OXIDISING CYCLOHEXANE

Air oxidises liquid cyclohexane in a complex reaction sequence discussed by Berezin (1). Present discussion uses the highly simplified scheme shown in Figure 1. The desired product is cyclohexanol, RHOH, although cyclohexanone, RO, is almost equally acceptable since mixtures are industrial commodities required for nylon manufacture. Unfortunately, the reactor outflow is a mixture containing acids resulting from rupture of the carbon ring. The acids react with cyclohexanol to form esters but these can be hydrolysed with caustic solution in an additional processing stage, releasing cyclohexanol and giving a separable solution of sodium salts for waste disposal.

* Department of Chemical Engineering, King's Buildings, University of Edinburgh.

PROCESS DESIGN

Reactor design must recognise some inherent constraints, notably :

1 A heat of reaction of -260 kJ/mol corresponding to a theoretical adiabatic temperature rise of about 700°C for complete conversion. Such intense heat generation cannot conceivably be removed by surface heat transfer while keeping the reactant liquid.

2 A heat of evaporation of 30 kJ/mol, sufficient to vapourise 6 mol of cyclohexane for every mol oxidised. A reactor with evaporative cooling is thus possible, operating at partial conversion with condensed cyclohexane being returned to the reactor. Partial conversion also limits formation of degradation products.

3 Cyclohexane vapour and air form explosive mixtures with a heat of combustion exceeding -1000 kJ/mol. Nitrogen emerging from the reactor should therefore be substantially free from unreacted oxygen. This also avoids vapour phase reactions.

A suitable reactor therefore comprises a well stirred pot under sufficient pressure to maintain cyclohexane in liquid phase at reacting temperature of 130-150°C ie not less than 5.5 bar. Total pressure is greater because of the partial pressure of nitrogen. Agitation is facilitated by sparging in air as finely divided bubbles which also increase the surface area for mass transfer. Some cyclohexanone is added as reaction initiator together with up to 0.05% cobalt naphthenate catalyst. (Loder. (2)).

Most of the cyclohexane is recovered from the nitrogen off-gas in a condensing reflux return system which may also include phase separation means to minimise water return to the reactor since this inhibits desired reactions (see Fig 3 Schematic flowsheet). Since design temperature for such a condensing system may be as high as 35°C, when cyclohexane vapour pressure is 0.2 bar, additional means of stripping cyclohexane from the off-gas are necessary. An absorption/desorption system using a non-volatile oil is suitable.

If hydrolysis is carried out in the presence of unreacted cyclohexane present in the reactor outflow, organic and aqueous layers separate readily and unwanted dioic salts may be passed to disposal as aqueous solution. Incineration is assumed since the organic content contributes some calorific value. The organic layer is purified by water washing and the separated layer used to dilute the caustic bought in as c.50% solution. Cyclohexanol partitions preferentially into the organic layer which is distilled to obtain cyclohexanol as bottom product. The overhead product of cyclohexane is returned to the reactor as the composition is not unduly critical.

CONVERSION AND SELECTIVITY

The process objective is to maximise cyclohexanol formation in preference to more highly oxidised products. Berezin et al (1) report "if oxidation exceeds 10-15% secondary products become more intense and reduce the yield of valuable products". They conclude that "the most effective method of increasing cyclohexanol and cyclohexanone yield known at present is to keep the oxidation conversion low".

Present discussion prefers the term "selectivity" to "yield". Conversions up to 12% are reported by Loder (2) with selectivities of 70 to 77% at conversions of 10.4 to 10.7% Alagy et al (3) report "selectivity less than molar 73% for conversion rates of about 5%". Schueler and Wolff (4) relate conversion and selectivity for various process modifications. In all cases, selectivity falls as conversion rises. Assuming sufficient retention time to convert hydroperoxide to cyclohexanol completely, 82% selectivity at 3.8% conversion falls to 60% at 11.4% Figure 2 shows how product concentration in the reactor outflow is assumed to vary with conversion in subsequent design calculations. Stoichiometric selectivity is assumed at zero conversion.

These considerations suggest that a "low conversion, high selectivity" plant should have high costs for recovering and recycling unreacted cyclohexane but relatively low costs per te of product for cyclohexane consumed. A "high conversion, low selectivity" design should show a contrary relationship.

BASE CASE DESIGN

Mass flows are most accurately known at 7.6% conversion and are used to design a "base case" plant following heuristic procedures from Happel and Jordan (5). Costs are from Sinnott (6), Hall et al (7) and elsewhere. Table 1 gives a summary specification for key plant items.

TABLE 1 - Summary Specification for Plant Items to produce 6.25 te/hr Product at 7.6% Conversion. (Base Case).

Item No.	Description/Duty/Special Features
101-A,B,C*	Cyclohexane storages
102*	Catalyst handling
201	Air filter
202	Air compressor + drive. 21 te/hr, 10 bar, 2-stage, 1.3 MW.
301-A,B*	Feed pumps
303-307	Reactors. 3.5m dia x 6.5m. 10 bar. SS. Agitated.
401-405	Off-gas condensers. SS/SS shell + tube 80m^2 ea.
407-411*	Reflux separators
501	Absorber. 1m dia x 10m. 1" packing. SS.
502,503*	Oil circulation tank and pump.
504-*	Stripper and ancillaries.
601,603*	In-line mixers.
602,604-5	Liquid/liquid separators. 2m dia x 7m. SS.
701	Distillation column. 2.7m dia x 12m. SS. 10 sieve trays.
702	Condenser/cooler. SS/SS shell + tube. 400 m^2.
703	Reboiler. SS. U-tube. 500m^2.
705-A,B*	Reflux pumps.
706*	Reflux drum.
707	Cyclohexane recycle tank. Fixed cone roof. 200m^3.
708-A,B,C,D*	Product tanks.
800	Existing incinerator charged at operating cost + return.

* Cost of these items is fixed irrespective of conversion.

Table 2 summarises the delivered costs of those items varying with conversion, also the design parameter which determines their capacity and exponent determining their cost in the standard expression :

$$Cost_2 = Cost_1 \left(\frac{Capacity_2}{Capacity_1}\right)^n$$

Oxidation section costs rise with conversion as more air is consumed in the highly oxidised by-products. Cyclohexane recovery costs fall. Total item costs rise significantly below 7% conversion. Figure 4 shows these variations.

TABLE 2 - Plant Item Costs and their Variation.

Item No.	Description	Cost £000 (mid-1979)	Capacity proportional to	Cost Exponent n
201	Air filter	24	Air	0.67
202	Compressor + drive	276	Air	0.8
301	Feed pumps	20	Feed	0.67
303-7	Reactors	350	Air	0.9
401-5	Condensers	150	Heat load	0.67
501	Absorber	35	Air	0.67
602-5	Separators	65	Total flow	0.67
701	Distillation column + plates	125	Recycle	0.67
702	Condenser	80	Recycle	0.67
703	Reboiler	100	Recycle	0.67
707	Recycle tank	120	Recycle	1.0
Total Plant Item Cost		1345		

These costs are converted to fixed capital costs by multiplying by :

(a) a cost inflator for chemical plant construction, currently estimated at 119 with mid-1979 being 86.5.

(b) overall Lang factors recorded by Bridgwater (8) totalling 4.4.

Working capital for 4 hours recycle stock is finally added. Total variable capital is tabulated in Table 3 for later use.

TABLE 3 - Costs vary with Conversion.

Conversion %	Capital Cost £000	Operating Cost £/te Product
11.4	10,040	760.2
10.0	10,045	738.0
7.6	10,145	697.4
6.0	10,397	668.6
5.0	10,763	661.6
3.8	11,604	657.8
3.0	12,688	673.3

Capital costs show a minimum of £10 million at 11.4% conversion, increasing at lower conversions. Any such increase must be justified by reduced operating cost.

OPERATING COSTS

The most important operating costs are :
Cyclohexane which rises with conversion because of decreasing selectivity.
Caustic usage
Waste Disposal charges for the 20% solution of dioic acid sodium salts. These also rise with conversion. The latter is charged at a commercial rate assuming incineration or other disposal facilities are provided by central facilities or by a contractor. Discharge to effluent treatment is unlikely to be acceptable because of high organic content and consequent BOD.
Electricity for air compression (lowest cost).
Steam and Cooling water for cyclohexane recovery.

Values for the base case are shown in Table 4.

TABLE 4 - Operating Costs for Base Case Design.

Raw Materials		Usage t/t Product	Price £/te	£/te Product
Cyclohexane		1.29	400	516
Caustic as 100% NaOH		0.42	150	63
Catalyst		-	-	-
Sub-total				579
Variable Expenses	Units	Usage	£/unit	£/te
Steam (inc. CW)	te	0.2 t/t b/u	13	61
Electricity	MW	1.3	35	7
Waste disposal	te act. wt.	4.99	10	50
Operating Cost				697

Figure 5 shows how operating costs vary with conversion at equal output, with a minimum of £658 /te product around 4% conversion.

RECONCILING CAPITAL AND OPERATING COSTS

The cases to be reconciled are :

Low Capital, High Operating Cost.
Minimum Capital £10.04 million at 11.4% conversion.
Operating Cost £760 /te Product.

High Capital, Low Operating Cost.
Minimum Operating Cost £658 /te Product at 3.8% conversion.
Capital Cost £11.6 million.

An extra £1.56 million capital brings operating cost advantages of £102 /te Product worth over £5 million a year at the plant's flowsheet capacity of 50 ktpa. The relation between capital and operating cost is not linear however as is shown in Figure 6. Marginal increases in capital are therefore best evaluated on an incremental basis. This may be done using net present value and DCFR techniques described by Allen (9).

However, Pike and Dobbins (10) report that pay-back calculation is the industrially preferred method of evaluation in currently prevailing economic circumstances. Answers obtained by all methods are dependent on the size of the increment being evaluated. In the present case, capital costs begin to rise rapidly at conversions below 5% due to the rapid increase in recycle. A graphical construction indicates the effect on pay-back. For a two year pay-back, extra capital spending of £1 million requires £10 /te operating cost saving at flowsheet production. A line with this gradient is shown in Figure 6 drawn tangentially to the cost curve. The point of contact indicates that capital expenditure of £10.8 million will pay for a plant with an operating cost of £662 /te. These conditions correspond to 5% conversion. A similar construction for a four year pay-back indicates a conversion slightly above 4% as optimal.

It is interesting to observe that Nypro's former Phase 2 Flixborough plant was apparently designed to operate at 6% conversion, whereas the former Phase 1 plant was based on hydrogenating phenol. The plant was commissioned in 1972 so that economic design studies would be carried out before the general price inflation initiated by OPEC in 1973/4. Cost differences between the two routes are marginal and fluctuating. It is therefore important to compare both routes at their optimal performance level to gain accurate insights. While a knowledge of capital/operating cost relationships is essential for this purpose, there is of course still the need to evaluate and compare profitability of the total project in both cases. Total costs are required for this rather than selecting only those costs which are variable with operating conditions, as in previous discussions.

FLEXIBILITY AND OPERATION

A calculated economic optimum depends on relative levels of capital and operating cost elements, on plant size and product demand and on design and forecasting uncertainties. Potential design errors arise from conversion/selectivity relationships, data assumptions, plant item performance and cost estimation. Calculations assume that item costs are continuously variable with capacity but since replicate reactors are required, variation will be stepwise. The largest errors usually concern product prices and demand and raw material and energy prices. Some provision for these uncertainties is desirable in the design.

Sections 600 (separation of hydrolysed liquors) and 700 (recovery of unreacted cyclohexane by distillation) are likely to be the most critical. The cost of section 600 is minor and over-design is tolerable. The capacity of section 700 could be increased by raising its pressure and design provision for this is desirable. Total extra capital for these uncertainties is about £250,000.

Once the plant is built, the capital is spent and irrespective of design assumptions, operations should seek to maximise the margin between sales revenue and production cost. The capacity of each section should be determined experimentally to establish what flexibility in operating conditions is available. The cost/conversion curve should also be redetermined using currently valid financial and technical data. If product demand is at design level, the plant should be operated to give lowest achievable production cost ie if other design assumptions are still valid as close to 3.8% conversion as practicable.

If demand should fall, increased selectivity at lower conversion may pay for increased recycle cost. Thus at 39 ktpa production rate, a 50 ktpa plant can save £148,000 a year by reducing conversion from design level of 5% down to the economic mimimum of 3.8%. Alternatively, market elasticity might allow demand to be stimulated by a drop in sales price. Margin per tonne would then be smaller but some compensation might then be obtained from increased volume. Sales into marginal export markets might also achieve this. Since the product under discussion is merely the start of a chain of intermediates culminating in synthetic fibre, this may seem unlikely but should be considered at the time.

CONCLUSIONS

The economics of processes with incomplete reactions and recycles should be examined to determine relationships between capital and operating costs. This requires a "base case" design from which costs varying under other conditions can be scaled. Starting with conditions which give lowest capital cost, operating savings resulting from incremental capital additions can be evaluated using pay-back criteria. The optimum occurs at the point of contact between a pay-back line of appropriate gradient and the curve relating capital and operating costs. Alternatively conventional step-wise calculations will indicate when pay-back extends beyond an acceptable time. Some minor extra capital is advisable to provide for uncertainties. Once the plant is built, the only economic objective is to maximise operating margin and reduce operating cost.

REFERENCES

1 Berezin,I.V., Denisov,E.T. and Emanuel,N.M., 1966, "The Oxidation of Cyclohexane" translated by Allen,K.A., Pergamon Press, Oxford.

2 Loder,D.J., 1940, U.S. Patent 2 223 494. "Production of Cyclic Alcohols and Ketones."

3 Schueler,A.P., and Wolff,F.A., 1958, U.S. Patent 2 825 742. "Liquid Phase Oxidation of Cyclohexane."

4 Alagy,J., Asselineau,L., Busson, C., Cha,B, and Sandler,H., 1968, Hydrocarbon Processing 47 (12) 131.

5 Happel,J., and Jordan,D.G., 1975, "Chemical Process Economics", Marcel Dekker, New York.

6 Sinnott,R.K., 1983 "An Introduction to Chemical Engineering Design" Pergamon Press, Oxford.

7 Hall,R.S., Matley,J., and McNaughton,K.J., 1982, Chem. Engg. 5 Apr.80.

8 Bridgwater,A.V., 1978, "Step Counting Methods for Preliminary Capital Cost Estimation". Transactions of the Fifth International Cost Engineering Congress, Utrecht, The Netherlands.

9 Allen,D.H., 1980, "A Guide to the Economic Evaluation of Projects" 2nd edition, The Institution of Chemical Engineers, Rugby.

10 Pike,R. and Dobbins,R., 1986, "Investment Decisions and Financial Strategy", Philip Allan, Oxford.

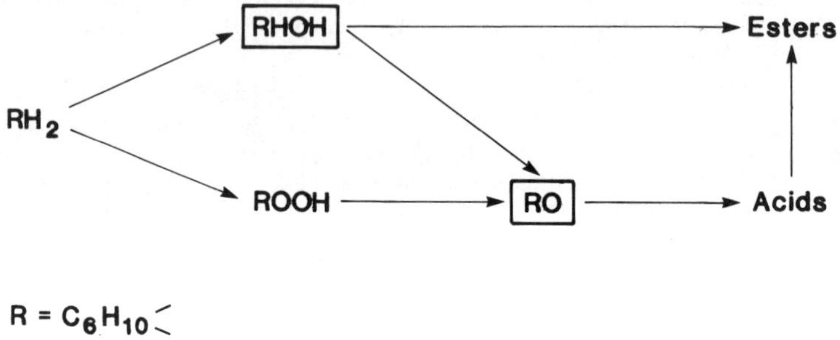

Figure 1 Simplified reaction scheme

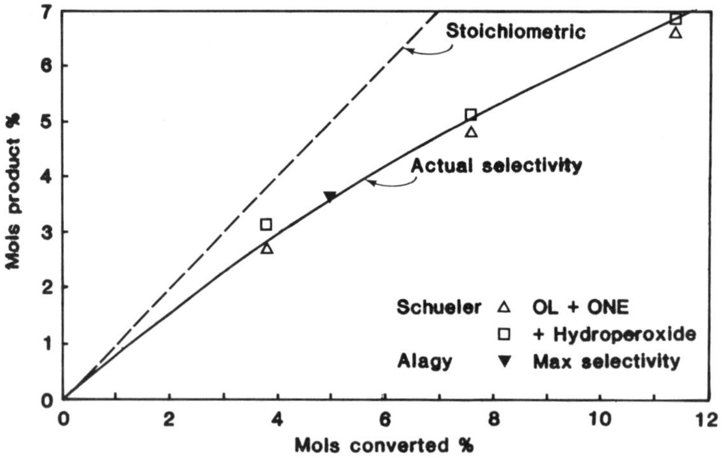

Figure 2 Selectivity and conversion

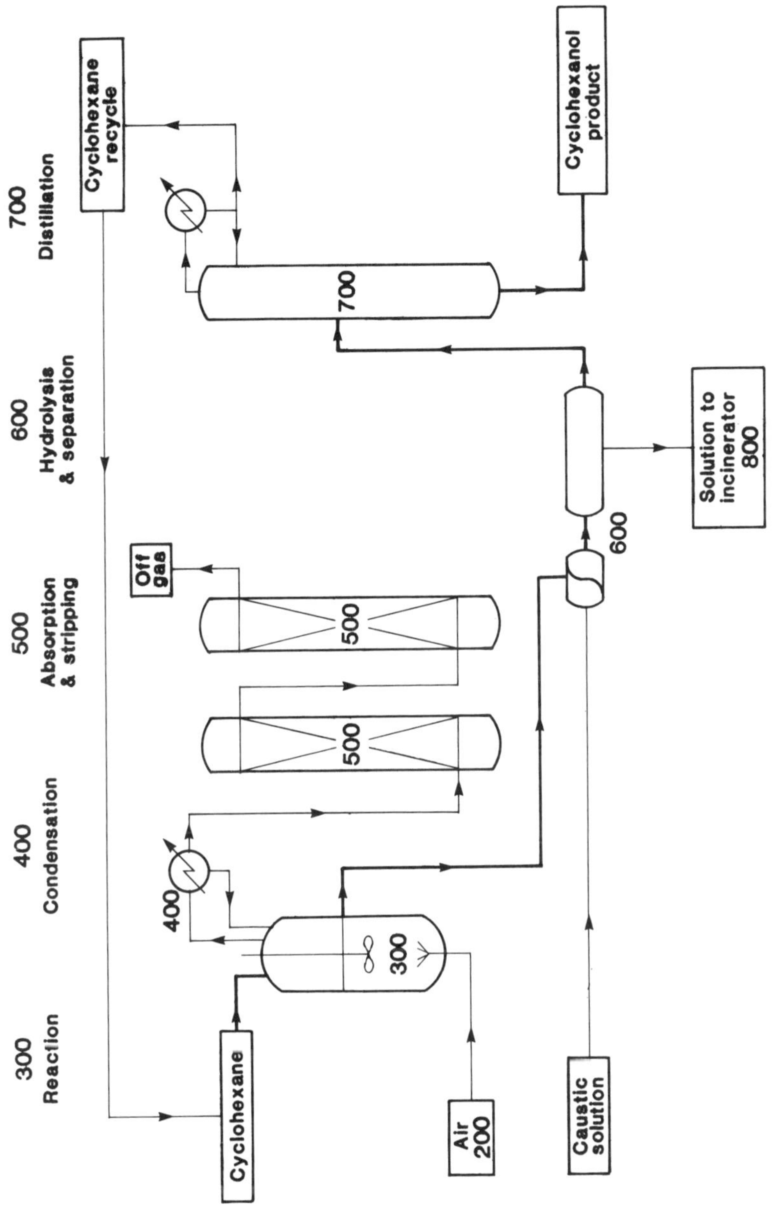

Figure 3 Schematic flowsheet for cyclohexanol

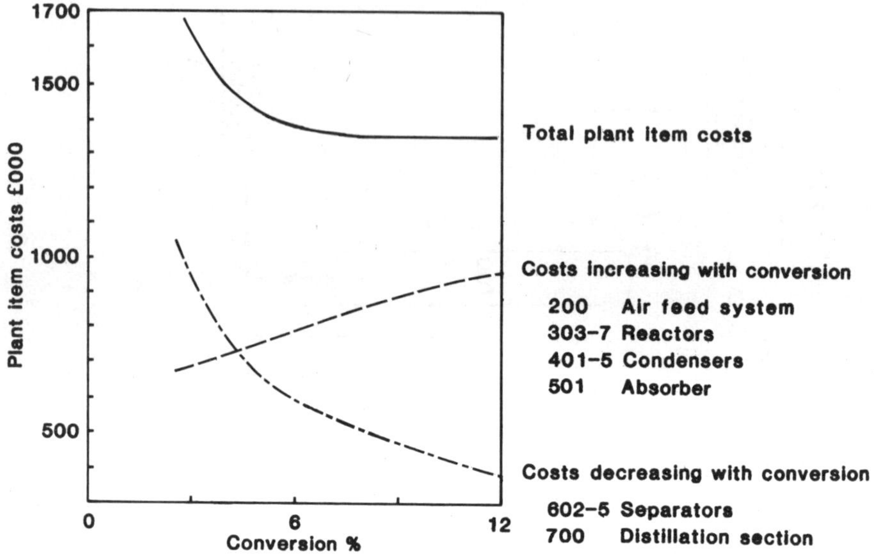

Figure 4 Plant item costs

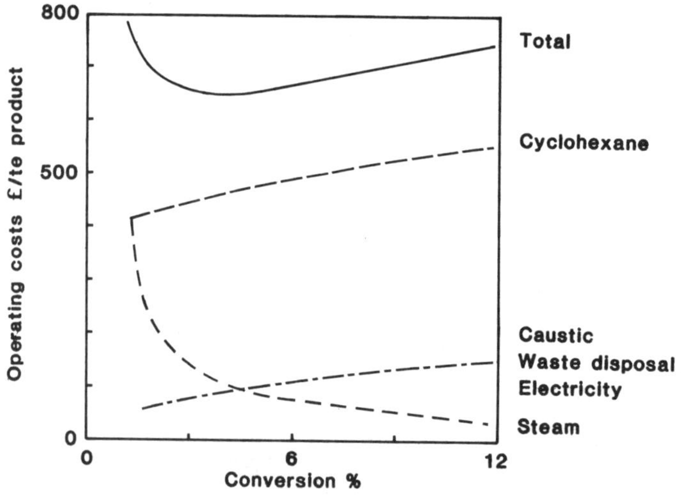

Figure 5 Operating costs

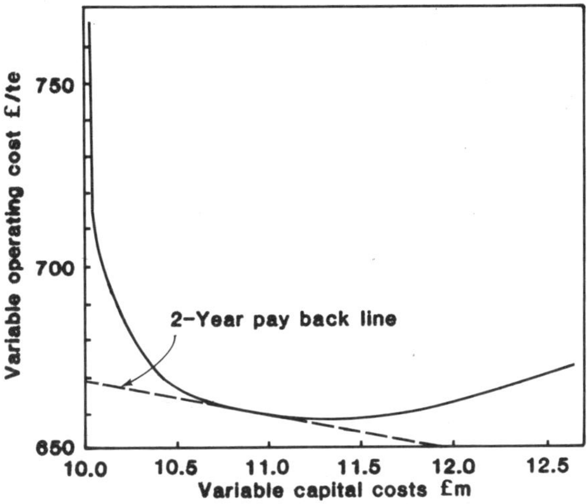

Figure 6 Capital vs operating costs

OPTIMISATION OF BATCH PLANT DESIGN

J.S.Grist*

The basic design of batch plants and particularly multi-product plants is frequently approached in a non-systematic manner which results in much redesign at a later and more expensive stage as the latent problems of cost or technical complexity reveal themselves. A systematic method is proposed for identifying the true requirements, establishing a minimum capital cost plant and verifying its workability.

INTRODUCTION

The increasing tendency to build batch process plants to produce more than one product has led designers to search for methodical ways to select the optimum design in terms of operability and capital cost. Each exercise is very individual and the course of such a systematic approach would be influenced from an early stage by the particular processes and the special circumstances of the project. The aim of this paper is to suggest in general terms a method of arriving at the optimum design. To do this a hypothetical plant is considered which is required to operate four processes having very different equipment requirements and producing very different tonnages. The numerical exercise is used as a vehicle to discuss the other aspects of batch plant optimisation.

WHAT IS BATCH PLANT OPTIMISATION?

For process plants the optimum is usually interpreted as being the best compromise in a complex situation between many constraints and several possible parameters which could usefully be maximised or minimised. For a large continuous plant built on an open site, optimisation may be very much flowsheet orientated with conversions or stream recoveries affected by a calculated change in temperature or pressure. The batch plant is not very amenable to this kind of process optimisation but its great versatility and frequent location within a process building provides many alternative meanings to the word optimum and many constraints which may need to be considered.

* JOHN BROWN Engineers & Constructors Limited,
1 Buckingham Street, Portsmouth, Hants, PO1 1HN.

Primarily, optimisation in this context is about achieving good plant operability and control with acceptable utility consumption costs at a minimum capital cost.

Constraints may include the reuse of an existing building, a predetermined capital budget, client's preferred proprietary equipment or limitations imposed on scale-up by lack of information.

To attempt a systematic approach to optimisation, a clear appreciation must be gained of the relative importance given to different and often conflicting parameters which are to be considered. Many of these will be both commercial and political and as such are outside the scope of this paper. They nevertheless have to be answered at a very early stage as they are fundamental to the design. A good vehicle for gathering this information is a well structured questionnaire.

OPTIMISATION OF A HYPOTHETICAL PLANT

Having established the ground rules, the task of optimising can commence against the agreed technical and non-technical criteria. The mass of technical data has now to be assessed and manipulated to provide the process design which comes nearest to satisfying the criteria. The first objective has got to be the reduction of the capital cost of the main plant items to a demonstrable minimum before considerations of operability can be addressed.

Since such topics are difficult to discuss in the abstract, a hypothetical design basis has been used to illustrate the pursuit of various optimisation strategies. For the purpose of this paper we can consider that the plant is required to make four products for which production data at the current scale of operation has been provided as indicated in Table 1 below.

TABLE 1 - PRODUCT OPERATING DATA

PRODUCT	TONNES/ ANNUM	TONNES/ BATCH	BX/ ANNUM	CYCLE HOURS	HOURS/ ANNUM
A	1000	5	200	40	8000
B	500	3	167	30	5000
C	100	2	50	30	1500
D	100	0.5	200	25	5000
			TOTAL HOURS		19500

The tonnes/batch and cycle times we can presume are based upon current methods of operation whilst the tonnes/annum figures and other derived figures are for the proposed new plant.

For the design of a completely new plant with no equipment constraints and particularly when using data from laboratory or pilot scale developments, it is preferable to first draw up independent flowsheets for each process. Short cut methods such as using one 'typical' flowsheet or working from equipment lists can result in incorrect assumptions being made and the requirement for additional equipment being missed. Attempts to pre-judge optimising reductions can cause confusion and incorrect solutions and should therefore be avoided. The flowsheets should, if possible, be scaled for the required annual capacity but otherwise at the capacity for which production data is available with each stage employing separate equipment. The resulting flowsheets provide the base case from which optimisation can commence, with each rearrangement being tested to see whether it is indeed an improvement.

To demonstrate the possibilities for optimisation without recourse to detailed flowsheets, some simplifying assumptions will be made about the potentially modular nature of most batch plants. Whilst it is true that the process flowsheet for a batch plant is invariably individual to the product it usually follows a general pattern of reaction, purification and isolation with remarkably few novel combinations.

Table 2 below represents the list of key equipment items for a typical production module which could produce one intermediate stage of a complex synthesis.

TABLE 2 - TYPICAL MODULE EQUIPMENT SCHEDULE

1	REACTOR
2	WASTE FILTER
3	WASH VESSEL AND RECEIVER
4	BATCH STILL AND RECEIVER
5	CRYSTALLISER
6	PRODUCT FILTER
7	BATCH DRYER

A dedicated plant could contain several such modules which for simplicity we can consider would vary in only two features, capacity and materials of construction. For the four products considered above the complete process for each product could be represented by a sequence of such modules suitably scaled to maintain the flow of batches through the plant at the given annual production rates as illustrated in Table 3 below.

TABLE 3 - SCHEDULE OF MODULE REQUIREMENTS PER PRODUCT

PRODUCT	MODULE 1	MODULE 2	MODULE 3	MODULE 4	MODULE 5
A	GLMS 2000L	316 SS 4000L	GLMS 10000L		
B	316 SS 1500L	316 SS 8000L			
C	HAST 500L	316 SS 1000L	GLMS 5000L		
D	316 SS 1000L	316 SS 2000L	GLMS 500L	GLMS 500L	316 SS 2000L

where GLMS = Glass Lined Mild Steel
316 SS = Stainless Steel
HAST = Hastelloy

Before considering possibilities of optimisation we must first review the plant as given, to establish an approximate method for estimating changes in capital cost resulting from attempted optimisations. If as a datum a 5000L stainless steel module is taken as costing £1.00M and if a multiplier of 1.5 is used for glass lined equipment and 3.0 for Hastelloy equipment then by using the 0.6 power rule for other sizes it can be shown that the above modules represent a total investment of £10.75M (see Table 4). This cost would be for the process modules only and would exclude all ancillary equipment, utilities and other offsites. This then provides the datum for optimising the process plant since alternative regimes can be similarly costed and compared.

COMMENCING PLANT OPTIMISATION

Referring to Table 1 and assuming an 8000 hour operational year, it can be seen that to employ the given batch sizes would require three plants operating simultaneously for products A, B and D, with product C being manufactured on plants for B or D as appropriate. Retaining such small scale operations may be favoured in circumstances where scale-up is potentially dangerous (e.g. munitions or the nuclear industry) or where certain materials are so expensive that there is a risk of unacceptable losses within the equipment. The general tendency, however, would be to try to simplify the operation and simultaneously reduce the capital cost of the plant by combinations of scale-up and campaign operation.

To demonstrate the outcome of three simple strategies a Spreadsheet computer utility has been used to calculate the resulting module sizes and costs which are discussed below.

Strategy 1

Scale up all four processes by a factor of 19500/8000 (see Table 1), eliminate modules which are sufficiently similar, and design one all-purpose plant which can handle each process in turn.

Result

The capital cost rose initially to £18.34M (Table 5) but was then reduced by the elimination of five duplicate modules or modules which could reasonably be considered duplicate on grounds of adequacy of materials of construction and capacity to £13.15M (Table 6).

Comment

This is a suitable method when it is important to maintain batch identity and advantageous to be able to produce finished product at short notice. The plant has been increased in scale by a factor of 2.44 which could create difficulties in obtaining propriatary equipment for the larger capacity stages.

The plant now produces one product and a maximum of four intermediates at any time, thereby simplifying the plant operation. The large variance in size of the equipment modules prevented further reductions and cost savings. As a consequence the plant modules are under-utilised and the capital cost £2.42M greater than the base case and therefore not a suitable means of optimisation in this case.

Strategy 2

As for Strategy 1 but scale up each process by a different factor in an endeavour to harmonise module capacities whilst maintaining batch identify for each product.

In this instance an arbitrary decision was taken to scale up processes B, C and D to increase the capacity of the largest module to 10000L to match STEP 3 of process A. Scales of 1.25, 2.0 and 5.0 were used respectively.

Result

The capital cost at this point has risen to £16.09M (Table 7). A second scale factor of 1.72 was then applied to all four processes to reduce the total operating time to 8000 hours. The capital cost increased further to £22.27M (Table 8). Finally by the elimination of seven modules the cost was reduced to £10.95M (Table 9).

Comment

This is an improvement upon Strategy 1 in that the number of modules has been reduced from 13 to 6 and the capital cost to £10.95M which is effectively the same as the base case.

Based upon the number of modules the building size should be about halved and the plant will be better utilised than in Strategy 1. Nevertheless, if the capital cost cannot be decreased any operational benefits over the base case will be difficult to sell.

Strategy 3

Break each process down into campaigns which result in a storable (usually solid) intermediate or product and consider scaling each campaign independently to maximise the utilisation of key equipment items.

Result

Since the plant is now a general purpose production facility and not sized for any particular product, there is great benefit in standardising the module sizes. The exception will be the Hastelloy module which is prohibitively expensive to scale up. This will stay as a small scale module. By inspection of Table 9 the mean module capacity was determined as 8800L. Anticipating some further economy, the scaling of each module to 80001 was investigated. The results based upon an 8000 hour operating year were:-

Glass Units	13563 hrs on line	2 Units
Stainless Units	13250 hrs on line	2 Units
Hastelloy Units	1500 hrs on line	1 Unit

The total capital cost was £7.35M (Table 10).

Comment

Moving from dedicated plants to campaign operation can present some operational problems if the batch identify is to be maintained because batches of intermediate may need to be bulked together to form the feed stock for the next scaled-up stage. In addition, from the commencement of operations it may be several months before finished product is produced since each intermediate will be made in separate campaigns each lasting for several weeks.

The proposed capacity of 8000L was about right. In theory the capacities could be decreased until 16000 hours on line was achieved (8000 hours per unit). However, in reality some capacity will be lost due to plant cleaning time between campaigns.

By contrast the size of the Hastelloy could easily be reduced but since it is only 500L capacity, scope for reduction is limited. The reduction of the plant to four standardised modules plus a small Hastelloy module at a capital cost reduction of £3.4M could be considered an optimisation.

Strategy 4

A mixture of the above three strategies. The above strategies are aimed at producing a plant consisting of a few scaled-up modules which can either run each complete process in turn or campaign produce one intermediate per module. Other strategies could be to run, say, Product A at its original capacity and apply the above strategies to the remaining three products.

OPTIMISING REAL PLANTS

The Strategies 1 to 4 above assume that the modules are sufficiently similar to be used on different products. Also since the modules are assumed to be self-contained there is no problem with the double booking of ancillary equipment. In most real optimisation studies, however, these modules are broken down to individual plant items and reformed as new modules using different combinations of equipment. It is then uncertain whether the plant is capable of producing material to the batch size and cycle times assumed.

A method is therefore required to verify that the plant is capable of producing the specified annual tonnages. Initially, production bar charts will need to be produced for each campaign to determine the batch cycle time and single batch time so that the duration of the campaign can be calculated. A matrix of equipment usages versus campaigns will also be required to identify campaigns which cannot run together because of equipment limitations.

PROVING THE REVISED PLANT

When proposing a sequence of campaigns, consideration must be given to the order of manufacture, availability of warehoused intermediates, allowances for shutdown and cleaning between campaigns and the allocation of ancillary equipment to process streams. All have to be considered when confirming that the plant will make the annual tonnage of all the products and intermediates.

Producing a week by week production schedule for the plant whilst bearing in mind all the constraints mentioned above can be a very time consuming trial and error operation which lends itself to computer assistance. Such a program was written for a plant to run two campaigns simultaneously from a menu of twelve. It required the designer to input, on a weekly basis, pairs of products to be made.

The program first checked that the campaign pairing was feasible with regard to the use of shared equipment and the availability of warehoused intermediate. It then checked that the products were not about to be over produced. Any such fault in the input was flagged to the designer who then made a new selection.

After a successful entry the program calculated and displayed the new stock levels in the warehouse of intermediates and products. The program then moved on to the next weeks production. At the end of the run the program printed the proposed schedule for the year and the quantities of products and unused intermediates in the warehouse. Even with such assistance it was possible to approach the end of the year and have left incompatible campaigns which were nevertheless needed to successfully meet the years production requirements. However, after a couple of attempts the more subtle rules of the game were appreciated and a successful run achieved. Once achieved, of course, the answer was obvious from the start!

In the event that a successful campaign pattern could not be found to produce the annual tonnages then a search was made to discover the source of the problem. This could be an equipment bottleneck prohibiting the necessary parallel operation of two campaigns. Alternatively, it may have been the use of too many campaigns of short duration which would reduce the annual productive hours due to the need for frequent plant cleaning. After alterations the program would be re-run to confirm the suitability of the modified plant.

DETAILED DESIGN

Having established a workable plant the next stage of design is to firm up the flowsheet details and devise an effective plant layout. It is preferable that these two activities are progressed simultaneously so that layout implications can be reflected in the flowsheet. The key issues to be considered here are the transportation and charging of solids, the flow and filtration of slurries and the segregation of plant areas for reasons of safety or cross contamination.

Batch reactors tend to be multi-functional by nature and this is particularly true of plants which have been condensed by an optimisation exercise. The temptation at this point is to equip all reactors with all features (e.g. solids charging system, distillation, etc.) in order to achieve a totally multi-purpose plant. However, when these possibilities are reviewed against the limitations of the plant layout (e.g. the need for short steep slurry lines and the desirability of creating a central filtration and drying area) some degree of equipment specialisation usually emerges as the practical solution.

Thus those reactors close to the filtration equipment will be employed as crystallisers and the others for raw materials receipt and reaction. Pursuing a totally multi-purpose plant design can become very expensive relative to the minimum adequate plant and can make the design very complex. It is therefore very desirable to initially design for the minimum requirement to provide a datum against which to measure the need and cost of additional features.

SPREADSHEET CALCULATION OF CAPITAL COSTS AND EQUIPMENT MODULE SIZES

TABLE 4

BASIS OF OPTIMISATION STUDY

PRODUCT	A	B	C	D				
TPA	1000	500	100	100				
BX/ANNUM	200	167	50	200				
CYCLE.HR.	40	30	30	25				
T/BX	5	3	2	0.5				
TPH	0.125	0.1	0.067	0.02				
HRS/ANNUM	8000	5000	1500	5000				
TOTAL HRS.	19500				CAPITAL COST IN £M			
		MODULE CAPACITIES			A	B	C	D
MODULE 1	2000	1500	500	1000	0.87	0.49	0.75	0.38
MODULE 2	4000	8000	1000	2000	0.87	1.33	0.38	0.58
MODULE 3	10000		5000	500	2.27	0.00	1.50	0.38
MODULE 4				500	0.00		0.00	0.38
MODULE 5				2000	0.00			0.58
					4.01	1.81	2.63	2.29
SCALE	1	1	1	1	TOTAL COST £M			10.75

TABLE 5

STRATEGY 1   SCALED BY 19500/800

PRODUCT	A	B	C	D				
TPA	1000	500	100	100				
BX/ANNUM	82	68	21	82				
CYCLE.HR.	40	30	30	25				
T/BX	12.19	7.313	4.875	1.219				
TPH	0.305	0.224	0.163	0.049				
HRS/ANNUM	3282	2051	615	2051				
TOTAL HRS.	8000				CAPITAL COST IN £M			
		MODULE CAPACITIES			A	B	C	D
MODULE 1	4875	3656	1219	2438	1.48	0.83	1.29	0.65
MODULE 2	9750	19500	2438	4875	1.49	2.26	0.65	0.98
MODULE 3	24375		12188	1219	3.88	0.00	2.56	0.64
MODULE 4				1219	0.00		0.00	0.64
MODULE 5				4875	0.00			0.98
					6.85	3.09	4.50	3.91
SCALE	2.438	2.438	2.438	2.438	TOTAL COST £M			18.34

SPREADSHEET CALCULATION OF CAPITAL COSTS AND EQUIPMENT MODULE SIZES

TABLE 6

STRATEGY 1 DUPLICATE UNITS DELETED

PRODUCT	A	B	C	D				
TPA	1000	500	100	100				
BX/ANNUM	82	68	21	82				
CYCLE.HR.	40	30	30	25				
T/BX	12.19	7.313	4.875	1.219				
TPH	0.305	0.244	0.163	0.049				
HRS/ANNUM	3282	2051	615	2051				
TOTAL HRS.	8000				CAPITAL COST IN £M			
		MODULE CAPACITIES			A	B	C	D
MODULE 1	4875	3656	1219	0	1.48	0.83	1.29	0.00
MODULE 2	9750	0	0	0	1.49	0.00	0.00	0.00
MODULE 3	24375		12188	0	3.88	0.00	2.56	0.00
MODULE 4				1219	0.00		0.00	0.64
MODULE 5				4875	0.00			0.98
					6.85	0.83	3.85	1.63
SCALE	2.438	2.438	2.438	2.438	TOTAL COST £M			13.15

TABLE 7

STRATEGY 2 PLANT SCALED TO 10000L

PRODUCT	A	B	C	D				
TPA	1000	500	100	100				
BX/ANNUM	200	133	25	40				
CYCLE.HR.	40	30	30	25				
T/BX	5	3.75	4	2.5				
TPH	0.125	0.125	0.133	0.1				
HRS/ANNUM	8000	4000	750	1000				
TOTAL HRS.	13750				CAPITAL COST IN £M			
		MODULE CAPACITIES			A	B	C	D
MODULE 1	2000	1875	1000	5000	0.87	0.56	1.14	1.00
MODULE 2	4000	10000	2000	10000	0.87	1.52	0.58	1.52
MODULE 3	10000		10000	2500	2.27	0.00	2.27	0.99
MODULE 4				2500	0.00		0.00	0.99
MODULE 5				10000	0.00			1.52
					4.01	2.07	3.99	6.01
SCALE	1	1.25	2	5	TOTAL COST £M			16.09

SPREADSHEET CALCULATION OF CAPITAL COSTS AND EQUIPMENT MODULE SIZES

TABLE 8

STRATEGY 1 PLANT SCALED TO 10000L - FURTHER SCALED FOR 8000 HRS

PRODUCT	A	B	C	D				
TPA	1000	500	100	100				
BX/ANNUM	116	78	15	23				
CYCLE.HR.	40	30	30	25				
T/BX	8.595	6.446	6.876	4.298				
TPH	0.215	0.215	0.229	0.172				
HRS/ANNUM	4654	2327	436	582				
TOTAL HRS.	7999				CAPITAL COST IN £M			
	MODULE CAPACITIES				A	B	C	D
MODULE 1	3438	3223	1719	8595	1.2	0.77	1.58	1.38
MODULE 2	6876	0	0	0	1.21	2.10	0.80	2.10
MODULE 3	17190		17190	4298	3.15	0.00	3.15	1.37
MODULE 4				4298	0.00		0.00	1.37
MODULE 5				17190	0.00			2.10
					5.56	2.87	5.53	8.32
SCALE	1.719	2.149	3.438	8.595	TOTAL COST £M			22.27

TABLE 9

STRATEGY 2 DUPLICATE UNITS DELETED

PRODUCT	A	B	C	D				
TPA	1000	500	100	100				
BX/ANNUM	116	78	15	23				
CYCLE.HR.	40	30	30	25				
T/BX	8.595	6.446	6.876	4.298				
TPH	0.215	0.215	0.229	0.172				
HRS/ANNUM	4654	2327	436	582				
TOTAL HRS.	13750				CAPITAL COST IN £M			
	MODULE CAPACITIES				A	B	C	D
MODULE 1	0	0	1719	8595	0.00	0.00	1.58	1.38
MODULE 2	0	0	0	17190	0.00	0.00	0.00	2.10
MODULE 3	17190		0	4298	3.15	0.00	0.00	1.37
MODULE 4				4298	0.00		0.00	1.37
MODULE 5				0	0.00			0.00
					3.15	0.00	1.58	6.22
SCALE	1.719	2.149	3.438	8.595	TOTAL COST £M			10.95

SPREADSHEET CALCULATION OF CAPITAL COSTS AND EQUIPMENT MODULE SIZES

TABLE 10

STRATEGY 3 MODIFICATION OF MODULE SIZES TO STANDARDISE AND OPERATE ON A CAMPAIGN BASIS

A1 = PRODUCT A - MODULE 1

EQUIPMENT SCALED FOR 8000L

PRODUCT	A1	A2	A3	B1	B2	C1	C2
TPA	1000	1000	1000	500	500	100	100
BX/ANNUM	50	100	250	31.25	166.67	50	6.25
CYCLE.HR.	40	40	40	30	30	30	30
TNS/BX	20	10	4	16	3	2	16
HRS/ANNUM	2000	4000	10000	937.5	5000	1500	187.5
CAPACITY	8000	8000	8000	8000	8000	500	8000
MATERIAL	GLMS	316SS	GLMS	316SS	316SS	HAST	316SS
SCALE	4	2	0.8	5.3333	1	1	8

PRODUCT	C3	D1	D2	D3	D4	D5
TPA	100	100	100	100	100	100
BX/ANNUM	31.25	25	50	12.5	12.5	50
CYCLE.HR.	30	25	25	25	25	25
TNS/BX	3.2	4	2	8	8	2
HRS/ANNUM	937.5	625	1250	312.5	312.5	1250
CAPACITY	8000	8000	8000	8000	8000	8000
MATERIAL	GLMS	316SS	316SS	GLMS	GLMS	316SS
SCALE	1.6	8	4	16	16	4

TOTAL HRS 28313

TOTAL HRS (GLMS) = 13563 = 2*8000L or £2.64M
TOTAL HRS (316SS) = 13250 = 2*8000L or £3.96M
TOTAL HRS (HAST) = 1500 = 1*1500L or £0.75M

TOTAL COST £7.35M

HAZARDS ASSESSMENT AT INITIAL LAYOUT STAGE

R.J. Aldridge*, J. Madden* and J.C. Mecklenburgh**

It is essential to obtain an understanding of possible major hazards at the earliest possible stage of design to ensure that hazard problems are not built in to the design. The paper describes a C.A.D. system based on the methods in Process Plant Layout for assessment of hazards related to layout features which enables rapid understanding and rectification of the likely hazards before project sanction or planning permission is sought.

INTRODUCTION

Plant layout is the single most important decision in plant engineering once the major process design parameters are established. A "good" plant layout will be safe, economic and easily operated; a "bad" layout may well prevent the plant functioning in any effective way. There is considerable benefit to be gained by optimising plant layout but there is also considerable difficulty. The layout designer is required to optimise the effect of a number of major and often conflicting constraints:

Safety - the plant must not endanger the plant operators or the surrounding community,

Economics - layout is a significant factor in construction and operating costs.

Operability - Operating or access difficulties have an adverse effect on operation and, consequently, on plant efficiency, and plant availability,

Time - the plant layout is required very quickly at the start of a project to define and initiate major design work such as Civils, Electrical, Equipment and Piping,

* Isopipe Ltd., Nottingham
** Late of Chemical Engineering Department, University of Nottingham.

Information - the information available to the layout designer is limited and often uncertain, but the layout must be definitive and remain essentially unchanged throughout the project.

In practice, there is no reliable numerate means of expressing the safety, economic and operability constraints. Furthermore the additional time and information constraints prevent the development of any true objective design optimisation procedures. The best available technique is that of convergence towards an acceptable layout, the degree of convergence often being limited by the demands of the project schedule.

Traditional design procedures used probable capital cost as the main independent variable (tempered by judgement of safety) to measure progress in converging to an acceptable layout. Despite the reliance on capital cost as the only objective measure, the large number of plants which have been designed and operated safely over many years suggest that reasonable optimisation was achieved, albeit by the use of intuition and experience rather than formal methods.

HAZARD ASSESSMENT IN LAYOUT

A radical change to the established methods was made necessary by the major incidents at Seveso (1) and Flixborough (2), followed by even more severe incidents at Mexico City (3) and Bhopal (4). In all these incidents, release of fluids from the plant produced ignition or toxic effects on other plant areas and the communities around the plants. The increased social concern about these incidents and the process industry at large is natural and understandable; the public which is affected by plant incidents knows little of process technology and always fears the unknown. The reporting and analysis of incidents in the media is not always accurate and is not always friendly towards the Industry.

Social concern has resulted in a growing degree of legislation and regulation (5,6) intended to ensure that the safety of plant operators and the public is considered at all stages of design but particularly those concerned with plant layout and location. Investigation of the likelihood and consequences of plant releases is now a major concern in developing and optimising the plant layout and in justifying the layout to planning and regulatory bodies (6).

The designer must postulate the releases which are feasible, assisted by reliability data, where it is available. The likelihood of a given release occurring spontaneously is effectively independent of layout and design proceeds, therefore, on the basis of a set of release cases whose consequential effects on the layout are to be investigated. Unacceptable effects indicate layout modifications should be made and further iterations of the release-consequence cycle enable the designer to move towards an optimised layout where the plant layout uses a realistic minimum space and releases have acceptable consequences. This case is likely to be a minimum capital cost case for the postulated releases, developed by safety and cost constraints along a design path whose major steps can be audited by designers and regulatory authorities.

ASSESSMENT OF RELEASES AND CONSEQUENCES

A number of methods of investigating the consequences of plant releases have been reported in recent years. Most of these methods consider one or more links in the consequence chain of a proposed plant failure which could lead to a release:

Fluid behaviour during and after release,
Rate and duration of release,
Explosive, ignition and toxic properties of released material in air
Severity of any explosion, fire or toxic effects,
Impact of the release on the plant or the surrounding community.

From the various release and consequence models published, those of Mecklenburgh (7), which originate from the work of the I.Chem.E 'Plant Layout' working party, were selected as the basis for the current project. This work encompasses a wide range of release and consequence models. The use of these models does not, however, imply rejection of other methods. Indeed for many release scenarios, more than one method may be needed to increase confidence in the design, and the selection of appropriate methods is an important part of hazard analysis studies. Whatever method is used, however, the same questions arise and the general procedure below usually follows:-

Type of Release

The designer must identify those areas of the plant where failure might occur, e.g.:-

- gasket failure
- pipe breakage
- pump seal failure
- vessel weld split
- total vessel failure, e.g. storage tank collapse

depending on his judgment of the behaviour of the plant equipment under all possible process conditions. By considering the process conditions and the nature of the plant item, the designer must specify feasible modes of failure for each item. Failure may result in a catastrophic instantaneous release of the entire vessel contents or a continuous release through a smaller opening of defined area. Inevitably, a lengthy list of possible releases, containing both instantaneous and continuous cases will be produced.

Released Fluid

The process material which is in the plant item at the time of failure will be released. This will not always be of known constant composition, or at constant temperature and pressure. From the process design data the range of compositions, temperatures and pressures can be obtained for different plant operating states. A release could occur under any set of operating conditions and thus a number of different release scenarios need to be considered for each of the predicted modes of vessel failure.

The rate and amount of release from a given plant item release case will also depend on such factors as the plant volume, and the physical properties of the process fluids. These must be defined to calculate release rates and volumes.

Behaviour of Released Fluid

When a release occurs, the fluid will reach atmospheric pressure almost instantaneously by flashing or expanding and it will then begin to cool down (or heat up) to atmospheric temperature. These effects will produce a gas, a vapour or a mixture of liquid and vapour. The process pressure and the size of the opening will determine whether a cloud, a jet or a pool (or a combination of these) will be formed.

If the fluid is flammable, it may ignite spontaneously on release or the gas/vapour part may drift to an external ignition source. Ignition may result in a fire, a fireball or an explosion either at the release point or in the vapour cloud drifting away from the release. The severity of fire or explosion is related to the physical properties and the quantity of fluid released.

If the fluid is toxic and does not ignite, a cloud will drift away from the release point under the influence of wind and weather. The cloud will gradually dilute as it disperses and mixes with air but will be hazardous until the toxic concentration in the cloud has fallen to a tolerable level.

Cloud behaviour (of both flammable and toxic clouds) is related to the physical properties of the released fluid, the wind speed and other meteorological conditions in and around the plant.

Clearly, any incident of fire, explosion or toxicity is most severe in the immediate vicinity and the effects of the incident are less severe at increasing distance. Calculations must, therefore, allow for distance effects in assessing the consequences of any release.

General Calculation Methods

Calculation methods such as Ref. 7 provide models for quantifying release effects such as jets, pools, clouds, etc. By reasoning through the behaviour of a released fluid the designer can specify an appropriate model to show the severity of the result of the release and to show the effect/distance relationship.

With these calculations available, attention can be given to the effects of a release. The affects upon people, plant items and buildings which are in the vicinity of the release must be assessed. How badly a given item (generally referred to as a target) is affected depends on its tolerance to physical effects such as explosion blast pressure, thermal radiation or toxic concentration. Data is published to indicate what levels various types of targets (e.g. Humans, Distillation Columns, Buildings) can sustain without death or total failure taking place.

The effect imposed on a target from any release should be compared with the tolerable value. Any case where the imposed effect exceeds the tolerable level must be improved, either by increasing the distance between the release and the target or by some other means. The layout designer can affect only the distance (by amending the layout if space is available); other action entails changing the process design and is outside the scope of the release model and this paper. It is obvious that the combination of many possible releases, many targets, much process and property data and many hazard calculations represent a formidable design problem. The complex data handling and calculation time may well inhibit the proper checking of a layout. A computer aided system is probably the only practical approach which will allow a proper study of the releases to be made and the subsequent layout actions to be investigated.

OUTLINE OF HAZARD CHECKING SYSTEM

A computer system has been developed (8), which enables hazard checking to become a useful tool for assisting in layout optimisation. Alternative layouts can be investigated quickly in response to hazard calculation data. The calculation methods suggested by Mecklenburgh (7) and other work present no inherent computing difficulty. However, to set up a hazard assessment of a layout requires the assembly of a considerable amount of data to define:-

- Process Temperatures, Pressures, Volumes and Compositions for equipment and pipes at Startup, Operation, Shutdown and excursion conditions,
- Physical properties of pure process materials and mixtures,
- Position and dimensions of equipment and other designated items (e.g. valves) from which releases are postulated,
- Windspeeds, atmospheric temperatures at the plant site,
- Effect of plant and surrounding topography

To facilitate the input, checking, use and modification of all this data, a modular system was used to hold the groups of data separately. Once input, the data held in the modules are accessed as required by calculation programs. Much of the data to be stored for a plant can be held in other CAD systems, e.g.

Process Data - Intelligent databased flowsheet packages, e.g. PEGS. (9)

Physical Properties - Databanks, e.g. PPDS (10).

Equipment Data - 3D modelling packages, e.g. PDMS (11).

Accordingly, the hazard checking system is designed to automate data transfer from or to such CAD systems and eliminate unnecessary transcription work.

The need for alternative calculation methods has been noted above and it is recognised that improved methods will emerge in the future. To permit the easy use of such alternative methods, the calculation programs are also modular and independent, so that changes to or addition of new calculation methods can be carried out by users without disturbing the integrity of the total system.

USING THE HAZARD CHECKING SYSTEM

When applied to a new design at the plant layout stage, the hazard checking system operates in 3 stages:

- Information Collection
- Calculation of Effects
- Reporting Hazards

Stage 1 - Information Collection

The designer assembles the information on process conditions and physical properties for entry to the data modules. Data from the proposed initial plant layout is also entered. This stage can be automated if other CAD systems are in use or data entry can be manual and interactive. All data is checked carefully to trap plausible but contradictory units or values.

A study of the process data and the layout is now carried out to define the release cases for each plant item. This operation is similar to a Hazop (12) study and requires considerable care, experience and judgement to ensure that all feasible release cases are identified for calculation. At this stage also, target items on the plant and in the surrounding area are identified to the system and their tolerance to blast, thermal radiation or toxic effects is specified. Plant items can act both as "releases" and "targets for other releases" to help identify possible initiation of consequence chains (ther so-called 'domino effect'). Finally, data on site weather and topography is entered to complete the data input.

Stage 2 - Calculation of Effects

With all data assembled, the calculation module are run. For each release specified, the appropriate physical properties and process data are identified so that the fluid behaviour can be deduced and suitable calculations performed. The calculations cover, typically,

- jets and jet fires
- pool and/or cloud formation
- flame height from pools or equipment
- cloud dispersion
- fireballs and vapour cloud explosions

The pessimistic assumption is made that any flammable release ignites or explodes and calculations are for this worst case. From each calculation, the property of interest is the hazard intensity (e.g. explosion overpressure, thermal radiation or toxic concentration) at the release point. With this basic value known, the module then examines in turn all defined targets. For each target, the "release" to "nearest point of target" distance is calculated to enable the intensity developed at the target to be determined, (i.e. allowing for the attenuating effect of release to target distance). The result is stored ready for output and the procedure repeated for every release and every target.

Further calculations may be made for flammable or toxic clouds to establish the distances from the release point to UFL, LFL and/or immediately dangerous and below danger level concentrations.

In addition, the designer can request for any release, calculations to show the hazard intensities at specified distances (and vice versa). A typical output resulting from such a request is shown in Fig. 1. The values marked with a * in the table are "distances/hazard intensities" for which specific calcualtions have been requested by the designer.

The calculation results corresponding to these specified values can be read from the table.

Stage 3 - Reporting Hazards

The system presents results in a form intended to assist the designer to identify each hazardous case and thereby assess what improvement can be made by modifying the layout. Two report formats are presented - Tabulated (Figs. 1 and 2) and Graphical (Fig. 3).

The Tabulated format (Fig. 2) lists all targets and shows for each target the intensity experienced from every release case. If the intensity at a target from any release scenario exceeds the permitted maximum (as specified in the Information Collection stage), a warning is printed to identify the corresponding release scenario and the excessive level of the hazard intensity. Unacceptable hazardous interactions between one target and one release are thus automatically highlighted.

With this consequence information the designer can judge what action might be taken to improve safety. Modifying the layout to change the release to target distances is easy to investigate by repositioning items and updating the data module. Effects of reduced inventories, lower pressure or temperature can be studied in a similar way. Release cases postulated can be modified by specifying higher quality equipment or by changing equipment types. The designers aim is a report showing no excessive hazard intensities as evidence of a reasonably safe layout.

The Graphical format (Fig. 3) provides a different insight into the consequences of releases, providing a means of showing the region within which a designer defined level of hazard intensity is exceeded as a result of given release scenarios. Examples might include heat flux or toxic concentrations. For a heat flux case arising from a large release of Propane, the designer might choose to see the region within which people would be exposed to an intolerable level of heat flux. Seeing this region superimposed on a plant layout drawing assists judgement of e.g. how many people are affected or whether plant roads would be rendered unusable. The effect of moving the Propane source or re-routing roads can be seen quickly as a guide to reducing the hazard by changing the layout. Similarly, for a release of chlorine, seeing the region in which an immediately dangerous concentration is present superimposed on a ground plan of the plant neighbourhood will quickly identify any areas of habitation in which people are at risk. Again, remedial changes to layout or plant siting can be investigated quickly.

CONCLUSIONS

The hazard checking system developed is a useful method of assisting designers to investigate plant releases and take remedial layout measures, without creating unrealistic workloads or imposing unacceptable delays in generating an acceptable layout at the initial project stages. It offers a vehicle for recording important stages in layout development and can assist designers and regulatory bodies to discuss proposed layouts and test their potential impact on the community.

The facility to add other calculation methods and interface to other in-house CAD systems provides designers with an extendable system to cope with growing hazard assessment techniques.

Further developments are possible in areas such as simultaneous multiple release cases, improving calculation methods or application of expert system technology and some work is currently under way.

REFERENCES

1. Health and Safety Executive, 'Seveso. A Translation of the Official Report', HMSO, London, 1980.

2. Parker, R.J. (Chrmn.), 'The Flixborough Disaster. Report of the Court of Inquiry', HMSO, London, 1975.

3. Redman, J., 'Pemex - the Forgotten Disaster', Chem. Engr. 418, 16, 1985.

4. Kharbanda, O., 'Another Bhopal? Never Again', Chem. Engr., 416, 40, 1985.

5. 'Notification of Installations Handling Hazardous Substances Regulations', HMSO, London, 1982.

6. Health and Safety Commission, 'Control of Industrial Major Accident Hazards. Draft Regulations and Guidance -Consultative Document', HMSO, London, 1983.

7. Mecklenburgh, J.C., 'Process Plant Layout', George Godwin/I.Chem.E., London, 1985.

8. Aldrige, R.J., 'Computerised Plant Layout for Hazards Containment', PhD Thesis, University of Nottingham, 1986.

9. Project Engineering Graphics System, CAD Centre Ltd., High Cross, Madingly Road, Cambridge.

10. Physical Property Data Service, Institution of Chemical Engineers, 165-171 Railway Terrace, Rugby.

11. Plant Design and Management System, CAD Centre Ltd., High Cross, Madingley Road, Cambridge.

12. CIA, 'A Guide to Hazard and Operability Studies' Chemical Industry Safety and Health Council of the Chemical Industries Association, London, 1977.

Dispersion Report for the Combined Initial and Secondary Clouds

Windspeed = 4.00 M/S
Volume of Cloud = 15272.3 CU.M.

Distance (M)	Concentration (V/V)	Dosage Time (S)	Time to Peak Concentration (S)
20.0 *	0.740740	15.0	5.0
40.0 *	0.565982	16.6	10.0
49.9	0.500000 *	17.4	12.5
60.0 *	0.443491	18.2	15.0
80.0 *	0.354844	19.9	20.0
100.0 *	0.288953	21.5	25.0
140.0	0.200000 *	24.8	35.0
150.0 *	0.183838	25.6	37.5
200.0 *	0.125065	29.6	50.0
232.6	0.100000 *	32.3	58.2
250.0 *	0.089415	33.7	62.5
300.0 *	0.066425	37.8	75.0
500.0 *	0.026098	54.1	125.0
571.4	0.020000 *	59.9	142.8
795.4	0.010000 *	78.1	198.9

Fig. 1 Typical tabular output following a detailed study of one specific release. The table shows the variation of concentration (and other dispersion parameters) with distance from a released vapour cloud. The table is built up from calculation requests made by the user (user-specified input is marked with a * in the table).

TARGET EQUIPMENT ITEM : INDUSTRIAL-SITE

CALCULATED FIREBALL HEAT FLUXES (KW/SQ.M) AT : IND-2

T180	T180	T190	T190
AN. 1C SO. 1	AN. 2 SO. 2	AN. 1C SO. 1	AN. 2 SO. 2
(9506.1,10960.7)	(9506.1,10960.7)	(9504.2,10951.0)	(9504.2,10951.0)
65.5840	134.1461	60.6152	123.9829

CALCULATED U.V.C.E. OVERPRESSURES (BAR) AT : INDUSTRIAL-SITE

T180	T180	T190	T190
AN. 1C SO. 1	AN. 2 SO. 2	AN. 1C SO. 1	AN. 2 SO. 2
(9506.1,10960.7)	(9506.1,10960.7)	(9504.2,10951.0)	(9504.2,10951.0)
0.0826	0.0959	0.0781	0.0906

TARGET EQUIPMENT ITEM : T190

CALCULATED END-ON HEAT FLUXES (KW/SQ.M.) AT : T190

T180	T180	T180	T180
AN. 1P SO. 1	AN. 3 SO. 3	AN. 4 SO. 4	AN. 5 SO. 5
(9506.1,10960.7)	(9506.1,10960.7)	(9506.1,10960.7)	(9506.1,10960.7)
************	************	************	************
T240	T240	T240	T250
AN. 2 SO. 2	AN. 3 SO. 3	AN. 4 SO. 4	AN. 1 SO. 1
(9195.3,11159.9)	(9191.0,11145.2)	(9185.9,11128.3)	(9268.3,11143.1)
0.2691	6.1648	************	12.3712
T250	T250	T260	T260
AN. 3 SO. 3	AN. 4 SO. 4	AN. 1 SO. 1	AN. 2 SO. 2
(9264.8,11128.0)	(9261.2,11111.0)	(9372.1,11119.1)	(9372.1,11119.1)
11.5109	************	39.9420	1.0519
T260	T270	T270	T270
AN. 4 SO. 4	AN. 1 SO. 1	AN. 2 SO. 2	AN. 3 SO. 3
(9366.7,11086.9)	(9446.5,11101.9)	(9446.5,11101.9)	(9443.7,11087.3)
************	110.0078	2.2670	126.1447

THE MAXIMUM TOLERABLE INTENSITY (10.0000) IS EXCEEDED BY THE FOLLOWING RELEASES :-

T250	ANALYSIS 1	SOURCE 1
T250	ANALYSIS 3	SOURCE 3
T260	ANALYSIS 1	SOURCE 1
T270	ANALYSIS 1	SOURCE 1
T270	ANALYSIS 3	SOURCE 3

Fig. 2 Typical tabular output from the inter-item hazard checking module, showing the consequences for both a nearby industrial site and an on-site storage tank (T190), from a range of on-site releases. Each calculation result is referenced to the specific mode of release ([SO]urce) and operating conditions ([AN]alysis) from which it results.

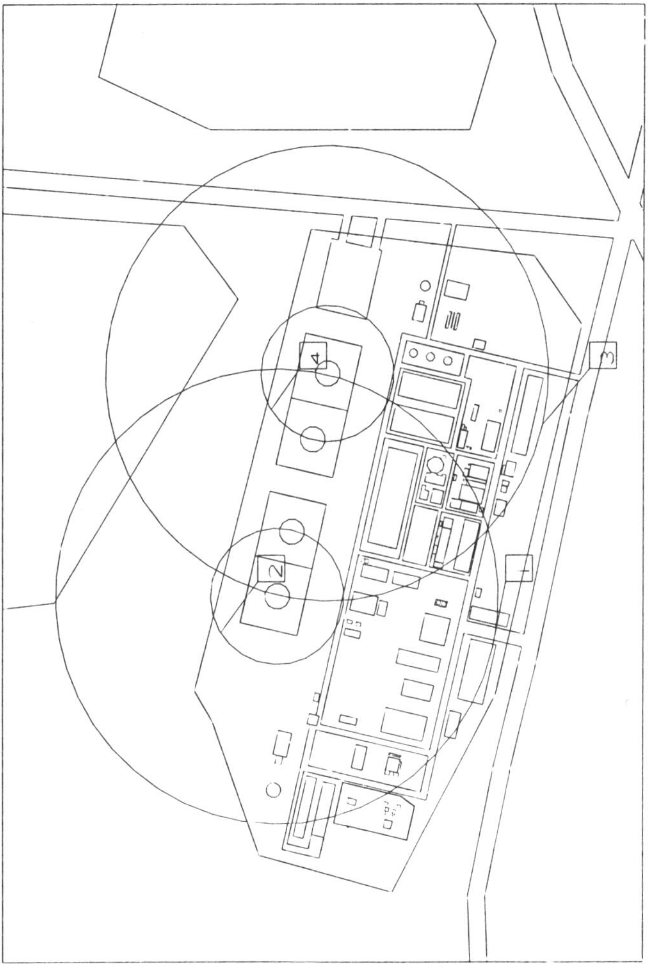

Fig. 3 Typical Graphical Output. The contours labelled 1-4 show the regions within which a pool fire heat flux in excess of 16 kW/m^2 might be expected following each of four alternative release scenarios from the tank farm area.

OPTIMISATION AND SAFETY

Trevor A Kletz*

Safety is usually considered as something apart from optimisation and as even opposed to it. Design changes to increase safety may well be necessary but increasing demands for ever higher standards are seen to result in expensive additions to plant cost, sometimes to the extent of making the whole process uneconomic.

This need not be the case: A design that has to be made safe by adding on expensive, complex protective equipment that requires testing and maintence is not an optimum design. In many cases we can avoid hazards by changes in design, but to do so we need to consider alternatives and carry out safety studies much earlier in design than is customary in many companies.

The thesis of this paper is illustrated by examples, including the design of plant for carrying out liquid phase oxidation processes, and the constraints that prevent development of inherently safer designs are discussed.

INTRODUCTION

Safety is usually considered as something apart from optimisation and as even opposed to it. Changes in design to increase safety may well be necessary but increasing demands for ever higher standards of safety are seen to result in expensive additions to plant cost which may make the whole design uneconomic.

It is the theme of this paper that this need not be the case: A design that has to be made safe by adding on to it expensive, complex protective equipment that has to be tested and maintained is not an optimum design and, in many cases we can avoid hazards by changes in design, in particular by reducing inventories of hazardous materials. Examples are given of what has been or might be done and the constraints that prevent more frequent use of this approach are discussed.

* University of Technology, Loughborough, Leics.

INTENSIFICATION

Many of the examples in this paper have not appeared in earlier papers[1] on the subject but the first example is a familiar one. However it is included again because it illustrates so well the essential features of the approach. Nitroglycerine (NG) used to be manufactured in a batch reactor containing about a tonne of material. The operator sat in front of the reactor, watching the temperature and controlling the addition rate of the reactants. If the reactor got too hot it would blow up, taking the operator with it. To make sure that he stayed awake he sat on one of the famous one-legged stools.

If we were asked to make this process safer, what would most of us do? We would add onto the reactor instruments for measuring temperature, pressure, flows, rate of temperature rise and so on and then use these measurements to operate valves which stopped flows, inceased cooling, opened vents and drains and so on. By the time we had finished the reactor would hardly be visible beneath the added-on protective equipment. However, when the NG engineers were asked to improve the process they did not proceed this way. They asked why the reactor had to contain so much material. The obvious answer is because the reaction is slow. But the chemical reaction is not slow. Once the molecules come together they react quickly. It is the chemical engineering - the mixing - that is slow. They therefore designed a small well-mixed reactor, holding only about a kilogram of material, that achieves about the same output as the batch reactor. The reactor resembles a laboratory water pump. The flow of acid through it sucks in the glycerine through a side-arm. Very rapid mixing occurs and by the time the mixture leaves the reactor reaction is complete. Similar changes have been made to the later stages of the plant where the NG is washed and separated.[2]

The modern NG process is an example of intensification, of achieving the same performance with a smaller inventory of material. Intensification is safer because there is less hazardous material present to leak out ("What you don't have, can't leak"). It is also cheaper in capital because the smaller inventory results in smaller vessels, pipes, structures, pumps, foundations etc. It is cheaper to operate because heat losses will be less. Safety and optimisation have gone together.

Intensified plants are said to be inherently safer as the increased safety is not obtained by adding on protective equipment which might fail but by reducing or eliminating the materials or circumstances that give rise to the hazard.

As another example, one where less success has been achieved, consider liquid phase hydrocarbon oxidation processes such as those used in the oxidation of cyclohexane (to cyclohexanol/cyclohexanone or KA mixture), cumene (to phenol and acetone) and p-xylene (to terephthalic acid). Reaction is slow and conversion is low so that large quantities of material are present in the reactors and most of this material has to be recovered and recycled. Many serious fires and explosions have occurred as the result of leaks from hydrocarbon oxidation plants including the explosion at Flixborough in 1974.[3] Obviously intensification of the reaction would result in a safer plant. Can reaction be made faster and conversion greater?

The reactions are not really slow. Once the molecules come together they react quickly. It is the mixing that is slow, the chemical engineering rather than the chemistry, so if the mixing can be improved, reaction volume can be reduced. Conversion is low because if more air (or oxygen) is used to increase conversion it is difficult to ensure that the air (or oxygen) is intimately mixed with the hydrocarbon and high local concentrations cause unwanted by-products to be formed. If a better method of mixing air (or oxygen) and hydrocarbon could be found then conversion could be increased.

Intensification would also result in a more economic plant. Most of the process material in hydrocarbon oxidation plants gets a free ride; only a small percentage is removed as product at each pass. What would we think of an airline in which only 5 percent of the passengers got out at the end of each trip and the rest travelled backwards and forwards? Yet as engineers we are quite prepared to design and operate similar processes. It is worth remembering that if we require an output of 20 000 tonnes per year it can all pass through a pipe 5 cms diameter, assuming a linear velocity of 1 m/sec which is not particularly fast. Any larger pipe in the process lines of a hydrocarbon oxidation plant is a reflection on our ability as engineers. The Flixborough plant contained 28 inch pipes (though this was partly because they had to carry gas and liquid flows).

Since Flixborough very few hydrocarbon oxidation plants have been built, for economic reasons. If a new plant was required one wonders if it would have a significantly lower inventory than existing plants or if existing designs would be reproduced, with minor improvements, because there was no time to undertake a programme of research on intensification. One company which started a programme on the intensification of a hydrocarbon oxidation process abandoned it, despite promising results, because they could see no hope of a new plant in the forseeable future. On the other hand a new stirred pot reactor for gas-liquid reactions has recently been described.[4] Instead of the gas entering the liquid through a sparger, a down-pumping impellor draws in gas from the gas phase above the liquid; any unreacted gas that escapes back into the gas phase is recirculated. The reactor was designed to increase efficiency and output but, compared with conventional reactors, less inventory is required for a given output.

Although most fires and explosions on hydrocarbon oxidation plants are the result of leaks, some occur because the atmosphere in the reactor enters the explosive range and there is an internal fire or explosion. The chance of such an incident can be reduced by operating further from the explosive region. A new catalyst makes this possible on o-xylene oxidation plants.[5] Similarly, on cumene oxidation plants it is now customary to operate much further away, than in earlier designs, from the temperature at which a runaway reaction occurs. Fewer by-products are produced and the change has been made mainly for economic reasons but the plant is now safer and requires less added-on safety features. It may no longer be necessary to keep a large storage tank half-full of water in which the reactor contents can be dumped if a runaway occurs.

If 20 000 tonnes/year needs only a 5 cms diameter pipe then 100 tonnes/year needs only a 1 cm diameter pipe. Yet how often do we consider a small continuous reactor for this output instead of a batch reactor?[6]

Here are some recent examples of successful intensification. In most cases safety was not the primary motivation. The designers were looking for a cheaper design and/or lower operating cost but the designs as well as being cheaper were also safer. Inherently safer designs are not just an isolated but desirable improvement but part of the total package of improvements the chemical industry needs to make.

The ICI Higee distillation process can reduce inventories by a factor of up to a thousand.[7]

In a new process for low density polyethylene, low pressure ethylene gas is reacted in a fluidised bed. In the older processes the ethylene is polymerised at high pressure (when it is a supercritical gas but has the same density as the liquid) or a flammable solvent is used.[8]

Some designs of plate heat exchangers contain low inventories. At one time their use was restricted by lack of suitable gasket materials but these difficulties now seem to have been overcome.[9] New designs of intensified heat exchangers have recently been described.[9a,9b]

A new centrifugal fluidised bed drier contains lower inventories, for a given output, than a conventional drier.[10]

Vortex mixing can reduce the inventory required for membrane separation.[11]

Membrane cells for chlorine manufacture, as well as being more economic, avoid the use of mercury and asbestos.[11a]

Storage

Operators and managers usually like to have as much intermediate storage as accountants and designers will let them have. But such storage is expensive in capital (fixed and working) and usually hazardous as intermediates are usually reactive chemicals, or they would not be used as intermediates. The methyl isocyanate (MIC) which leaked at Bhopal with the loss of over 2000 lives was an intermediate, not a product or raw material. It was convenient to store it but not essential. Now Union Carbide, the company concerned, and other companies are saying that they have eliminated, or will eliminate, all intermediate storage of MIC. The material will be used as it is formed and instead of up to 100 tonnes in a tank there will be only a few kilograms in a pipe. Stocks of other toxic intermediates are also being reduced.[12]

At one time chlorine manufacturers carried very large stocks, sometimes measured in thousands of tonnes. Now stocks are much lower. The change has been made possible by improvements in the process used for the manufacture of chlorine which make it possible to increase and decrease rate very quickly when demand changes.

Computer simulations based on variations in demand and the known pattern of shutdowns, planned and unplanned, in the producing units are often used to show that so much intermediate or product storage is necessary. However, these simulations take no account of the fact that if stocks are low people improvise, change plans and so on in order to maintain production in a way not foreseen at the design stage. "Grey hairs cost the company nothing". Holusha, reviewing a book on Japanese industry writes: "... the American practice of having buffer stocks of partly finished components all along the production line concealed problems. If a machine broke down, parts from a nearby buffer stock could keep the assembly line working until somebody got round to fixing the machine. But ... if the security of the buffers was removed, if the whole plant were in jeopardy of being shut down by one malfunctioning machine, the workers would be forced to tend their machines more carefully, so that all worked correctly all the time. The discipline this system imposed brought surprising improvements in productivity and quality."[13] In short, the need for large stocks is a self-fulfilling prophecy. In the process industries reduced stocks will not make operators treat their plants more carefully but they will encourage the maintenance organisation to get the plant back on line quickly after a shutdown.

SUBSTITUTION and ATTENUATION

If intensification is not practicable it may be possible to substitute a less hazardous material or process for a hazardous one. For example, the product made at Bhopal (carbaryl) is made from alpha-naphthol, methylamine and phosgene. At Bhopal the first two of these were reacted together to make MIC, which was then reacted with phosgene. In the alternative process, used by the Israeli company Makhteshim, alpha-naphthol and phosgene are reacted together to make a chloroformate ester which is then reacted with methylamine to give carbaryl. MIC is not formed at all (Figure 1).[14] Another example: Alpha-naphthol, required for the manufacture of carbaryl, is made (by both Union Carbide and Makhteshim) by a process (Figure 2) which avoids the formation of the traces of beta-naphthol which are formed in the old nitration-reduction method.[14]

In the manufacture of peroxides water is removed with a dehydrating agent. An investigation of one such process showed that if magnesium or aluminium sulphate is used as the agent the process is inherently protected against a rise in temperature as this causes the release of water by the agent, diluting the reactants and stopping the reaction.[15]

Other examples of substitution are the use of non-flammable heat transfer agents such as water or fluorinated hydrocarbons instead of flammable oils[16] and the use of non-flammable insulating and cooling liquids for transformers.[17] If we cannot use a non-flammable heat transfer oil it may be possible to use a higher-boiling and therefore less hazardous one.

Substitution has been used extensively in the manufacture of silicon semiconductors. Toxic gases such as phosphine, diborane and silane can be replaced by liquids such as trimethyl phosphite, trimethyl borate and tetraethylorthosilicate. Phosphorus oxychloride can replace phosgene and 1,1,1-trichloroethane can replace HCl gas. All these changes have technical as well as safety advantages.[17a]

The economic advantages of substitution are not as great or as obvious as those of intensification but they are still there. For example, when a fluorinated hydrocarbon was chosen as the refrigerant for a new ethylene liquefaction plant in place of propylene, it was found possible to make the plant more compact, as the fire risk was reduced, and the compressor could be obtained more quickly. Less fire insulation was also required.

Substitution is more effective than intensification in reducing fugitive emissions, the small, continuous emissions than produce concentrations of toxic materials in the working environment.

Another alternative to intensification is attenuation: handling a material in a safer form. An example is the transportation of organic peroxides in solution. This costs more, as the solvent has to be transported, but the alternative may be not transporting it at all, or transporting it in very small quantities or with extreme precautions.[18]

Nitration reactions have to take place, if possible, for economic reasons, at temperatures close to those at which decomposition occurs. They can be carried out safely at these temperatures if the reaction mixture is sufficiently dilute.[15] Good mixing can compensate for the effect of dilution.

SIMPLIFICATION

There can be no doubt that simplification is desirable on economic grounds. Complicated equipment costs money, to install, test and maintain. It also provides increased opportunities for error, leading to hazards and operating problems. By simplifying plants we can often eliminate hazardous operations or equipment. For example, leaks and fires often occur in the pump rooms of ships and men have often been overcome in them due to shortage of oxygen. One way of dealing with the problem is to use pumps submersed in the ships' tanks and eliminate the pumproom. This is cheaper as well as safer.[19] Another example of simplification achieved by combining two items of equipment is a combined fan and deduster.[6]

In simpler plants there are fewer points from which fugitive emissions can occur.

Liquid phase hydrocarbon oxidation processes can be simplified, and made safer, by eliminating off-gas recycle. Explosions have occurred in the recycle lines because too much oxygen was added. If there are no recycle lines this cannot occur and the advantages of off-gas recycle have been shown to be illusory.[20]

CONSTRAINTS

The reasons why progress towards inherently safer and simpler plants has been somewhat disappointing are discusssed in detail in reference 1. The principal constraints are time and company procedures. If a plant includes innovation, of any sort, it will probably take longer to design than a familiar plant and longer to come on line and come up to flowsheet output. There may be unforeseen difficulties. The output from the new plant is probably wanted as soon as possible, or a market opportunity will be lost. The company cannot risk the delay and decides to repeat the original design. For this reason Malpas[21] recommends that, during design, we should always consider the plant after next. During design we are conscious of all the things we would like to do but do not have time to do.

In adddition to shortage of time, a further constraint is that company procedures often do not call for detailed consideration of alternatives during the early stages of design or during the conceptual stage when the process is being chosen. In many companies safety studies take place late in design or even when construction is complete. Safety and loss prevention staff take part in relief and blowdown reviews, hazard and operability studies (hazops), model reviews, inspections of the finished plant and so on but by the time these studies are made all they can do is add on protective equipment to control the hazards that are found; they cannot eliminate them by a change in design. We need to examine designs critically during their early stages. In addition to the conventional hazop of the line diagram I suggest that we need two earlier studies, one at the conceptual stage and one at the flowsheet stage. For the flowsheet hazop the conventional guide words (none, more of, less of, part of, more than, other) should be suitable but for the conceptual hazop different guide words or questions are needed. There are some suggestions in reference 1, Chapter 4.

A further constraint is that many companies are now organised in business areas rather than functional departments (such as research, process design, etc) and as a result it is not clear whose job it is to innovate. The business is managed but not the technology and business-oriented managers may be umsympathic to technological problems. If the problem is recognised it can be overcome. ICI's research and technology director says, "We have a process by which scientists can talk about science strategy which cuts across business boundaries. We have seven science groups covering the whole of ICI's scientific ambitions. Each of these scientific areas might serve up to five or six different business strategies".[22]

Acknowlegements

Thanks are due to the many colleagues who suggested items for inclusion in this paper and to the Leverhulme Trust for financial support.

REFERENCES

1 T A Kletz, *Cheaper Safer Plants or Wealth and Safety at Work*, Institution of Chemical Engineers, 2nd edition, 1985.

2 N A R Bell, *Institution of Chemical Engineers Symposium Series No 34*, 1971, p 50.

3 T A Kletz, *Loss Prevention*, Vol 12, 1979, p 34.

4 L M Litz, *Chemical Engineering Progress*, Vol 81, No 11, Nov 1985, p 30.

5 T Sato et al, *Hydrocarbon Processing*, Vol 62, No 10, Oct 1983, p 107.

6 C Ramshaw, *The Chemical Engineer*, No 416, July/Aug 1985, p 30.

7 C Ramshaw, *The Chemical Engineer*, No 389, Feb 1983, p 13 (& No 388, Jan 1983, p 7).

8 C R Raufast, *Hydrocarbon Processing*, Vol 63, No 5, May 1984, p 105.

9 L Sennik, *The Chemical Engineer*, No 406, Aug/Sept 1984, p 41.

9a W T Cross and C Ramshaw, *Chem Eng Res Des*, Vol 64, July 1986, p 293.

9b T Johnson, *The Chemical Engineer*, No 431, Dec 1986, p 36.

10 *The Chemical Engineer*, No 416, July/Aug 1985, p 25.

11 D Williams (ed), *Research in Medical Engineering - A Review of Projects Supported by the Engineering Board of SERC*, Science and Engineering Research Council, 1985.

11a W N Brooks, *Chemistry in Britain*, Vol 22, No 12, Dec 1980, p 1095.

12 *Plant/Operations Progress*, Vol 5, No 2, April 1986, p A3.

13 J Holusha, *New York Times*, 2 April 1986 (in a Review of M A Cusumano, *The Japanese Automobile Industry*, Harvard University Press).

14 B M Reuben, private communication.

15 H G Gerrirsen & C M van't Land, *I&EC Process Design and Development*, Vol 24, 1985, p 893.

16 J Singh, *Heat Transfer Fluids and Systems for Process and Engineering Applications*, Dekker, 1985.

17 *Chemistry in Britain*, Vol 20, No 5, May 1984, p 392.

17a *Product Application Notes Nos 5, 6 & 15*, J C Schumacher Co, 580 Airport Road, Oceanside, California 92054.

18 D J Lewis, *Hazardous Cargo Bulletin*, Feb 1986, p 31.

19 K Fleming, *Hazardous Cargo Bulletin*, Jan 1984, p 9.

20 A Krzysztoforski et al, *Proceedings of the 5th International Symposium on Loss Prevention and Safety Promotion in the Process Industries*, Societe de Chimie Industrielle, 1986, Paper 45.

21 R Malpas in B Atkinson (ed), *Research and Innovation for the 1990's*, Institution of Chemical Engineers, 1986, p 28.

22 C Reece, quoted by R Stevenson, *Chemistry in Britain*, Vol 22, No 8, August 1986, p 695.

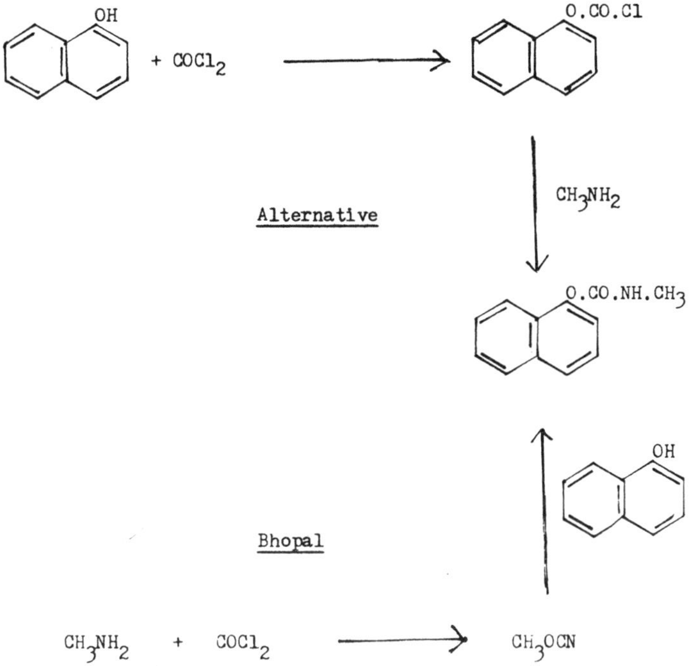

Figure 1 Routes to Carbaryl

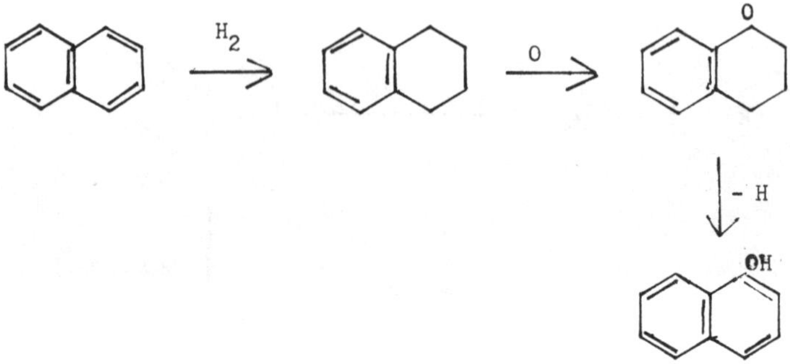

No trace of 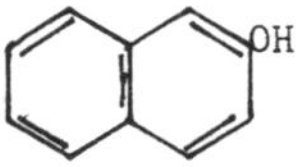

formed by this process.

Figure 2 Alpha-Naphthol from Naphthalene

DYNAMIC MODEL BASED OPTIMISATION IN THE DESIGN OF A BATCH PROCESS INVOLVING SIMULTANEOUS REACTION AND DISTILLATION

J.A. Wilson*

The objective of this paper is to demonstrate that with the aid of even a simple process model the economic optimisation of a batch process can be taken significantly further than is traditional at the predesign stage. The process used to demonstrate this is a complex one, itself developed with modelling assistance, in which chemical reaction and distillation steps are combined into an integrated system. The effects of prime design variables on the economic performance of the process around the optimum are highlighted. The approach used is a general one however and, with a suitable model, should be widely applicable.

INTRODUCTION

During recent years digital computer based aids have had considerable impact on the engineering of continuous process systems. In the design environment, aside from widely used draughting and data management systems, it is now quite commonplace to find plant item and even flowsheeting simulation software being used routinely. As a result designs can be both more precise and better posed and tested. There is also interest in the development of more approximate aids for use at the predesign stage to enable a measure of optimisation in both selecting between processes and narrowing down the choice of the most important process variables (refs. 1 and 2).

In strong contrast to this batch process engineering is by comparison poorly catered for in terms of computer aid. Paradoxically, the pressure to get batch processes up and running within a tight time scale is perhaps even stronger than in the case of continuous processes. As a result process engineering activities are curtailed and plant design often centres on direct scale-up of chemical laboratory equipment. With computer aid a more detailed economic and engineering analysis should be possible in the time available thus driving process decisions closer to an optimum at a very early stage. Even having defined the process however the inherent time varying behaviour of batch systems makes precise plant design difficult. To cope with this a conventional approach is to introduce sufficient flexibility, capacity or over design into the system to allow any shortcomings to be "fine tuned" out once the plant is built and on stream. Weaknesses in the design will often translate into lengthened cycle times, but provided sufficient flexibility is built into production schedules to

*Department of Chemical Engineering, University of Nottingham.

allow for this, the required production will still be achieved. Obviously, though, a real economic cost is paid for this "fine tuning" in the production situation and it is in an attempt to avoid this cost penalty that the work described here is directed. Some practical fine tuning will inevitably be necessary in starting up any new plant but the objective here is to side step avoidable penalties by moving directly into a region of efficient operation through a measure of optimisation early in the design evolution.

A prerequisite to the detailed analysis of any process is a precise definition of its characteristics and behaviour. Because in batch processing we are interested in the movement of process variables within the time scale of a batch cycle this translates into a need for a dynamic process model in a form that is suited to digital computer simulation. The effort entailed in producing such a model centres largely on the capture of sufficient experimental data (perhaps laboratory generated) to define the rates of process movement (e.g. reaction kinetics). By comparison the jobs of defining clearly the basic engineering chemistry and of its conversion into the form of a set of dynamic process equations is far less demanding than many people anticipate. Furthermore, as the availability of suitable applications orientated computer aids grows the development and use of such models should become progressively more easy. The main purpose here however is to make use of such a dynamic process model as an aid in process optimisation. No further attention will therefore be paid to model development other than to point out that in the case of the specific process studied an extensive process model has been developed as part of a SERC/DTI Teaching Company project between the Boots Company PLC and the Department of Chemical Engineering at Nottingham University (Ref. 3)

Thus, as a basis for optimisation there is a requirement for a simplified model that embodies sufficient features of the process to provide an explicit link between the prime design variables and economic performance whilst minimising computational demands. As a result the model can act as a vehicle providing a more precise visualisation of the sensitivity of performance to unconstrained design variable choices. Considered decisions can then be taken by a designer regarding the position of optimum choices in the face of constraints where necessary.

PROCESS DESCRIPTION

The process under study is a complex one in which a pharmaceutical intermediate compound is produced in the organic liquid phase of a two phase reaction mixture. Reaction rates relate primarily to the composition of the organic phase and the reaction scheme reduces to one of the parallel form

where for reasons of both simplicity and commercial confidentiality the main components are given pseudonyms as follows:

A the main reactant (volatile)
B the main product (volatile)
C an unwanted side product (involatile)
D a solvent/catalyst (involatile)

The reaction proceeds over a period of hours at elevated temperature, a further complication being the progressive decomposition of D to other involatile impurities.

Options for recovery of product B include evaporation of volatiles A and B from the reaction vessel, leaving the other involatile organic and inorganic phase components behind for separate discharge. Pure B can then be recovered in a subsequent distillation step leaving unreacted A for recycle. Fortuitously, in this particular reaction system, product B is the most volatile component present, giving rise to the possibility of its distillation from the reaction vessel whilst reaction is in progress. This has the advantageous effect of artificially boosting the concentration of reactant A in the organic phase, and thus of increasing the rate of formation of product. As a result a higher conversion to product B will be achieved in a fixed reaction time than if the reaction were conducted with no distillation. This feature is illustrated in Fig. 1 for the extreme cases of no distillation and perfect distillation of product B. Furthermore, no separation of B from unreacted A is necessary subsequent to the reaction step. At the close of reaction any remaining A can be rapidly stripped directly from the reactor. The whole process could thus be conducted in a single reactor with integrally mounted distillation column, rather than in separate reaction and distillation vessels. This simultaneous reaction/distillation approach therefore has clear capital cost and equipment utilisation advantages and forms the basis of this study.

PROCESS MODEL

The approximate model equations used to represent the process are presented for the reactions (Appendix I), the distillation column (Appendix II) and the economic performance indicators (Appendix III). A total of seven differential equations is involved (five relating to reacting components, and two relating to distillation aspects). The full set of equations can be simulated on a digital computer using standard numerical integration techniques. As a measure of computational loading, the profiles presented here were infact all generated using a fourth order Runge-Kutta integration procedure mounted on a BBC microcomputer. In general integration time steps of the order of one hour process time were used over a batch cycle of up to 40 hours. Thus only longer term movement in process variables is accounted for. Examination of short time scale movements, as for example might be needed in control system design, would be more appropriately done using a more detailed model, and consequently more computational resources. Relatively little shift in optimum design or operating conditions would be envisaged on the basis of such a more detailed approach however.

The approximate model equations used to represent the distillation column (Appendix II) embody the significant assumption that hold-up in the distillation column is negligible in comparison with quantity of material held in the still. As a result compositions throughout the distillation column respond to changes in still composition with no dynamic lag. The mole fraction of product in the overhead stream at any instant $y_b(t)$ is calculated on the basis of ideal binary distillation (i.e. no heat losses, fixed relative volatility, equimolal overflow). The algorithm employed for this is an iterative one based on use of Smoker's equation for a rectifying column.

OPTIMUM SEQUENCE STRATEGY

The conversion profiles presented in Fig. 1 represent the extremes of reaction with no distillation and reaction with simultaneous "perfect" distillation. "Perfect" here means that a column of infinite separating power is used such that pure product B is distilled immediately on its formation with x_b, the mole fraction of B in the reactor/still, being always zero. The cost penalty against this is obviously infinite in that an infinite size of column is implied. Any real column will only be capable of sustaining a distillate composition to specification using a reflux ratio above the minimum and with a non-zero mole fraction x_b in the reactor/still. Even for a fixed column design (i.e. equivalent to a fixed number of theoretical plates) a variety of reflux and boil-up strategies are possible. For example the optimum distillation strategy in boil-up and reflux ratio allows instantaneous distillate compositions below specification provided the total distillate collected is to specification. This is a computationally demanding problem to solve so here, because the prime objective is a rapid low computational overhead approach, a relaxation to the more easily handled strategy of always recovering distillate at the required specification is chosen. Although a suboptimal strategy this still involves profiling reflux ratio and boil-up and it lies far closer to the optimum than the more normal practical approaches (e.g. stepping reflux ratio at widely spaced predefined time intervals or perhaps even holding reflux ratio steady throughout). It thus represents a useful yardstick against which to compare design variable choices.

Profiling boil-up is a practice that is even less common in a production environment than is profiling reflux ratio, despite the increasing use of on-line computer sequence control systems capable of implementating such an approach. In this particular study however there is little alternative if best advantage is to be gained from the proposed integrated process, simply because product off-take rate, and hence boil-up rate, is driven by the varying rate of product formation.

Turning now to design aspects of the distillation column, any real distillation system will lead to conversion profiles lying between the two extremes of Fig. 1, the closeness of approach to the "perfect" distillation case being related to the separating power of the integrally mounted column. A column equivalent to a fixed number of theoretical plates will be unable to achieve distillate to specification at the very start of reaction, even under total reflux conditions, since there will be little or no product in the reactor to vaporise. A waiting period under total reflux is called for to allow the reactor/still product mole fraction x_b to rise above the minimum consistent with a specification distillate. Product draw-off can then be initiated at a finite reflux ratio. In order to use the full separating power of the column the distillate draw-off rate F(t) should be set to maintain a fixed mole fraction x_b in the reactor/still.

A strategy regarding reaction temperature T is also required. Here the suboptimal approach of holding T steady during the batch has been chosen. The optimum approach might be seen to involve time profiling T to compensate for falling reactant concentrations but here the added computational overhead of determining an optimum profile was considered disproportionate to its effects on plant design. Once design of the plant is defined, however, profiling might usefully be investigated, with the aid of the model presented, as an operating problem.

A steady reaction temperature can be achieved in practice by manipulating the system operating pressure throughout the reaction period, itself an important factor in that operating pressure has various cost implications including its effect on distillation column size and cost. Alternative charging strategies for catalyst/solvent D could also be considered although again in this study the suboptimal approach of charging all D at the start of the reaction phase is used.

Thus a variety of fairly involved dynamic optimisation problems are embedded in defining a global optimum operating strategy. To allow a reasonably close approach to the global optimum whilst keeping computational overheads to a reasonable level a suboptimal sequence strategy is chosen here. Having defined a broad choice of design variables and hence design of the plant on this basis some of the more sophisticated optimisations could be fruitfully investigated for the purpose of production sequencing (i.e. trimming around the operating optimum).

In summary then, for design purposes the selected suboptimal operating sequence involves charging reactant A and solvent/catalyst D (together with the second inorganic phase components) at a fixed reaction temperature. Product B is distilled to specification as early as possible from the column head whilst reaction proceeds in the reactor/still vessel. Reaction temperature is fixed by manipulation of system operating pressure. At the end of the chosen reaction period all remaining volatiles (i.e. unreacted A and any residual product B) are stripped quickly from the reactor (i.e. column on total draw-off) for recycle to the next batch.

SIMULATED BATCH BEHAVIOUR

The approximate process model equations can be integrated for a given batch charge and fixed column size under the described operating sequence and as a result time profiles of the main process variables can be plotted. Fig. 2 shows typical variations in the reactor organic phase mole fractions $x_a(t)$ and $x_b(t)$ of reactant A and product B respectively. The initial period of product build up in the reaction mixture is terminated by the start of distillate draw-off. The distillate rate F(t) is used to hold $x_b(t)$ fixed whilst reflux ratio is adjusted to keep the distillate composition $y_b(t)$ at specification. The effect of distillation in delaying the fall in $x_a(t)$, the reactant concentration, is quite evident though in the long term reaction will tail off as reactant A is consumed. The reaction could be terminated at any chosen point in time through the batch by simply switching to total off-take of volatiles from the reactor.

The necessary variations in distillate rate F(t), reflux ratio R(t), and hence boil-up rate F(R+1) are shown in Fig. 3. As can be seen, reflux ratio, distillate and boil-up rates must all fall progressively from maximum values at the start of distillation in order to recover acceptable product with fixed x_b. Total quantity of organic phase in the reactor M(t) and the overheads receiver ($\mu_a + \mu_b$) are shown in Fig. 4.

Optimisation under the set operating procedure described reduces to fixing reaction temperature, maximum reflux ratio, reactor residence time (i.e. time for switch over between product B and reactant A recycle cuts) and distillation column size (i.e. column diameter and number of plates or packed height).

OPTIMISING PRODUCTION PLANT DESIGN

The design objective is to define the reactor capacity and distillation column height and diameter sufficient to achieve a target annual tonnage of product. Integration of the process model equations, as already described, provides information on a reaction/distillation period of t hours duration. Appendix III shows how costs relating to a suitable production unit operating with a fixed reaction period of the same t hours can be calculated from the model (i.e. t hours reaction/distillation followed by stripping of volatiles before reactor discharge). The production unit's performance is given in terms of C_p the production cost contribution per unit product. Thus at each instant t during a dynamic simulation of the process the performance of a suitable production unit operating with a fixed reaction period of t can be calculated. As a result profiles of product cost contribution C_p against reactor residence time can be plotted for a variety of design variable choices.

Fig. 5 shows the product cost contribution C_p together with its materials and capital cost components (C_m and C_c) for a given number of column plates, reaction temperature and initial reflux ratio. Whilst the capital cost contribution C_c shows a minimum for shorter reaction times the progressive improvement in C_m the dominating materials cost leads to a minimum product cost C_p at much higher reaction times. The materials contribution is clearly asymptotic to a fixed cost per unit product when 100% conversion of reactant is achieved and cost of the initial reactor charge is shared over the maximum yield of product. The capital cost shows a minimum through its dependence on production batch size. A short reaction time implies use of a large reactor in order to achieve the required annual production at low conversion. Longer reaction times imply better conversions but the reactor size must again increase since fewer batches can be fitted into the production campaign.

Fig. 6 shows the variation in C_p with reaction temperature T for fixed column plates N_p and initial reflux ratio. The strong effect of temperature on performance is quite evident and an optimum temperature can be quickly located by trial and error over several simulation runs.

Fig. 7 shows the variation in C_p with N_p the number of column plates for fixed initial reflux ratio and reaction temperature. Clearly there is a disincentive to include more than twelve column plates.

The number of column plates N_p and the initial reflux ratio interact in their effects on process performance and in order to arrive at an optimum choice a series of simulation runs were completed at fixed ("optimum") reaction temperature. Values for N_p in the range 2 to 20 and initial reflux ratios in the range 1 to 10 were used. The minimum product cost C_p^* was noted for each combination. Fig. 8 shows the results in terms of contours of this minimum product cost contribution C_p^* plotted against N_p and initial reflux ratio. Whilst smoother contours would result from using a finer grid of calculated points the plot is adequate for the purpose of identifying the main trends. An optimum choice of about 12 column plates is indicated but initial reflux ratio shows no optimum in the range examined. This behaviour with initial reflux ratio arises primarily from the relatively low level effect of energy cost compared with materials cost. Clearly from a practical standpoint an upper constraint on initial reflux ratio must be applied. The cost penalty incurred by applying such a constraint is however quantifiable from Fig. 8 and thus a considered decision can be made as to where the constraint should be placed.

Having defined an optimum design on the basis described the model could be used to examine performance of that design more fully. For example factors like reaction temperature or catalyst addition profiling or the effects of recycle stream compositions could be investigated. In the case of the particular process considered here however the effects of such factors on the optimum design are likely to be slight.

CONCLUSIONS

On the basis of an approximate dynamic model of the batch process under study it is possible to generate a series of plots of process efficiency (in terms of product cost contribution per unit product) for a range of alternative process and design variable choices. As a result a quantified approach to optimisation is made possible.

The fact that reactant costs are dominant is expected and typical of many batch processes involving complex substances. It translates into the batch processing ethos of maximising reaction yield which of itself, whilst in essence correct, provides little insight into developing a considered design. The numerical approach presented demonstrates how a quantifiable link can be forged between design and economics that should enable more rational design decisions to be made.

The graphical representation of performance afforded by this model based approach facilitates manual convergence towards an optimum process and plant design. Because of the relative scale of their effect on process performance an optimum choice of reaction temperature can be made prior to and almost independently of distillation column design (i.e. number of plates and initial reflux ratio). Having thus defined the plant design, further but progressively more minor performance improvements will be possible with refinement of the selected operating sequence.

REFERENCES

1. Fisher, W.R., Doherty, M.F., Douglas, J.M., "Evaluating Significant Economic Trade-Offs for Process Design and Steady-State Control Optimization Problems".
A.I.Ch.E. Journal 31, 9 (September) 1985, 1538-1547.

2. Ingleby, S., Rossiter, A.P., Douglas, J.M., "Economic Evaluation of Integrated Reactor Systems".
Chem. Eng. Res. Des. 64, (July) 1986, 241-247.

3. Wilson, J.A., Smith, W., "Modelling Batch Processes - A Case Study" in "Control and Instrumentation - The Changing Scene".
Institute of Measurement and Control Symposium, Harrogate 1985, 113-118.

4. Strangio, V.A., Treybal, R.E., "Reflux-stages Relations for Distillation", Ind. Eng. Chem. Process Des. Develop. 13, 3, 1974, 279-285.

APPENDIX I: REACTOR MODEL

$r_a(t)$ the rate at which feed A reacts through to products B and C depends upon organic phase mole fractions according to:

$$r_a(t) = (k_b + k_c)\, x_a(t)\, x_d^n(t) \qquad (A1.1)$$

whilst $r_d(t)$, the rate of decay of D, is given by

$$r_d(t) = k_d\, x_d(t) \qquad (A1.2)$$

The reaction rate constants k_b, k_c and k_d all follow an Arrhenius relationship with reactor temperature T^oK:

$$\ln k_i = A_i + B_i/T \qquad (A1.3)$$

The equivalent rates for formation of B and C are

$$r_b(t) = k_b\, x_a(t)\, x_d^n(t) \qquad (A1.4)$$

and $$r_c(t) = k_c\, x_a(t)\, x_d^n(t) \qquad (A1.5)$$

Allowing for removal of a vapour stream from the reaction system at a molar flowrate F, and assuming a binary vapour mixture, molar balances on the principle reacting and volatile components give:

$$\dot{m}_a(t) = -r_a(t)\,M(t) - F(t)\,y_a(t) \qquad (A1.6)$$

$$\dot{m}_b(t) = r_b(t)\,M(t) - F(t)\,y_b(t) \qquad (A1.7)$$

$$\dot{m}_c(t) = r_c(t)\,M(t) \qquad (A1.8)$$

$$\dot{m}_d(t) = -r_d(t)\,M(t) \qquad (A1.9)$$

where x_i are the corresponding liquid phase mole fractions. Products from decay of component D appear on an equimolar basis and consequently the total number of moles of D and its decay products in the organic phase remains fixed throughout the batch. Hence an overall molar balance (addition of A1.6 to 9) gives

$$\dot{M}(t) = -F(t) \qquad (A1.10)$$

Because of the reaction stoichiometry loss of material from the reactor is due totally to vapour removal. A closed form equation set arises from using mole fractions as state variables such that

$$\dot{m}_i(t) = \frac{d}{dt}(M(t)x_i(t)) = x_i(t)\,\dot{M}(t) + M(t)\,\dot{x}_i(t) \qquad (A1.11)$$

Hence on substitution we have

$$\dot{x}_a(t) = -r_a(t) - F(t)(y_a(t)-x_a(t))/M(t) \qquad (A1.12)$$

$$\dot{x}_b(t) = +r_b(t) - F(t)(y_b(t)-x_b(t))/M(t) \qquad (A1.13)$$

$$\dot{x}_c(t) = +r_c(t) + F(t)\,x_c(t)/M(t) \qquad (A1.14)$$

$$\dot{x}_d(t) = -r_d(t) - F(t)\,x_d(t)/M(t) \qquad (A1.15)$$

Molar accumulation of condensate in the overheads receiver is given by

$$\dot{\mu}_b(t) = F(t)y_b(t) \qquad (A1.16)$$

and $$\dot{\mu}_a(t) = F(t)y_a(t) \qquad (A1.17)$$

Because only a binary distillation is involved equation A1.17 is redundant and $\mu_a(t)$ can instead be obtained from an overall balance on reactor and product receiver:

$$\mu_a(t) = M(o) - M(t) - \mu_b(t) \qquad \text{(A1.18)}$$

Equations A1.10 and A1.12 to A1.16 form a basic reactor and distillate receiver model, provided $y_b(t)$ can be defined. An approximate approach to this is given in Appendix II.

APPENDIX II: DISTILLATION MODEL

In order to minimise computational overheads, whilst retaining the effects of prime design variables on system performance, the distillation model is deliberately simplified to that of an ideal binary mixture under conditions of equimolal overflow. Liquid hold-up in the column is assumed negligibly small such that distillate composition y_b will move rapidly to a new steady state following changes in either reactor/still composition, boil-up rate or reflux ratio (i.e. a steady state column model).

The strategy chosen as a basis for optimisation demands that a fixed overhead product mole fraction y_b be maintained in the face of variable reactor/still composition $x_b(t)$ by choice of R(t) the reflux ratio and F(t) the distillate flowrate (which in combination fix boil-up rate at (R(t) + 1)F(t)). The column operating pressure P is chosen to maintain a fixed reactor/still bubble point temperature T^oK and can thus be calculated from

$$P(t) = p_a\, x_a(t) + p_b\, x_b(t) \qquad \text{(A2.1)}$$

Defining Reflux Ratio R(t)

In order to evaluate the necessary instantaneous reflux ratio R(t) an iterative scheme based on Smoker's equation for the rectifying section of a distillation column is employed. In the form quoted by Strangio (Ref. 4) this links distillate composition y_b, reflux ratio R(t) and the liquid mole fraction at which the operating and equilibrium lines intersect on a McCabe-Thiele diagram k_1 through the implicit scheme

$$M = R(t)/(R(t)+1) \qquad \text{(A2.2)}$$

$$C = 1 + (\alpha-1)k_1 \qquad \text{(A2.3)}$$

$$Y = \alpha/C \qquad \text{(A2.4)}$$

$$Z = y_b/(R(t)+1) \qquad \text{(A2.5)}$$

and $$f_1(k_1) = (M-Y)k_1 + Z = 0 \qquad \text{(A2.6)}$$

where α is the relative volatility of product B to reactant A. Given R(t) equation A2.6 is easily solved for k_1 using a Newton iteration starting from $k_1 = 0$ together with the analytic function derivative

$$\frac{df_1(k_1)}{dk_1} = M - Y/C \tag{A2.7}$$

For a given size of column (i.e. number of theoretical column plates N_p defined) the corresponding still mole fraction x_s is then given by

$$X = \frac{MC(\alpha-1)(y_b-k_1)}{(\alpha - MC^2)} \tag{A2.8}$$

$$Y = Y/(MC)^{N_p} \tag{A2.9}$$

$$\text{and } x_s = k_1 + (y_b - k_1)/(Y(1-X) + X) \tag{A2.10}$$

In order that this calculated still composition x_s corresponds with that actually present in the reactor vessel x_b, we must iterate to find a suitable reflux ratio R(t). This can again be achieved using a Newton iteration, this time incorporating a numerically evaluated function derivative:

$$\text{error function } f_2(R) = x_b - x_s = 0 \tag{A2.11}$$

$$\frac{df_2(R)}{dR} \approx \frac{f_2(R + \delta R) - f_2(R)}{\delta R} \tag{A2.12}$$

A reflux ratio increment δR of 1% of current value was found adequate in this application. Termination of the two nested iterations described was accepted when

$$f_1(k_1) \leqq 10^{-8} \tag{A2.13}$$

$$\text{and } f_2(R) \leqq 10^{-4} \tag{A2.14}$$

The result is a value for reflux ratio R(t) that forces distillate composition y_b to meet the required specification with closer than four figure correspondence between calculated and prevailing still compositions (respectively x_s and x_b).

Defining Distillate Flowrate F(t)

Having ensured that the overhead product is to specification we also need to ensure that product B is distilled at the same rate at which it is being formed by reaction from A. This holds composition x_b steady in the reactor/still whilst distillation of product is in progress. Thus distillate flowrate F(t) must be chosen to force $\dot{x}_b(t)$ in equation A1.13 to zero.

$$\therefore \; F(t) = r_b(t) M(t)/(y_b(t) - x_b(t)) \qquad \text{(A2.15)}$$

Clearly for a given reactor/still composition $x_b(t)$ the required overhead composition $y_b(t)$ can only be achieved if there are sufficient plates in the column. Thus distillate rate F(t) must be held at zero until reactor/still product composition $x_b(t)$ climbs to a level sufficient to allow product to be drawn from the column head at the required rate and specification. Here, a maximum feasible reflux ratio is enforced and product off-take is started only once the distillate specification can be reached at this maximum reflux ratio. In order to accurately fix the time at which distillation should start reduced integration time steps of as small as 0.1 hours were used during the phase of reaction without distillation.

Total Vapour Generated $\mu_v(t)$

In order to calculate the energy used in providing boil-up to the column during the whole of a batch we must integrate the instantaneous vapour flowrate over time. Thus the total quantity of vapour generated $\mu_v(t)$ is given by integrating

$$\dot{\mu}_v(t) = (R(t) + 1)\, F(t) \qquad \text{(A2.16)}$$

APPENDIX III : ECONOMIC MODEL

The basis for evaluation of economic performance used is the contribution to production cost of unit weight of pure product B. Cost components arising from process materials and services are included together with a capital cost component arising from the main plant items. Whilst other cost components could be included, perhaps allowing calculation of a realistic production cost, only those mentioned are directly affected by the design variables of interest. Thus the optimum choice of design variables arising from these cost components should also correspond closely with an optimum in actual production cost.

Materials Cost Component C_m

The cost of materials charged per unit weight of product can be calculated on the basis of a fixed initial reactor charge. Integration of the process model equations (Appendices I and II) for this base case batch gives $\mu_b(t)$ the quantity of product recovered in the distillate receiver t hours after charging. The materials cost per unit weight of product is then given from:

Cost of initial charge $= m_a c_a + m_d c_d + m_i c_i$ (A3.1)

Value of unconsumed reactant A at time t $= x_a(t)\ M(t)\ c_a$ (A3.2)

giving a net materials cost per unit weight of product of

$$C_m = [(m_a - x_a(t)M(t))c_a + m_d c_d + m_i c_i]/(\mu_b(t) mw_b) \quad (A3.3)$$

Clearly C_m will take a finite value only once some product has accumulated in the distillate receiver (i.e. $\mu_b(t) > 0$).

Energy Cost Component C_e

The main energy cost in the process arises in generating boil-up for the distillation column. The total quantity of vapour generated up to time t into the batch is $\mu_v(t)$ given from integration of model equation A2.16. The energy cost contribution per unit product is thus given as

$$C_e = \mu_v(t)\ c_v/(\mu_b(t) mw_b) \quad (A3.4)$$

An additional allowance could be included for energy used in stripping A remaining at the close of reaction for recycle.

Capital Cost Component C_c

Integration of the model equations for the fixed base case reactor charge gives $\mu_b(t)$ the quantity of product recovered over t hours of reaction. If a fixed annual production of W_p is required over an annual campaign of t_c hours then, allowing for a reactor turnaround time of t_t per batch, N_{bx} the number of production plant batches required per annum is

$$N_{bx} = t_c/(t + t_t) \quad (A3.5)$$

whilst W_{bx} the quantity of product required from each production batch is

$$W_{bx} = W_p/N_{bx} \quad (A3.6)$$

We can thus define a capacity factor K that relates this minimum size of production batch required (for a reaction time of t hours) to the size of the base case batch simulated by the model equations:

$$\text{Capacity factor } K = W_{bx}/(\mu_b(t) mw_b) \quad (A3.7)$$

The reactor/still capacity is fixed by the size of initial charge since reactor hold-up will deplete once distillation starts. Thus if W^o is the initial total weight of materials charged for a base case batch then W^o_{bx} the equivalent production plant initial charge weight is given as

$$W^o_{bx} = K W^o \qquad (A3.8)$$

The annualised capital cost of the reactor/still vessel and the distillation column can be averaged over the fixed annual production W_p to give a capital cost component per unit product. Since the reactor/still capacity is fixed by the size of initial charge W^o_{bx} its capital cost component can be taken in the form

$$C_r = k_r(W^o_{bx})^{0.667} \qquad (A3.9)$$

The equivalent column capital cost component can be expressed as

$$C_{dc} = k_{dc}(N_p)^{0.802} (d)^{1.066} \qquad (A3.10)$$

where N_p is the number of theoretical plates and d is the column diameter (Ref. 1). Whilst N_p is a prime design variable the column diameter d is fixed by the maximum vapour flowrate allowed by the column internals. Because the process demands draw-off of product B at the rate of its appearance by reaction the column vapour flowrate will always be at a maximum at the very start of product draw-off. Allowing for pressure and temperature conditions in the column the diameter d^o necessary for the base case batch can thus be calculated from the model using a relationship of the form

$$d^o = k_d(F(R+1)T/P)^{0.5} \qquad (A3.11)$$

where F and R are the distillate rate and reflux ratio at the start of product draw-off. The corresponding column diameter for the production plant is then given as

$$d = d^o K^{0.5} \qquad (A3.12)$$

The capital cost component C_c can thus be calculated as

$$C_c = C_r + C_{dc} \qquad (A3.13)$$

Contribution to Product Cost

Using the fixed base case initial charge the model allows calculation of cost components arising from materials C_m, energy C_e and both reactor C_r and column C_{dc} capital cost. The objective function for optimisation is Cp, the contribution to product cost, which is given by the sum of these components

$$C_p = C_m + C_e + C_r + C_{dc} = C_m + C_e + C_c \qquad (A3.14)$$

NOMENCLATURE

c_i	cost of reaction component i (£/kmol)
C_j	production cost component j (£/kg)
C_j^*	minimum value of C_j (£/kg)
d	production plant distillation column diameter (m)
d^o	base case batch distillation column diameter (m)
f(.)	a function
F(t)	distillate flowrate (kmol/h)
k_i	reaction rate constant for component i
k_j	a constant
K	production plant capacity factor over base case batch
$m_i(t)$	moles of reaction component i (kmol)
mw_i	molecular weight of component i
M(t)	moles of reactor organic phase (kmol)
N_{bx}	number of production batches per year
N_p	number of theoretical plates in column
p_i	vapour pressure of component i at T^oK (bar a)
P(t)	system operating pressure (bar a)
$r_i(t)$	reaction rate for component i (kmol/h)
R(t)	reflux ratio
t	reaction time (since start of batch) (h)
t_c	production campaign length (h)
t_t	reactor turnaround time (h)
T	reactor/still temperature (^oK)
W^o	weight of base case batch initial charge (kg)
W_{bx}	weight of product per production batch (kg)

W^{o}_{bx}	weight of production batch initial charge (kg)
W_p	weight of product required per annum (kg)
$x_i(t)$	mole fraction of component i in reactor
x_s	calculated still mole fraction
$y_i(t)$	mole fraction of component i in distillate
α	relative volatility of product B to reactant A
δR	increment in reflux ratio
μ_i	moles of overhead component i (kmol)

<u>i-subscripts</u>

a	reactant A
b	product B
c	side product C
d	solvent/catalyst D
i	inorganic phase
v	vapour

<u>j-subscripts</u>

c	capital cost
dc	distillation column
e	energy
m	process material contribution
p	production cost
r	reactor vessel

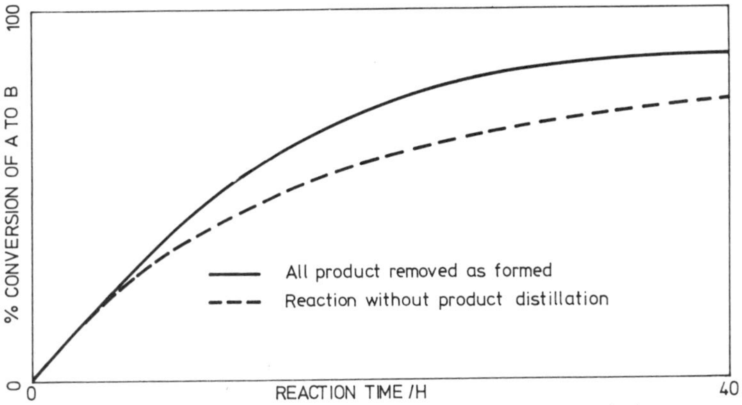

Fig. 1: Idealised Effect of Distillation on Reactant Conversion to Product

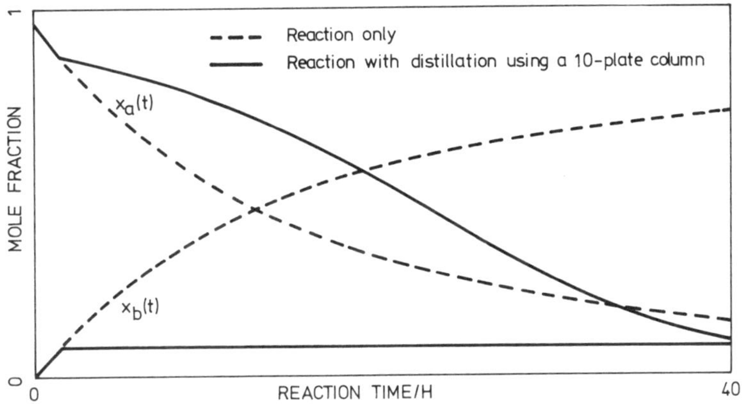

Fig. 2: Effect of Product Distillation on Reactor Compositions

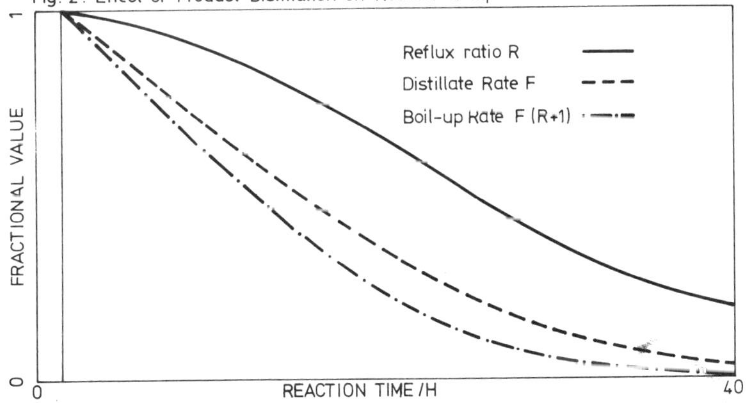

Fig. 3: Time Profiles in Reflux Ratio, Distillate and Boil-up Rates Corresponding to Fig. 2 (expressed as fraction of maximum value required)

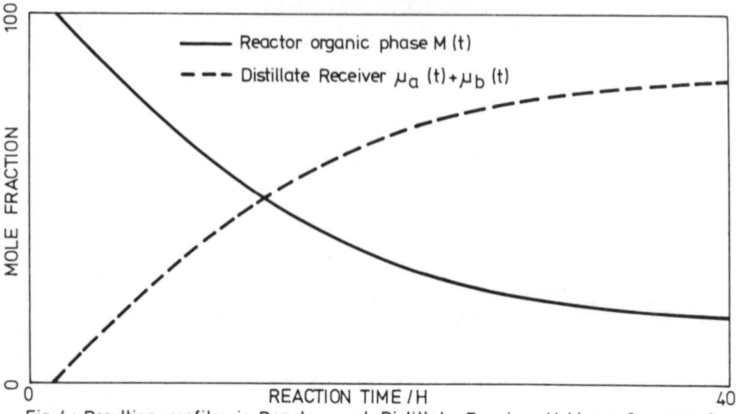

Fig. 4: Resulting profiles in Reactor and Distillate Receiver Hold-up Corresponding to Fig. 2

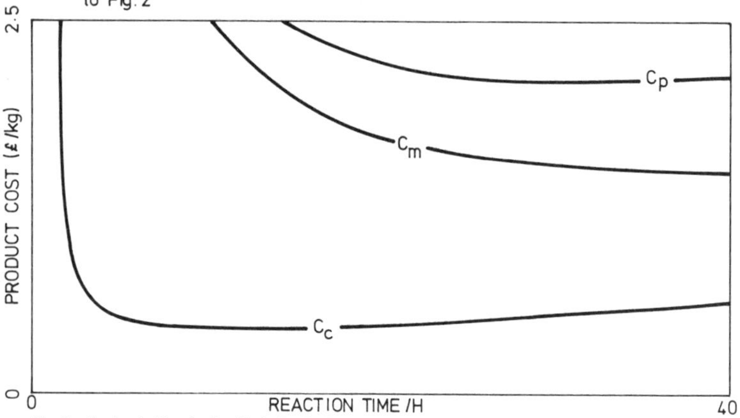

Fig. 5: Typical Plant Capital Cost C_c and Materials Cost C_m Contributions to Product Cost C_p

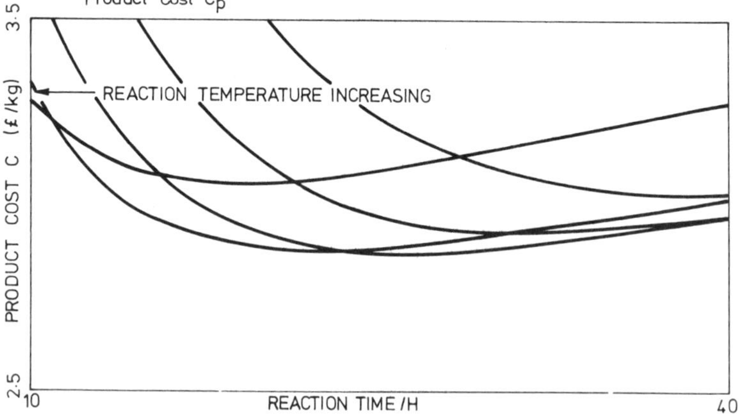

Fig. 6: Effects of Reaction Temperature on Product Cost C_p (N_p and initial R fixed)

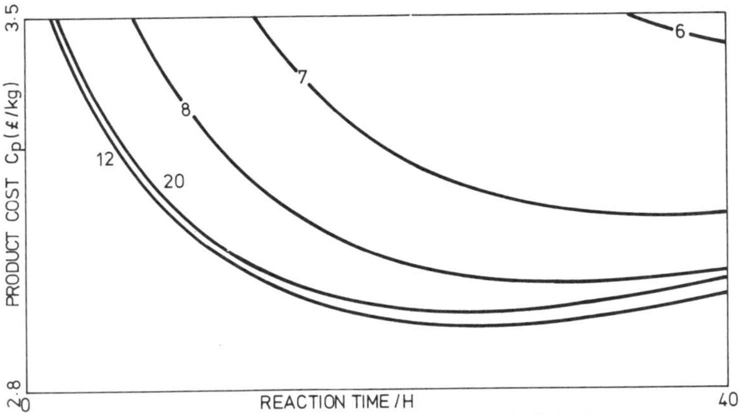

Fig.7: Effect of Number of Column Plates N_p on Product Cost C_p

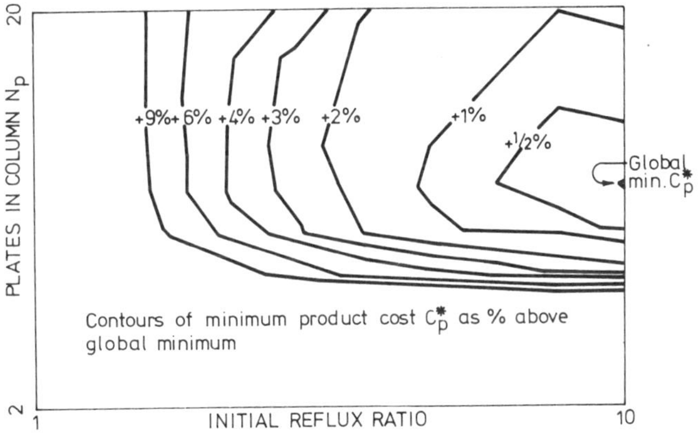

Fig. 8: Combined Effects of Column Plates and Initial Reflux Ratio on Minimum Product Cost C_p^*

INTEGRATED CONTROLLED RANDOM SEARCH: APPLICATION TO A WASTEWATER TREATMENT PLANT MODEL

J. Rodriguez Banga* & J.J. Casares Long*

An integrated procedure,ICRS, has been developed for the solution of constrained optimization problems. The method combines an equation-sequencing and design variables specification algorithm with a random search technique, CRS,which itself contains a Newton-Raphson algorithm to cope with cases where a full sequence of equality constraints is not attainable. The application of ICRS to a fairly complex model demonstrates the capabilities of the proposed procedure.

BACKGROUND

Constrained optimization has been, to a great extent, restricted to theoretically sound algorithms that have yet to prove their reliability and efficiency for solving real life problems such as the mathematical models of typical chemical and industrial processes although steady progress, in methods and software, has been made (1). One of the major problems associated with constrained optimization is that there is no way to choose between infeasible approximate solutions and therefore methods that provide feasible points at all times are to be preferred even if the algorithm fails to find a local optimum (2). The fact that there are various approaches to the general constrained problem from theoretical point of view, proves the limited application of any one in spite of a sound background (3). A detailed survey of existing general optimixation programs can be found in reference 4.

Deterministic algorithms can be all placed in one of two groups, a)Linear approximation techniques of which the sequential quadratic programming methods are the most popular (5) and, b)Transformation methods, where the method of multipliers has shown a certain degree of robustness for small sized problems (6,7).

Direct search methods have received comparatively little attention as capable techniques, for the solution of the general non-linear programming problem. Nevertheless, random search methods have become increasingly popular, specially amongst chemical engineers because they have shown to be reliable and have a

* Departamento de Ingenieria Quimica, Universidad de Santiago, Santiago de Compostela, Spain.

gcod convergence rate (8,9). Random search techniques are competitive when functions are difficult to compute, there is limited computer memory and, with multimodal objetive functions there exists a good probability of hitting the global minimum.

From the first review on random search in 1963, (10), various approaches have been adopted towards designing intelligent random procedures. Basically, all methods can be classified into two broad groups, those that have tried to introduce a degree of randomization into well established direct search unconstrained minimization techniques, and pure stochastic methods based on probability functions only.

Randomized direct search optimization techniques include the Complex method developed by Box (11) following the ideas of the Simplex method of Spendley and, was able to accommodate inequality constraints only. The hyperpolyhedron of serach points is built up using k points ($k \geq n+1$), the initial feasible point x^0 and k-1 vertices obtained by the formula,

$$x_i^j = l_i + r_i(u_i - l_i) \qquad i=1,2,\ldots,n \quad j=1,2,\ldots,n,\ldots,k-1$$

where r_i is a pseudo random deviate uniformly distributed over the interval (0,1) and u_i and l_i are the upper and lower bounds of the i-th variable respectively. The basic procedure has been refined later (12) and computer codes are available in the literature (13). Also, a randomized pattern search was devised by Lawrence & Steiglitz (14) based on Hooke & Jeeves method. The idea was to choose at random search directions during the exploratory search. Beltrami & Indusi (15) refined the method as to handle constraints by means of augmented penalty functions.

Pure random search methods derive from the original method of Brooks (16) where random search points were selected from a probability over the entire feasible region and the point with the smallest value of the objective function is taken as an approximation for the optimum. Although being a non-learning and time-consuming technique for high dimensional problems, it has been used to oooose a starting point for sequential search algorithms (17). A highly directional random search method was proposed by Matyas (18) and later refined to handle constraints (19). Convergence proofs of the updated methods have been obtained lately (20,21).

A sampling procedure with interval reduction was first presented by Luus & Jaakola (22), in 1973. The search region was periodically reduced to approximate the solution. Search points are computed through the expression,

$$x_n^{j+1} = x_n^j + l_n \cdot r_n^j$$

where x_n^j is the n-th independent variable of the current best value of the objective function, l_n is a pseudo-random deviate from a uniform distribution over the interval (-0.5,0.5) and, r_n^j is the interval size of the n-th variable at the j-th iteration. The range was decreased after each iteration by an empirical formula

$$r_n^j = (1-\varepsilon) r_n^{j-1}$$

with ε taking a fixed value (typically 0.05). Various modifications on the initial value of x and the initial search region, have been introduced later (23,24). A microcomputer implementation of this algorithm has been presented by

Ong (25).

Adaptive random search has been proposed by Gall (26) for multifactor optimization and search points are selected in accordance with the formula

$$x_n^{j+1} = x_n^{j} + R_n (2\theta - 1)^q$$

where q is an odd integer, R_n is the range of variable x_n and θ is a random deviate uniformly distributed between 0 and 1. The procedure was adapted to solve constrained optimization problems by Heuckroth et al (27). A number of refinements were incorporated to the basic procedure to improve its efficiency regarding the exponent q, systematic range reduction and skewing. To further improve the algorithm, a learning parameter was added by Martin & Gaddy (28). Similar techniques have been proposed for problems arising in medical and other engineering areas (29,30).

INTEGRATED CONTROL RANDOM SEARCH : A BRIEF DESCRIPTION

Integrated controlled random search, ICRS, is an integrated technique with a structuring and ordering algorithm to deal with the equality constraints of the mathematical model to be optimized plus the basic features of the controlled random search algorithm (31) complemented by a Newton-Raphson procedure for those cases with blocks of non-linear equations.

The method operates as follows:

Stage 1.-A tearing algorithm (32,33), initiates the optimization process, see Figure 1, in order to reduce/eliminate the number of equality constraints to be solved simultaneously in stage 2. The procedure will provide a set of variables (independent variables) which need to be specified such that the remaining variables (dependent variables) can be computed either seriatim or by means of a Newton-Raphson algorithm. Simultaneously, the number of iterative variables is minimized.

As a result of stage 1, the best independent variables are determined and the order of the equality constraints for the computation of the dependent variables is given. At this point the user is requested to explicit the dependent variables for those equations that are to be solved sequentially and the explicited equations are added automatically at their correct position in the computer code.

Stage 2.-The random search procedure is initialized by generating a feasible point from a uniform probability distribution over each of the independent variables. Subsequently, feasible search points are generated by means of a normal probability function. All basic features of the algorithm such as step size and stopping criterion are kept as in the original version. The major refinement to this stage is that if some of the equalities can not be solved seriatim, then a Newton-Raphson procedure is used to solve the block(s) of non-linear systems for the corresponding variables everytime a feasible set of independent variables is generated. Although time consuming, it has been tested with a number of problems and proved to be the simplest way, so far of handling this fairly uncommon situation with chemical process models.

ICRS APPLICATION TO A WASTEWATER TREATMENT PLANT

Controlled random search has been applied successfully in the past to a number of chemical process models and mathematical problems (31,34). The simple structure of the method was not suitable for attempting more complex processes such as the wastewater treatment plant model developed by Mishra et al (35,36) from the early work of Naito & Takamatsu (37). The objective is to minimize the total construction cost of a system consisting of a trickling filter, an activated sludge aeration vessel and a secondary clarifier as shown in Figure 2. The model equations are those reported by Himmelblau (38) and Wang & Luus (39), and are shown in Figure 3. The model includes 53 variables of which 13 are prespecified and 31 equality constraints plus a number of inequalities. Therefore, 9 variables will be chosen by the algorithm in stage 1 as independent to be used as such in the CRS code.

The set of independent variables and the sequence of the equality constraints obtained by the equation sequencing and design variables algorithm are given in Figure 4.

The results of the optimization process shown in Table 1 refer to two different approaches. Cases 1A and 1B correspond to a search from a single starting point generated by the uniform distribution. In cases 2A and 2B the starting point for the second phase of the search algorithm is the best (minimum value of objective function) of a set of 10 feasible points from the uniform distribution. For the dependent variables at the optimum, x_2^r (cell concentration of the recycle stream) is the one showing unreasonably high values ($>10^7$ ppm) and therefore bounds should be introduced to keep it under control. An upper bound of 10^5 ppm is adopted as in references 35,38 and 39.

The optimum values obtained by ICRS are compared in Table 2 with those recorded in the literature. Wang & Luus (39) and Mishra et al (35) solve the problem using a direct search method whilst Himmelblau´s solution (38) is atained with a deterministic algorithm.

Execution times on an IBM PC varied from two hours without a mathematical coprocessor to less than ten minutes when an 8087 coprocessor was used.

CONCLUSION

ICRS proves to be a flexible and robust procedure for constrained optimization of a medium size highly nonlinear problem. More complex models with non-sequential constraints are being tested. It is certainly, a major advantage that such an algorithm can be run successfully on IBM PC computers which are widely available.

REFERENCES

(1) Gill,P.E. et al,"Trends in nonlinear programming software".European Journal of Operational Research, Vol 72, pp 141-149,1984.

(2) Dennis,J.E.,"A user´s guide to nonlinear optimization algorithms", Proceedings of the IEEE, Vol 72,pp 1765-1776,1984.

(3) Gill P.E.,Murray W. and Wright M.H.,"Practical Optimization", Academic Press,1981.

(4) Waren,A.D. and Lasdon,L.S.,"The status of nonlinear programming software", Operations Research, Vol 27,pp 431-456,1979.

(5) Powell,M.J.D.,"Variable metric methods for constrained optimization",in A.Backem et al(eds),"Mathematical Programming :The State of the Art", Springer Verlag,1983.

(6) Pierre D.A.and Lowe M.J.,"Matehmatical Programming via Augmented Lagrangians",Addison-Wesley,1975.

(7) Newell,J.S. and Himmelblau D.M.,"A new method for nonlinearly constrained optimization",AIChE Journal, Vol 21,pp 479-486,1975.

(8) Solis,F.J. and Wets,R.,"Minimization by random search techniques", Mathematics of Operations Research, Vol 6,pp 19-30,1981.

(9) Reklaitis,G.V.,Ravindran,A. and Ragsdell,K.M.,"Engineering Optimization", John Wiley & Sons,Inc.,1983.

(10) Karnopp,D.C.,"Random search techniques for optimization problems", Automatica, Vol 1,pp 111-121,1963.

(11) Box,M.J.,"A new method of constrained optimization and a comparision with other methods",The Computer Journal, Vol 8,pp 42-52,1965.

(12) Ghani,S.N.,"An improved complex method of function minimization",Computer Aided Design, Vol 4,pp 71-78,1972.

(13) Richardson,J.A. and Kuester,J.L., Algorithm 454:The Complex method for constrained optimization",Communications ACM, Vol 16,pp 487-489,1973.

(14) Lawrence,J.P. and Steiglitz,K.,"Randomized pattern search",IEEE Transactions on Computers, Vol C-21,pp 382-385,1972.

(15) Beltrami,E.J. and Indusi,J.P.,"An adaptive random search algorithm for constrained minimization",IEEE Transactions on Computers, Vol C-21,pp 1004-1007,1972.

(16) Brooks,S.H.,"Discussion of random methods for locating surface maxima", Operations Research, Vol 6,pp 244-251,1958.

(17) Devroye,L.P.,"Progressive global search of continuous functions", Mathematical Programming, Vol 15,pp 330-342,1978.

(18) Matyas,J.,"Random optimization",Automation and RemoteControl, Vol 26,pp 244-250,1965.

(19) Baba,N.,"Convergence of a random optimization method for constrained optimization problems",J.O.T.A., Vol 33,pp 451-461,1981.

(20) Dorea,C.C.Y.,"Expected number of steps of a random optimization method", J.O.T.A., Vol 39,pp 165-171,1983.

(21) Dorea,C.C.Y.,"Limiting distribution for random optimization methods",SIAM J. Control and Optimization, Vol 24,pp 76-82,1986.

(22) Luus,R. and Jaakola,T.H.I.,"Optimization by direct search and systematic reduction of the size of search region",AIChE Journal, Vol 19,pp 760-766,1973.

(23) Wang,B. and Luus,R.,"Reliability of optimization procedures for obtaining global optimum",AIChE Journal, Vol 24,pp 619-626,1978.

(24) Lou,Q.,"Modified random search algorithm and its application in chemical engineering",International Chemical Engineering, Vol 25,pp 730-737,1985.

(25) Ong,S.L.,"Microcomputer application for constrained nonlinear optimization program",Industrial Engineering, Vol 17,pp 24-27,1985.

(26) Gall,D.A.,"Recent Advances in Optimization Techniques",John Wiley & Sons,Inc.,1966.

(27) Heuckroth,M.W.,Gaddy,J.L. and Gaines,L.D.,"An examination of the adaptive random search technique",AIChE Journal, Vol 22,pp 744-750,1976.

(28) Martin,D.L. and Gaddy,J.L.,"Process optimization with the adaptive randomly directed search",AIChE Symposium Series, Vol 78,pp 99-107,1982.

(29) Bekey,G.A. and Masri,S.F.,"Random search techniques for optimization of nonlinear systems with many parameters",Mathematics and Computers in Simulation, Vol XXV,pp 210-213,1983.

(30) Pronzato,L. et al, A general purpose global optimiser:Implementation and applications",Mathematics and Computers in Simulation, Vol XXVI,pp 412-422,1984.

(31) Goulcher,R. and Casares Long,J.J.,"The solution of steady-state chemical engineering optimization problems using a random search technique", Computers and Chemical Engineering, Vol 2,pp 33-36,1978.

(32) Ramirez,W.F. and Vestal,C.R.,"Algorithms for structuring design calculations",Chemical Engineering Science, Vol 27,pp 2243-2254,1972.

(33) Book,N.L.and Ramirez,W.F.,"The selection of design variables in systems of algebraic equations',AIChE Journal, Vol 22,pp 55-66,1976.

(34) Goulcher,R.W. et al,"A random search method for the solution of constrained nonlinear optimization problems",Anales de Química, Vol 74,pp 1310-1313,1978.

(35) Mishra,P.N.,Fan,L.T. and Ericksson,L.E.,"Biological wastewater treatment system design, Part I", Canadian J. of Chemical Engineering, Vol 51,pp 694-701,1973.

(36) Mishra,P.N.,Fan,L.T. and Ericksson,L.E.,"Application of mathematical optimization techniques in CAD of wastewater treatment systems", AIChE Symposium Series, Vol 71,pp 136-153,1976.

(37) Naito,M.,Takamatsu,T. and Fan,L.T.,"Optimization of the activated sludge process", Water Research, Vol 3,pp 433-443,1969.

(38) Himmelblau,D.M.,"Optimal design via structural parameters and nonlinear programming", Engineering Optimization, Vol 2,pp 17-27,1976.

(39) Wang,B.C. and Luus,R.,"Optimization of non-unimodal systems",International Journal for Numerical Methods in Engineering, Vol 11,pp 1235-1250,1977.

		CASE 1A	CASE 2A	CASE 1B	CASE 2B
Objective function C^T_{min}		157983.4	157902.8	171902.8	170901.8
INDEPENDENT VARIABLES AT MINIMUM	q^{AS}	3.778	8.918	6.393	5.052
	α_{12}	0.094	0.0089	0.015	0.104
	α_{11}	0.237	0.031	0.015	0.012
	α_{20}	0.0037	0.701	0.976	0.985
	α_{21}	0.762	0.941	0.204	0.0118
	α_{22}	0.111	0.658	0.504	0.363
	x_1^{TF}	90.081	92.393	47.874	14.579
	x_1^1	90.081	92.387	47.866	14.574
	x_2^{TF}	1603.875	1522.213	3791.790	7306.536

A refers to search with x_2^r unbounded;B refers to search with $x_2^r \leq 10^5$

Table 1.- Output of second stage of integrated algorithm.

ALGORITHM		$C^T_{minimun}$	Comments
G.R.G. (38)		243841	$0<x_2^r<10^5$
Simplex Pattern Search (35)		215095	$0<x_2^r<10^5$
Modified Random Search (39)		155000 approx.	x_2^r unbounded
		166409	$0<x_2^r<9.10^4$
ICRS	Case 1A	157983	x_2^r unbounded
	2A	157902	
ICRS	Case 1B	171548	$0<x_2^r<10^5$
	2B	170902	

Table 2.- Minumun construction cost of wastewater treatment plant; Comparison of results

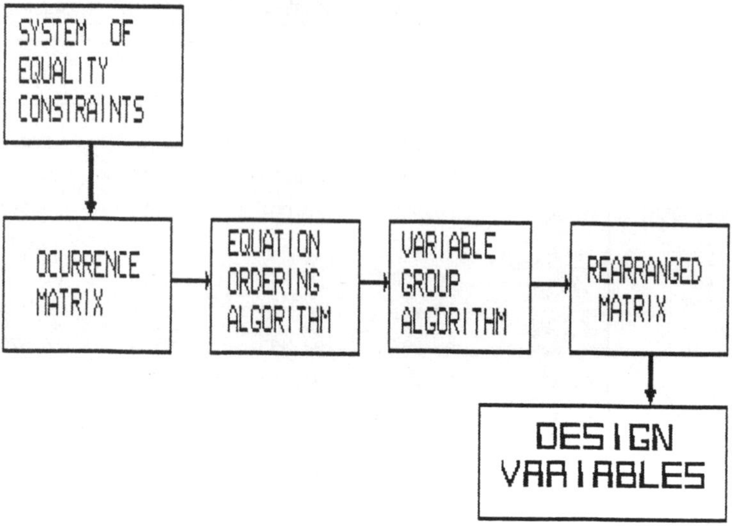

Figure 1.- Block diagram for equation sequencing and design variable selection.

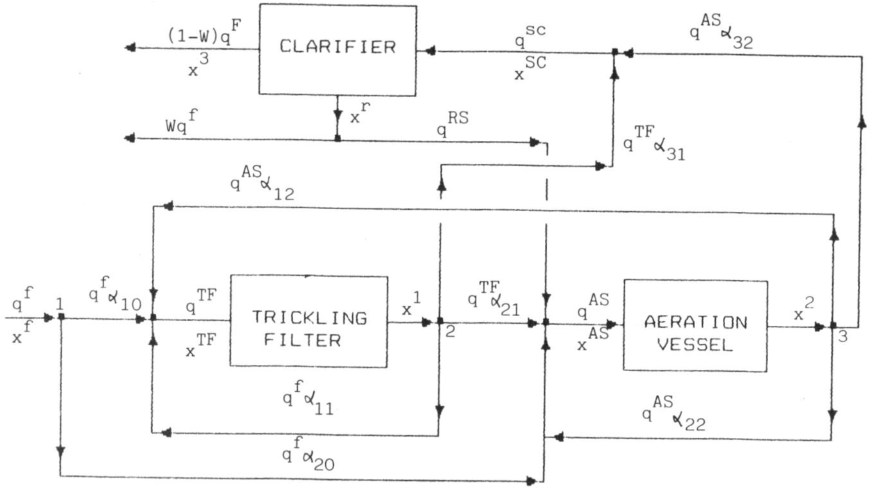

Figure 2.- Basic layout for wastewater treatment plant

OBJECTIVE FUNCTION

$$C^T = 1.4\,(C^{TF} + C^{AS} + C^{SC})$$

EQUALITY CONSTRAINTS

1. $q^{TF} = (q^f \alpha_{10} + q^{AS}\,\alpha_{12}) / (1 - \alpha_{11})$

2. $\alpha_{20} = 1 - \alpha_{10}$

3. $\alpha_{31} = 1 - \alpha_{21} - \alpha_{11}$

4. $\alpha_{32} = (q^{SC} - \alpha_{31}\, q^{TF}) / q^{AS}$

5. $\alpha_{22} = 1 - \alpha_{32} - \alpha_{12}$

6. $r_1 = q^{AS}\,(1 - \alpha_{11})\,(1 - \alpha_{22}) - q^{AS}\alpha_{12}\alpha_{21} - [(1 - \alpha_{11})\alpha_{20} + \alpha_{10}\alpha_{21}]q^f$

7. $r_2 = q^{AS}\alpha_{12}\alpha_{21} + q^{AS}(1 - \alpha_{11})\alpha_{22} + [(1 - \alpha_{11})\alpha_{20} + \alpha_{10}\alpha_{21}]q^f$

8. $r = r_1/r_2$

9. $q^{RS} = r(q^f\alpha_{20} + q^{TF}\alpha_{21} + q^{AS}\alpha_{22})$

10. $x_1{}^2 = (q^{TF}x_1{}^{TF} - q^f\alpha_{10}\, x_1{}^f - q^{TF}\alpha_{11}x_1{}^1)/(q^{AS}\alpha_{12})$

11. $x_1{}^{SC} = (\alpha_{31}q^{TF}x_1{}^1 + q^{AS}\alpha_{32}x_1{}^2)/q^{SC}$

12. $x_1{}^{SC} = (\alpha_{21}q^{TF}x_1{}^1 + \alpha_{22}q^{AS}x_1{}^2 + q^{RS}x_1{}^{SC} + \alpha_{20}q^f x_1{}^f)/q^{AS}$

13. $x_2{}^1 = x_2{}^{TF} + Y(x_1{}^{TF} - x_1{}^1)$

14. $x_2{}^2 = (q^{TF}x_2{}^{TF} - q^f\alpha_{10}x_2{}^f - q^{TF}\alpha_{11}x_2{}^1)/(q^{AS}\alpha_{12})$

15. $V^{AS} = q^{AS}(x_1{}^{AS} - x_1{}^2)Y(K + x_1{}^2)/(kx_1{}^2x_2{}^2)$

16. $x_1{}^2 = x_2{}^2 - \dfrac{V^{AS}}{q^{AS}}\left[\dfrac{k\,x_1{}^2\,x_2{}^2}{K + x_1{}^2} - k_D\,x_2{}^2\right]$

17. $x_2{}^r = (q^{AS}x_2{}^{AS} - \alpha_{21}q^{TF}x_2{}^1 - \alpha_{22}q^{AS}x_2{}^2 - \alpha_{20}q^f x_2{}^f)/q^{RS}$

18. $x_1{}^3 = x_1{}^{SC}$

19. $x_2{}^{SC} = (\alpha_{31}q^{TF}x_2{}^1 + q^{AS}\alpha_{32}x_2{}^2)/q^{SC}$

20. $x_2{}^3 = (Cd - x_1{}^3)/m$

21. $T^{SC} = -\ln\left[\dfrac{x_2{}^3}{p(x_2{}^{SC})\beta}\right] \Big/ K_5$

22. $D^{TF} = -20^n \ln\left[\dfrac{x_1{}^1}{x_1{}^{TF}}\right] \Big/ (K^{TF} A_p{}^{TF})$

23. $V^{SC} = T^{SC}(832.67*160.544q^{SC})/24$

24. $A^{SC} = V^{SC}/D^{SC}$

25. $x_1{}^r = x_1{}^{SC}$

26. $A^{TF} = 2178q^{TF}$

27. $V^{TF} = A^{TF}D^{TF}$

28. $W = (q^{SC}x_2{}^{SC} - q^{RS}x_2{}^r - q^f x_2{}^3)/(q^f x_2{}^r - q^f x_2{}^3)$

29. $C^{TF} = 25(V^{TF})^{.827} + 3{,}650 + 1{,}150q^{TF}\alpha_{11}$

30. $C^{AS} = V^{AS}[175{,}000 + 36{,}500(V^{AS})^{-.818}] + 3{,}650 + 1{,}150q^{AS}\alpha_{12}$

31. $C^{SC} = 12.6\,A^{SC} + 5{,}350\,[A^{SC}/1{,}000] - 0.126 + 3{,}650 + 1{,}150q^{RS}$

INEQUALITY CONSTRAINTS

$0 \le q^{AS} \le 10$	$0 \le \alpha_{12} \le 1$	$0 \le \alpha_{11} \le 1$
$0 \le \alpha_{20} \le 1$	$0 \le \alpha_{21} \le 1$	$0 \le \alpha_{22} \le 1$
$0 \le x_1{}^{TF} \le 100$	$0 \le x_1{}^1 \le 100$	$0 \le x_2{}^{TF} \le 10{,}000$

Figure 3.- Model to be optimized

INDEPENDENT VARIABLES

q^{AS}, α_{12}, α_{11}, α_{20}, α_{21}, α_{22}, x_1^{TF}, x_1^1, x_2^{TF}

SEQUENTIAL ORDER FOR THE SOLUTION OF CONSTRAINTS

2, 7, 6, 5, 3, 1, 10, 8, 4, 13, 11, 9, 18, 14, 12, 20, 19, 15, 21, 16, 26, 23, 22, 17, 28, 27, 25, 24, 31, 30, 29

Figure 4.- Output of first stage of integrated algorithm

EXPLOSION VENTING OF WEAKLY FLAMMABLE MIXTURES: IMPLICATIONS FOR EXISTING PLANT MODIFICATIONS

D. G.Robson*

Most fixed bed catalytic partial oxidation processes designed in the 1960's started with an initial design premise of operation below the lower flammable limit. This required the compression and handling of additional diluent air or inert gases throughout the total process train. In the 1980's, recognising the inherent operating costs of diluent handling, design of similar new units moved to deliberate flammable operation, higher reactor design pressures and positive explosion venting. Even in the 1960's, however, it was often thought prudent to add explosion relief on partial oxidation reactors in case control inadvertently allowed operation to move into the flammable region. In these circumstances, a detailed examination of the reactor mechanical design, the existing venting arrangements, and the explosive nature of the mixture handled can allow operation to move closer to 1980 design standards without incurring the capital costs of replacement reaction systems. This paper will describe the details of such an investigation of the partial oxidation of o-xylene to phthalic anhydride. Work on computer simulation has enabled a full description of the built-up back pressure behind the explosion venting of weakly flammable mixtures.

INTRODUCTION

The key input for a detailed mechanical analysis of the effect of an explosion on a vessel is a pressure time diagram. Not only is the maximum pressure reached significant but also the rate of pressure rise towards that maximum. Very rapid stressing can increase the "effective" maximum by a factor of two when analysing the response of a steel structure. Pressure time diagrams for o-xylene air mixtures are derived below and the mechanical response of the top head of the reactor shown in Fig.1 examined.

DESIGN METHODS

The standard method used by Phthalic Anhydride reactor manufacturers for the calculation of an explosion venting area for the top heads of reactors working in the explosive region is based on the work of Heinrich (1)

* Engineering and Technical Department, B.P. Chemicals

The basic vent sizing equation is:

$$F = \frac{V_L^{1/3} \times V^{2/3} \times \left(\frac{dpex}{dt}\right) Pred\ V_L}{a \times (2RT/M)^{0.5} \times (Pred\ (Pred-Po))^{0.5}} \qquad (1)$$

It enables the sizing of vent area, F, for an acceptable maximum pressure, Pred, ie. a pressure reduced by venting below that of a completely contained explosion. The equation requires explosion bomb testwork to find the maximum rate of pressure rise during an explosion from various o-xylene air composition. (Table 1, below).

The Heinrich equation gives an extremely conservative value for venting area for a fixed reduced explosion pressure. Conversely, for existing reactor heads with fixed vent areas, the equation gives a high prediction of reduced explosion pressure.

In particular, the equation gives a high prediction of explosion pressure for a 50 g/Nm^3 case because design is always based on the maximum rate of pressure rise during explosion testwork in a standard bomb. This maximum for o-xylene occurs at a concentration of 110 g/Nm^3. The measured rate of pressure rise at 110 g/Nm^3 is 430 bar/s whereas that at 50 g/Nm^3 is 28 bar/s.

The mixture control systems as set up on the best PA plants are designed to hold rigorously o-xylene concentration at set point. The system normally includes an independent trip system. A pessimistic view on running with a set point of 50 g/Nm^3 is that control would not allow a deviation beyond 58g/Nm^3 without shutdown. This extreme upper figure has been taken as a design basis in all the work below, relating to a 50 g/Nm^3 design case.

Heinrich's work was primarily concerned with venting dust explosions and hence with low set pressures and nearly incompressible flow situations. The equation is based on a venting balance between the incompressible volume flow from an explosion vent and the volume generation rate at the surface of a spherical flame whose volume equals that of the reactor head. It only addresses that static balance at maximum pressure and gives no information on the rate of pressure rise towards that maximum value.

In addition, the shape of most PA reactor heads, severely restricts the possible spherical flame surface and at a relief set pressure of 0.8 barg, compressibility is certainly a significant factor in calculating the relief flow from the vent with full built up backpressure.

Since equation (1) was judged to give an extremely conservative design, particularly when working just inside the flammable limit, a more detailed calculation involving a computer simulation was set up.

CALCULATIONS BASED ON BOMB TESTWORK

A basic requirement for a computer simulation of a vented explosion is a description of the rate of growth of the flame front surface area and the additional volume of gas created by combustion.

A convenient source of the basic data for this in our particular case is the bomb testwork carried out on o-xylene/air mixtures for the Heinrich equation. This is used to derive mixture flame speeds suitable for further computer modelling.

A programme has been written to describe the growth of a spherical flame growing at an assumed constant flame speed in the explosion bomb. The additional volume of combustion products at the flame front is related simply to the ratio of the final to the initial pressure in the bomb. (N.B. bomb flame temperatures are included implicitly when this is later used for simulation).

ie. $$\frac{dp}{dt} = 4\pi r^2 \times V_F \times (E-1) \times P\ \text{initial}/V_{\text{initial}} \qquad (2)$$

and $$\frac{dr}{dt} = V_F \text{ where } E = \text{final bomb pressure}/_{\text{initial pressure}}$$

The above equations when used to match closed bomb testwork give the velocity of the flame relative to the original point of ignition i.e. it includes the flow velocity of unburnt gas compressed ahead of the flame front.

Matching to bomb experiments lasting 100 milliseconds has been done on a millisecond basis. With a simple time-step method V_F will be overestimated depending on the number of steps chosen but this is a safe situation regarding later simulations.

A match on the test results has been made by trial and error. The flame speed was varied until dp/dt max and Pmax/P initial agreed with the measured value.

Table 1 Explosion Bomb Testwork and Calculated Flame Speed

Concentration (g/Nm^3)	dp/dt max bar/s	P max/ P initial	Flame Speed (m/s)
50	28	4.3	0.38
60	200	5.45	2.05
70	290	6.5	2.51
90	397	7.3	2.83
110	431	7.6	3.04
130	404	7.4	2.83

The flame speed at 50 g/Nm^3 is substantially laminar whereas, at higher concentrations, combustion is vigorous enough to impart a significant turbulent character to the flame front. Fig. 2 gives expansion ratio and flame velocity plotted against o-xylene concentration.

COMPUTER SIMULATION OF A VENTED EXPLOSION

The computer modelling of an explosion and its subsequent venting to atmosphere is well covered in the literature. In particular, papers by Chappell (2), Rust (3), Palmer (4), Pasman (5) and Chippett (6) have been used during the setting up of the programme detailed below.

Both the flame speed, the expansion ratio derived from closed bomb testwork and the flame surface area are required to describe the rate of internal volume generation due to combustion.

If P<1.8 bara $\frac{dp}{dt} = 4\pi r^2 \times V_F \times (E-1) \times Pi/V_{head}$ (3)

The programme follows the growth of the flame surface within the reactor head at millisecond intervals. The growing flame surface is taken as spherical and smooth except when confined by the reactor walls.

A knowledge at any instant of flame speed, expansion ratio and flame front surface area completely defines internal volume generation. Up to the point of venting, the pressure in the reactor head is directly proportional to the extra volume accumulated therein.

The set pressure for explosion relief is taken as a minimum at 0.8 barg and hence flow through the vent is described by a simple sonic flow equation. At the same time, the flame speed is multiplied by a factor of five to allow for further turbulence due to venting. Combustion continues until all the o-xylene is burnt or otherwise vented.

If P>1.8 bara and subsequently $\frac{dp}{dt} = 5\,[4\pi r^2 \times V_F \times (E-1)]\,\frac{Pi}{V_{head}} - Fv$ (4)

where Fv = 122.98 x A x P (see Ref.4) and $\frac{dr}{dt} = V_F$ (5)

The assumption of a fivefold increase in flame speed due to turbulence in the head following venting was verified independently (Appendix 1).

A typical output from the programme is given in Figure 3. It describes the situation where, with a set point of 50 g/Nm^3, errors in control have let o-xylene concentration increase to 58 g/Nm^3 with coincident ignition in the reactor head.

Fig. 4 is a collection of results to show the effect of o-xylene concentration, where both maximum reduced explosion pressure and rate of pressure rise increase with increasing concentration.

EXPLOSIBLE DUCT SYSTEMS

In the work above, ignition is assumed to occur only in the reactor top head (Fig.1). The existing o-xylene injection system and feed ductwork would be used to add only 40 g/Nm^3 of o-xylene (i.e. a non-flammable mixture) with a further addition of 10 g/Nm^3 at a small secondary injection vessel at the reactor head. Alternatively, all the o-xylene could be added at the secondary vessel. In either case, however, the ducting would run non-flammable up to the second vessel.

A significantly cheaper feed system could be realised by using existing o-xylene addition pots and allowing the feed ducting to run entirely in a flammable condition. With a flammable duct system, however, and with pipe L/D ratios greater than 40, an ignition at the feed vessel would lead to a flame front accelerating down the feed ducting and subsequently expanding into the reactor head ie. the phenomenon of pressure piling.

The recognised method of controlling explosions in ductwork is by venting at regular intervals so as to minimise the self-acceleration of the flame front. The analysis of such a vented feed line is described below.

DEVELOPMENT OF FLAME FRONT SURFACE AREA DURING PIPELINE PROPAGATION - THE METHANE/AIR SYSTEM

Two graphs describing the results of experiments on the propagation of a flame in a methane-air mixture in a straight 400mm pipe have been taken from Bartknect (7). They plot flame speed against distance. For the purpose of analysis the two graphs have been combined, using one to extend the plot of the other to longer lengths of pipe and into extremely turbulent regions of flow, as detailed in Fig. 5.

Fig. 5 has been verified independently by some unpublished work by Steen & Schampel (8).

A computer programme has been written to match the observed flame velocity in Fig. 5 at each increment of pipe diameter along the length of a 400mm pipe. The computer searches for a dimensionless flame surface area (ie. flame front surface/pipe cross-sectional area) which on burning at a calculated turbulent burning velocity through the next increment of length will match the graphed flame speed at the next reference point.

The programme uses a formula from Lee (9) to derive the turbulent burning velocity at the flame surface from the laminar value for methane (0.366 m/s) viz.

$$V_{turbulent} = 0.1761 \times V_{Laminar} \times (D)^{0.256} \times (Re)^{0.238} \qquad (6)$$

- where D is the pipe diameter and Re the Reynolds number based on the flame speed. Both the flame front Reynolds number and the turbulent burning velocity are recalculated throughout the computer run on a millisecond basis as the flow in the pipe increases.

The expansion of gas (7.33 vol/vol) on combustion at the flame front is taken into account when calculating the internal pressure at the pipe exit. Venting to atmosphere of unburnt material ahead of the flame front is described by subsonic/sonic flow equations and is the second part of the volume balance equation for internal pressure rise.

The flame speed is the sum of the turbulent burning velocity multiplied by the flame surface area plus the physical displacement due to the build up of combustion products behind the flame front.

At very high superficial mass flows in the pipe when sonic conditions exist at the exit an appreciable pressure drop occurs in the pipe. The pressure at the ignition source is thus calculated using the Fanno equation for compressible flow with a friction factor of 0.0035.

The output from the programme is given in Fig.6. and a listing with explanation given in Appendix 2.

LAMINAR FLAME VELOCITY FOR O-XYLENE AT 58 g/Nm^3

To use the work described above on methane-air systems for o-xylene air mixtures of differing concentrations the laminar flame velocity for o-xylene in air must be calculated. A previous analysis of the BAM o-xylene testwork has given Fig.7 - a graph of maximum test apparatus flame speed and expansion ratio against o-xylene concentration. For calculation it has been assumed that laminar burning velocities can be scaled using the BAM results.

The stoichiometric flame velocity for o-xylene has been calculated from that of methane using the reciprocal of the quenching diameter, thus since d_q methane = 3.10mm and d_q o-xylene = 3.20mm.

$$V_L \text{ (o-xylene)} = 0.366 \times \frac{3.20 \text{ mm}}{3.10 \text{ mm}} = 0.38 \text{ m/s}$$

and scaling from the BAM testwork gives:-

$$V_L \text{ (58 gm/Nm}^3\text{)} = 0.38 \times \frac{\text{V (BAM at 58 gm/Nm}^3\text{)}}{\text{V (BAM at 92 gm/Nm}^3\text{)}}$$

$$V_{\text{LAMINAR}} = 0.38 \times \frac{1.76}{2.95} = \underline{\underline{0.24}} \text{ m/s}$$

The laminar burning velocity calculated above is subsequently used with the expansion ratio from the BAM graph in the simulation of a duct/head explosion with subsequent venting.

DEVELOPMENT OF FLAME SURFACE AREA DURING PROPAGATION ALONG THE PA PLANT FEED DUCTING AND SUBSEQUENT VENTING

The surface area versus distance characteristic of the methane/air system, as defined in the programme above, is stored on disc and then recalled and used to describe the growth of flame surface area as a flame propagates from the o-xylene injection vessel on the PA plant towards and into the reactor head space.

Since the duct normally runs at a velocity of 24 m/s the initial flame surface area for the methane-air mixture calculated at this velocity is assumed to exist at the source of ignition.

The longest duct on an existing PA plant has a straight length of 46.2 metres but two bends increase the equivalent frictional resistance to 81.6 pipe diameters. The flame surface area in the feed ducting is thus assumed to increase from that at 24 m/s to a value equivalent to 81.6 pipe diameters of propagation from this point (see fig.6).

This additional information on the growth of flame surface area in the feed ducting has been incorporated into a reactor head venting programme. Additional venting in the feed ducting has also been incorporated and allows for the venting of burnt gas from behind the flame front using venting equations derived from Wilson's thesis (10) i.e. from his graphs of venting down after maximum explosion pressure.

When the explosion vents open, the flame front normally accelerates and enters the reactor head at a high turbulent flame speed. It has been assumed that there will be no further increase in flame speed on entering the head but only that the basic flame surface will grow spherically until confined by the reactor head.

Fig.7. gives the pressure in the reactor head and duct system against the position of the flame front in the ducting together with other significant parameters. The position of the flame front is calculated each millisecond and thus the additional frictional turbulence at a bend can be incorporated as the flame front sweeps out this position in the duct.

The relative areas for venting of burnt/unburnt gas is likewise determined by the position of the flame front.

For the example given in Fig.7 the maximum explosion pressure with four 600mm vents in the ducting is 1.36 barg.

Despite the use of four vents behind the flame front the acceleration in the pipe is such that the flame enters the head at high velocity. The expanding flame front then creates an extremely rapid pressure pulse.

Note that a number of the assumptions made in the duct propagation model have not yet been fully verified. More work in particular is required to verify the representation of bends as their equivalent frictional resistance of straight ducting. The use of Lees equation at such high velocities is also questionable.

A number of simplifying assumptions have been required in the work above and due regard to these must be taken when applying the results.

MECHANICAL CRITERIA

A review of the mechanical design of a particular reactor has been completed. This involved a finite element analysis of a representative reactor section using a given pressure - time curves as input to the analysis. The maximum allowable pressure within the head for pressure pulse rise times of 200-420 milliseconds has been agreed as in the range of 1.86 barg to 1.96 barg. This criteria has been applied for all the cases of ignition within the reactor headspace.

When considering ignition propagating within the feed ducting and expanding into the reactor head, the pressure pulse rise within the head is extremely rapid and a dynamic factor of 2 has thus been taken. The maximum allowable pressure within the head then drops to 1.4 barg. The following table illustrates the results of the calculations described above.

Table 2 Summary of Results

Case	Max Pressure (barg)	Allowable Pressure (barg)	Acceptability
$50g/Nm^3$ head	0.8	2.8	Yes
$58g/Nm^3$ head	1.3	1.96	Yes
$68g/Nm^3$ head	1.8	1.86	Yes
$110g/Nm^3$ head	2.2	1.86	No
$58g/Nm^3$ head and duct	1.4	1.4	No

REFERENCES

1. H.J. Heinrich, Chem Ing Techn 38/3, (1966). 1125

2. Chappell, W.G. "Pressure Time Diagram for Explosion Vented Space", 1977 AIChE Loss Prevention Symposium, pp 76-86.

3. Rust, E.A. "Explosion venting for low pressure equipment", Chemical Engineering, Nov. 5, 1979.

4. Palmer K.N. "Relief Venting of Dust Explosions". Loss Prevention Symposium, 1974.

5. Pasman, H.J. et al, "Design of Pressure Relief Vents" Loss Prevention Symposium, 1974.

6. Chippett, S. "Modeling of Vented Deflagrations", Combustion and Flame 55 127-140 (1984).

7. W. Bartknecht "Explosions" Chapter 3 Figures 1-61, 1-62.

8. Steen & Schampel, PTB Braunsweig Feb 2, 1977. Unpublished Note, "Increased flame velocities, transition from explosion to detonation in methane/air mixtures".

9. F.P. Lees "Loss Prevention in the Process Industries" pg 250 Section 16.5.3. "Flames in Pipes".

10. M.J.G. Wilson Ph.D. Thesis Imperial College. 1955.

NOMENCLATURE

r = flame radius m

1 Nm^3 = one cubic metre at Normal Temperature and Pressure

V_L (m^3) = volume of test vessel.

V (m^3) = volume of vessel to be protected.

(dpex/dt) pred. V_L 4 (bar, s^{-1}) = maximum rate of explosion pressure rise, in the test vessel, at the reduced explosion pressure, for gas - air mixture in question.

a = discharge coefficient, assumed to be 0.8
M (kg. mol^{-1}) = molecular mass of explosion gases discharged.
T (K) = initial absolute temperature of gas mixture
R (J) = gas constant (taken as 8.31).
pred (bar) = highest pressure to be permitted in vessel to be protected.
Po (bar) = environmental pressure into which relief takes place.

APPENDIX 1

FLAME VELOCITY ENHANCEMENT DURING VENTING

Two of the pressure-time curves from Wilson's thesis (10) have been examined in detail.

1. Closed Vessel Explosion - 2.7% Pentane

The pressure-time curve for a 2.7% pentane explosion in a closed 60 cubic ft test vessel was taken and pressure-time values input into a computer programme in 10 millisecond steps. The programme then followed a spherical flame growing within the vessel and searched for a velocity for the next 10 millisecond step that would give the next measured pressure on the pressure-time curve. The results are given in Fig.8.

The flame radius - velocity information generated by the programme as it matched the experimental plot was stored on disc as an input to a second programme which matched a vented 2.7% pentane explosion.

2. Vented Explosion - 2.7% Pentane

The pressure-time curve for 2.7% pentane explosion in the same test vessel, but with a 6" vent, was fed into a second programme in 10 millisecond steps as for the work above. The second programme again searched for a flame velocity to match the next pressure point 10 milliseconds ahead, though also now accounting for the volume of material vented. The volume vented per unit time was derived from the measured rate of venting down after the end of the explosion and this rate converted to the venting of unburnt gas for the period up to the maximum explosion pressure (- using a multiplier of 0.76 taken from Wilson's thesis).

The velocities required to match the vented curve were compared with those from the closed vessel explosion at the same radius by examining the disc record set up in the first programme. The ratio was output as "flame factor", Fig.9.

3. Enhancement during Vented Explosions

From the two curves examined above, I conclude that an enhancement of up to 3.8 times has been demonstrated immediately after the vent opened followed by a period when the effect of the vent is apparently to delay the growth of the flame and then finally followed by enhancement of up to 3.3 times at the end of the explosion.

Taking a factor of 5 times the maximum velocity derived from closed vessel testwork as has been done for the PA explosion work thus seems to have taken a safety factor both in the basic flame velocity (ie by taking the maximum value) and in the factor used for the enhanced velocity.

APPENDIX 2. PROGRAM TO MATCH EXPERIMENTAL RESULTS FOR PIPELINE FLAMES (L/D>40)

```
                         GOTO 260
20            [SUBROUTINE FOR PIPE DELTA PRESSURE CALCULATION (LINES 30 TO 80) ]
30 P1=P:RHO2=1.204*(P1^.71)  [DISCHARGE PRESSURE/GAS DENSITY FOR GAMMA AIR=1.4 ]
45 P1=P+DELP:RHO1=1.204*(P1^.71)                [INLET PRESSURE/GAS DENSITY]
60 DELP=4*.0035*((F3*RHO2/A)*(F3*RHO2/A))*.0012/(RHO1+RHO2)+(F3*RHO2/A)*
      (F3*RHO2/A)*9.999999E-06*(1/RHO2-1/RHO1)  [FANNO FLOW EQUATION FOR VENT]
70 IF DELP-(P1-P)>.01 THEN GOTO 45 ELSE P3=P+DELP         [CONVERGENCE CRITERIA]
                         RETURN
90  [SUBROUTINE FOR PIPE VENT FLOW CALCULATION WITH DISCHARGE PRESSURE P BARA ]
100 IF P<1.89 THEN F3=.62*A*SQR(580300!*P*((1/P)^.71)*(1-(1/P)^.29)) ELSE
    F3=.62*A*SQR(96716.66*P*((1/P)^.71))       [COMPRESSIBLE/INCOMPRESSIBLE FLOW]
                         RETURN
150                    [INPUT LINE FOR AN INCREMENT IN S (SURFACE AREA FACTOR)]
160 IF PV<60 THEN INC=.01 ELSE INC=.1  [AT HIGHER VELOCITIES TAKE LARGER STEPS ]
170 S=S+INC:DIST=DISTFIX:T=TFIX:RE=REFIX:DISTFAX=DISTFIX    [NOTE NEW VALUE OF S]
    UT=UTFIX:V3=V3FIX:F3=F3FIX:P=PFIX:GOTO 500        [RETRIEVE STARTING VALUES]
180                    [INPUT LINE FOR A DECREMENT IN S (SURFACE AREA FACTOR)]
190 IF PV<60 THEN INC=.01 ELSE INC=.1                    [SEE LINE 160 ABOVE]
200 S=S-INC/2:DIST=DISTFIX:DISTFAX=DISTFIX:T=TFIX:RE=REFIX
    UT=UTFIX:V3=V3FIX:F3=F3FIX:P=PFIX:GOTO 500            [SEE LINE 170 ABOVE]
210               [SUBROUTINE FOR LINEAR INTERPOLATION TO THE NEXT MATCH POINT]
220 I=0                                                  [INITIATE DATA CLIMB]
230 I=I+1                                                    [CLIMB DATA FILE]
240 IF DIST/D<DDAT(I+1) THEN PVDAT=VDAT(I)+(VDAT(I+1)-VDAT(I))* [CRITERIA AND
    (DIST/D-DDAT(I))/(DDAT(I+1)-DDAT(I)) ELSE GOTO 230     [LINEAR INTERPOLATION]
                         RETURN
270                     [READ IN EXPERIMENTAL DATA TO BE MATCHED  (LINE 310)]
280 FOR I=1 TO 20:READ DDAT(I),VDAT(I):NEXT I     [OBSERVED DISTANCE & VELOCITY]
310 DATA 0,0.366,10,27,20,42,30,60,40,80,50,110,55,135,60,160,65,190,70,220,
75,260,80,320,85,400,90,510,95,680,100,850,105,1060,110,1280,115,1450,120,1650
320                       [SET UP INITIAL SYSTEM PARAMETERS  (LINES 330,460)]
340 PRINT"INPUT PIPE DIAMETER-M,LAMINAR FLAME VELOCITY-M/S
          ,EXPANSION RATIO":INPUT D,UL,E
450 A=3.1416*D*D/4:V2=A*120*D:P(1)=1:V3=V2  [PIPE AREA,INITIAL PIPE VOLUME/    ]
460 DIST=.000001:DIST1=1:S=.175:P=1         [ PRESSURE,START DISTANCE,SA FACTOR]
500 DISTFIX=DIST:TFIX=T:REFIX=RE:UTFIX=UT:V3FIX=V3 [ INITIALISE KEY PARAMETERS ]
    F3FIX=F3:PFIX=P:DISTFAX=DISTFIX             [FOR SEARCH IN NEXT MILLISEC]
510 IF T=0 THEN UT=UL:IF T=0 THEN DISTFAX=0     [SET VALUES AT T=0 MILLISECS]
                         GOSUB 90
530 DIST=DIST+.001*UT*S+.001*F3/A [FLAMEFRONT POSITION DUE TO BURNING (UT*S)   ]
540 PV=(DIST-DISTFAX)/.001        [AND VENTING (F3 M3/S),VELOCITY=DISTANCE/TIME]
550 DISTFAX=DIST                  [HOLD NEW POSITION AS VARIABLE  DISTFAX      ]
560 RE=1.204*(P^.71)*PV*D/.0000179       [CALCULATE FLAMEFRONT REYNOLDS NUMBER ]
570 UT=.1761*UL*((D*100)^.256)*(RE^.238) [CALCULATE TURBULENT FLAMESPEED   M/S ]
580 F5=S*UT*A                            [F5=VOLUME OF GAS BURNT/SECOND (M3/S)]
590 F4=F5*(E-1)-F3          [F4=VOLUME INCREASE CONTAINED WITHIN PIPE/SECOND (M3/S)]
600 V3=V3+F4*.001           [V3=VOLUME CONTAINED WITHIN PIPE (M3) AFTER 1 MILLISEC ]
610 P=V3/V2:T=T+1   [P=PIPE PRESSURE (BARA),PROPORTIONAL TO V3 (INITIAL 1 BARA)]
630 IF DIST/D<DIST1 THEN GOTO 510  [CONTINUE UNTIL FLAME HAS MOVED 1 DIAMETER  ]
                         GOSUB 210
660 IF PVDAT<10 THEN DIFF=PVDAT/50 ELSE DIFF=PVDAT/100    [SET MATCHING CRITERIA]
670 IF ABS(PVDAT-PV)<DIFF THEN GOTO 695                             [CHECK MATCH]
680 IF PVDAT-PV>DIFF THEN GOTO 150                [INCREASE SURFACE AREA FACTOR]
690 IF PVDAT-PV<DIFF THEN GOTO 180                [DECREASE SURFACE AREA FACTOR]
                         GOSUB 20
710 D(DIST1)=DIST/D:V(DIST1)=PV:S(DIST1)=S:RE(DIST1)=RE
    P(DIST1)=P:P3(DIST1)=P3                     [STORE CALCULATED VALUES FOR OUTPUT]
760 IF DIST/D>109 THEN GOTO 790 [BEYOND 109 DIAMETERS GOTO PRINTOUT OF RESULTS ]
770 DIST1=DIST1+1             [INCREMENT DISTANCE FOR NEXT MATCH (DIAMETERS) ]
                         GOTO 500
780 D(0)=0:V(0)       [INITIAL VALUES FOR OUTPUT OF DISTANCE D(0) & VELOCITY V(0)]
                      PRINTOUT OF RESULTS FOLLOWS
```

Figure 1. GENERAL ARRANGEMENT OF REACTION SYSTEM

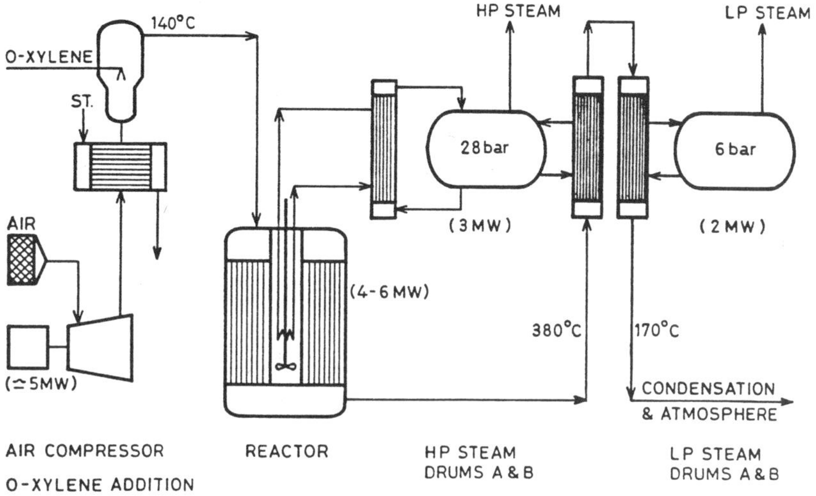

FIG 2. FLAME SPEED/EXPANSION RATIO

FROM BOMB TESTWORK

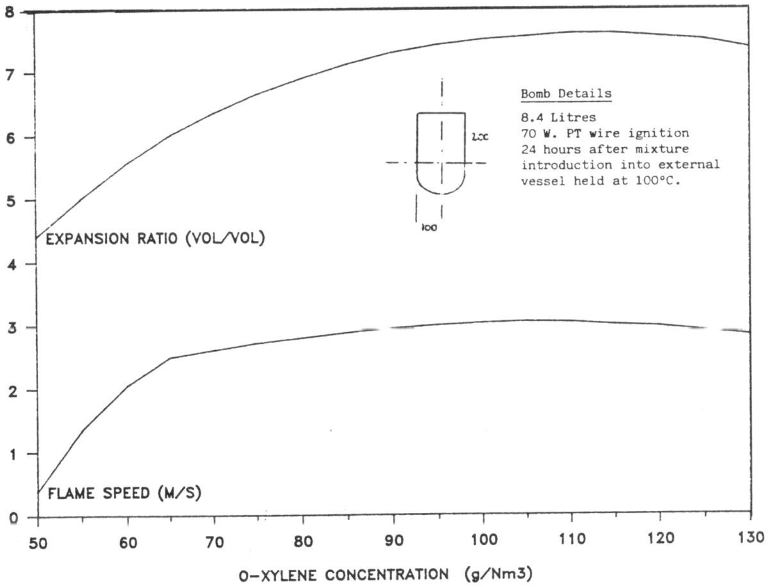

F I G U R E 3 VENTING OF REACTOR HEAD

INLET O-XYLENE CONCENTRATION	=	58 GM/NM3
CONFINED FLAME VELOCITY	=	1.76 M/SEC
VOLUME EXPANSION RATIO	=	5.3 VOL/VOL
INITIAL HEAD WORKING PRESSURE	=	1.3 BAR ABS
HEAD VOLUME	=	16.2 M3
BURSTING DISC SET PRESSURE	=	1.8 BAR ABS
BURSTING DISC RELIEF AREA	=	.59 M2

HEAD PRESSURE BAR ABS	PERCENT FLAMMABLES VENTED	OUTFLOW THROUGH EXP VENTS M3/SEC	EXPANSION RATE FOR EXPLOSION M3/SEC	TIME AFTER IGNITION MILLISECS	
1.30	0.00	0.00	0.04	10	
1.31	0.00	0.00	3.63	100	
1.36	0.00	0.00	14.52	200	
1.51	0.00	0.00	29.56	300	
1.53	0.00	0.00	30.67	310	
1.55	0.00	0.00	31.81	320	
1.57	0.00	0.00	32.99	330	
1.59	0.00	0.00	34.21	340	
1.61	0.00	0.00	35.46	350	
1.64	0.00	0.00	36.75	360	
1.66	0.00	0.00	38.08	370	
1.69	0.00	0.00	39.44	380	
1.71	0.00	0.00	40.84	390	
1.74	0.00	0.00	42.27	400	
1.77	0.00	0.00	43.74	410	
1.80	0.00	0.00	45.25	420	
1.80	0.00	6.53	45.40	421	VENTS OPEN
1.92	4.33	76.46	264.09	431	
2.01	17.73	146.02	308.75	441	
2.11	36.71	152.96	357.95	451	
2.23	57.75	161.86	411.69	461	
2.33	0.00	169.32	0.00	468	COMBUSTION ENDS
2.20	0.00	159.71	0.00	478	
2.08	0.00	150.66	0.00	488	
1.85	0.00	134.05	0.00	508	
1.74	0.00	126.44	0.00	518	
1.55	0.00	112.51	0.00	538	
1.46	0.00	106.13	0.00	548	
1.30	0.00	94.43	0.00	568	

MAXIMUM EXPLOSION PRESSURE = 2.333567 BAR ABS AT 467 MILLISECONDS

FIG 4.PRESSURE VS.TIME DIAGRAMS

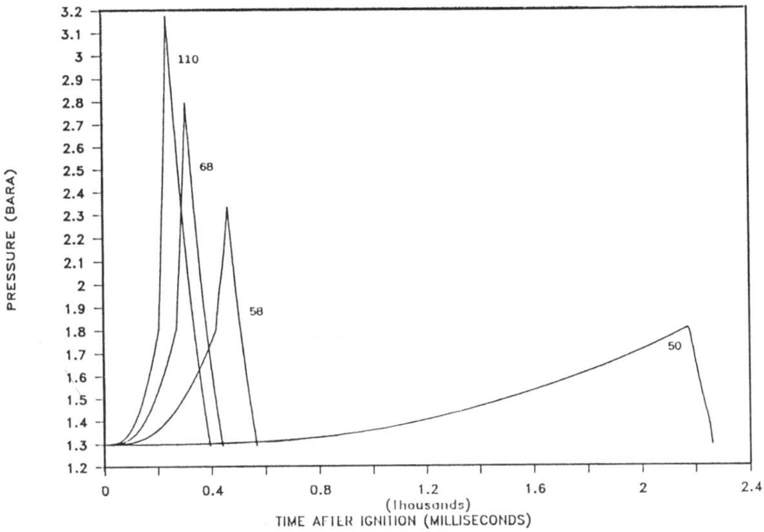

FIG 5.DATA ON METHANE-AIR DEFLAGRATIONS

400 MM PIPE STATIC/TURBULENT VELOCITIES

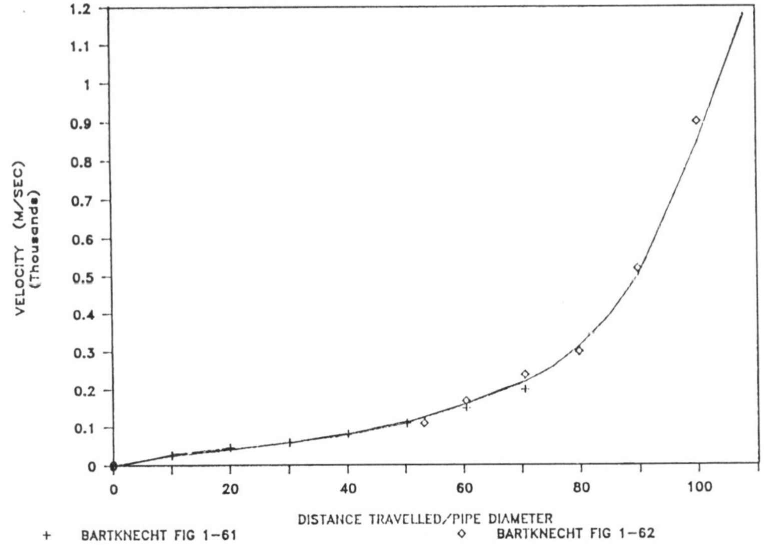

FIG 6.FLAME SURFACE FROM FIG 5.DATA

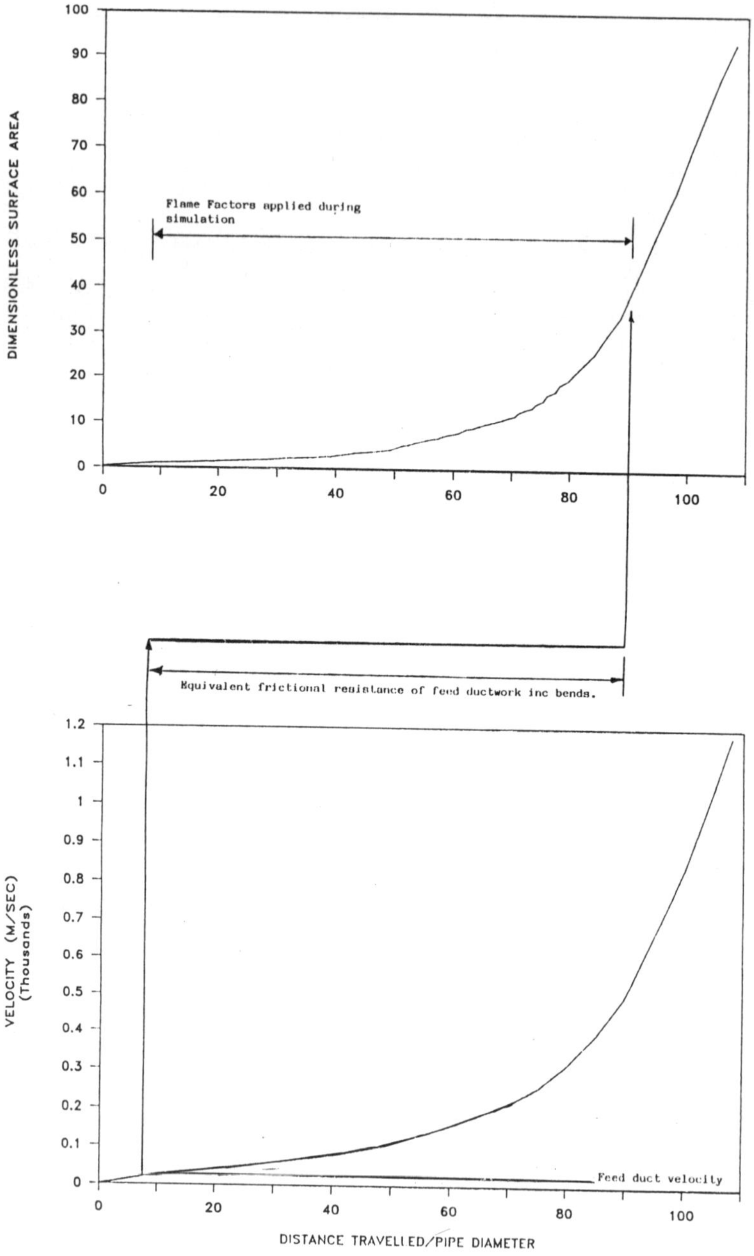

FIGURE 7 VENTING OF REACTOR FEED DUCTING PLUS HEAD

INLET O-XYLENE CONCENTRATION = 58 GM/NM3
LAMINAR FLAME VELOCITY = .24 M/SEC
VOLUME EXPANSION RATIO = 5.29 VOL/VOL
INITIAL HEAD WORKING PRESSURE = 1.3 BAR ABS
HEAD VOLUME +DUCT = 39.99 M3
BURSTING DISC SET PRESSURE = 1.8 BAR ABS
INITIAL DUCTING FLOW VELOCITY = 24 M/S
DUCTING DIAMETER = .81 M
DUCTING LENGTH = 46.17 M
VENT AREA M2 = .2827,.2827,.2827,.2827,.59
POSITION M = 0, 24, 34, 44,48

HEAD PRESSURE BAR ABS	PIPE VEL/ DISTANCE ALONG DUCTING		% BURNT/ FLAME VELOCITY M/SEC		OUTFLOW DUCT/HEAD EXP VENTS M3/SEC		EXPANSION RATE FOR EXPLOSION M3/SEC	TIME AFTER IGNITION MILLISECS	
1.31	42.2	1	0.1	5	0	0	10.07	25	
1.32	42.4	3	0.4	5	0	0	10.50	72	
1.34	44.5	5	0.7	5	0	0	12.05	118	
1.36	45.6	7	1.1	6	0	0	13.06	163	
1.38	47.4	9	1.4	7	0	0	14.64	206	
1.40	49.2	11	1.8	7	0	0	16.27	248	
1.42	50.6	13	2.2	8	0	0	18.19	287	
1.44	52.8	15	2.7	9	0	0	19.81	326	
1.47	53.7	17	3.2	10	0	0	21.11	364	
1.50	57.2	19	3.7	11	0	0	24.43	400	
1.52	57.6	21	4.3	12	0	0	25.65	434	
1.55	59.6	23	4.9	13	0	0	28.14	468	
1.58	61.4	25	5.5	14	0	0	30.67	501	
1.60	63.3	26	5.9	15	0	0	33.21	518	BEND
1.62	80.7	27	6.2	22	0	0	49.59	532	BEND
1.67	119.8	29	7.2	38	0	0	85.06	551	
1.71	131.5	31	8.2	45	0	0	99.59	567	
1.76	139.4	33	9.3	53	0	0	116.62	581	
1.80	151.0	34	10.2	57	429	178	126.37	591	VENTS OPEN
1.56	136.5	37	11.7	61	391	152	135.15	608	
1.34	93.7	39	13.3	60	356	122	133.61	626	
1.18	169.9	41	14.7	85	226	92	187.30	644	BEND
1.17	279.7	42	15.2	138	213	89	304.79	648	BEND
1.18	523.4	43	15.8	225	221	91	497.28	651	
1.20	529.3	46	17.0	276	332	66	610.69	655	
1.22	559.3	46	17.7	280	351	68	2479.77	657	HEAD BURNS
1.28	0.0	46	19.1	280	445	76	5256.97	658	
1.70	0.0	46	28.6	280	552	114	2743.00	660	
1.84	0.0	46	32.6	280	580	122	2743.00	662	
1.97	0.0	46	36.9	280	608	126	2743.00	664	
2.10	0.0	46	41.3	280	624	127	2743.00	666	
2.23	0.0	46	46.0	280	637	128	2743.00	668	
2.29	0.0	46	48.4	280	643	128	2743.00	669	
2.36	0.0	46	50.8	0	649	129	0.00	670	FLAME ENDS
2.11	0.0	46	50.8	0	625	127	0.00	680	
1.65	0.0	46	50.8	0	540	110	0.00	700	

MAXIMUM EXPLOSION PRESSURE = 2.358709 BAR ABS AT 669 MILLISECONDS

FIG 8.WILSONS CLOSED VESSEL TESTWORK

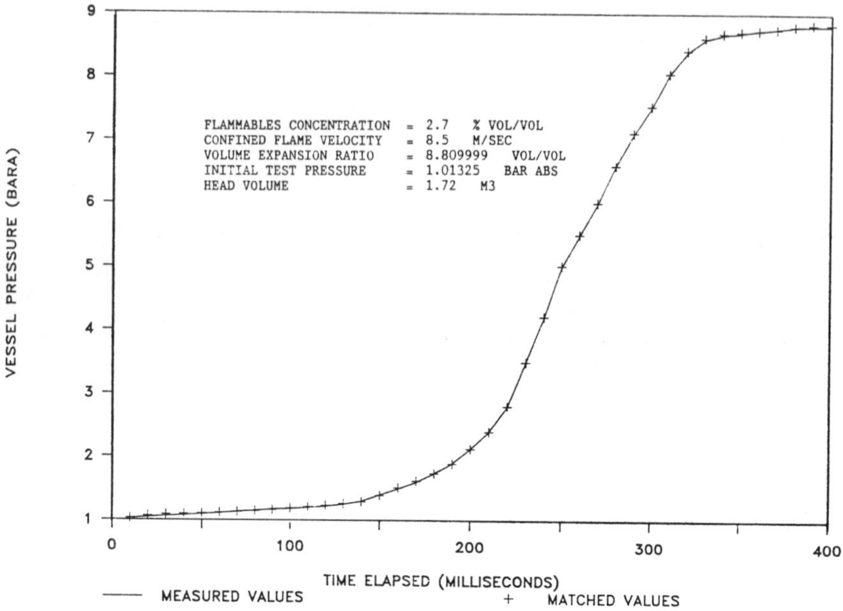

TEST PRESSURE BAR ABS	RATE PRESSURE RISE BAR/S	FLAME RADIUS/ VELOCITY MM/M-S		EXPANSION RATE FOR EXPLOSION M3/SEC	TIME AFTER IGNITION MILLISECS
1.01	0.0	1	8.5	0.00	0
1.06	5.3	137	5.2	1.14	20
1.09	1.0	161	0.7	0.23	40
1.11	1.2	175	0.7	0.27	60
1.14	1.7	190	0.8	0.36	80
1.18	2.2	208	0.9	0.49	100
1.23	2.6	226	0.9	0.57	120
1.29	3.4	245	1.0	0.75	140
1.51	13.0	297	2.6	2.83	160
1.74	13.0	337	2.0	2.82	180
2.13	24.9	388	2.9	5.40	200
2.80	43.4	454	3.7	9.43	220
4.22	77.0	551	4.4	16.73	240
5.51	51.4	617	2.3	11.17	260
6.60	58.7	638	2.5	12.77	280
7.54	42.2	637	1.8	9.17	300
8.42	36.3	637	1.5	7.89	320
8.69	7.0	635	0.3	1.52	340
8.73	2.3	635	0.1	0.51	360
8.80	4.7	635	0.2	1.01	380
8.83	0.0	635	0.0	0.00	400

MAXIMUM EXPLOSION PRESSURE = 8.850982 BAR ABS AT 389 MILLISECONDS
MAXIMUM RATE OF PRESSURE RISE = 98.22226

FIG 9. WILSON'S VENTED EXPERIMENT

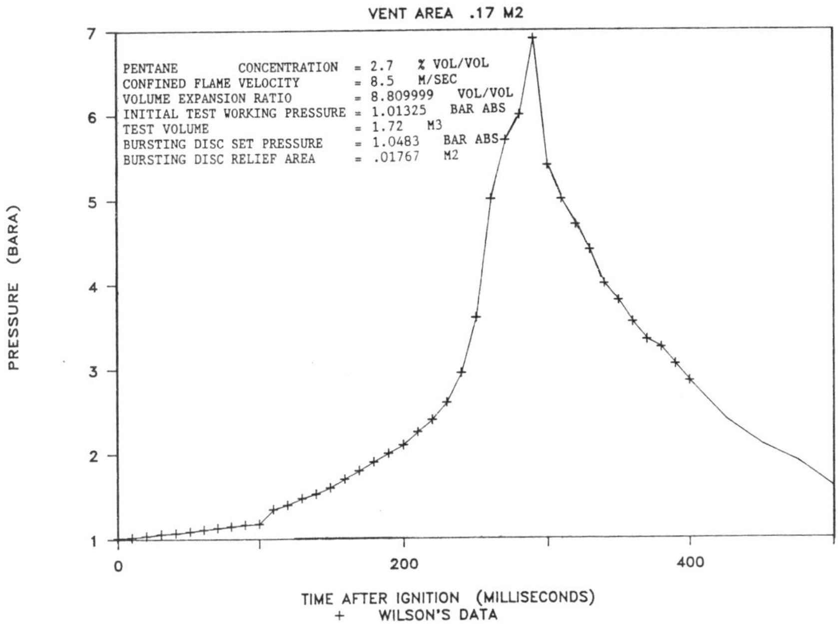

TEST PRESSURE BAR ABS	FRACTION OF UNBURNT GAS/FLAME FACTOR	OUTFLOW THROUGH EXP VENTS M3/SEC	EXPANSION RATE FOR EXPLOSION M3/SEC	TIME AFTER IGNITION MILLISECS	
1.03	1.00 1.00	0.00	0.78	10	
1.05	0.99 1.00	3.12	1.75	14	VENTS OPEN
1.06	0.99 1.87	3.14	1.65	20	
1.08	0.99 3.79	3.24	1.00	30	
1.11	0.98 2.11	4.42	1.06	50	
1.13	0.97 2.06	4.55	1.22	60	
1.17	0.96 0.79	3.63	1.12	80	
1.20	0.95 0.33	3.77	0.82	100	
1.42	0.91 0.95	4.71	2.60	120	
1.55	0.88 0.61	5.29	2.32	140	
1.69	0.85 0.59	5.89	3.32	160	
1.91	0.80 0.57	6.85	4.18	180	
2.12	0.76 0.39	7.34	4.17	200	
2.40	0.70 0.37	7.49	5.78	220	
2.95	0.60 0.63	7.76	10.74	240	
3.59	0.50 1.00	8.26	18.59	250	
5.00	0.29 3.27	9.61	36.74	260	
5.71	0.16 1.98	10.24	22.30	270	
6.19	0.00 0.00	14.01	0.00	287	COMBUSTION ENDS
6.03	0.00 0.00	13.90	0.00	290	
1.02	0.00 0.00	3.92	0.00	650	

MAXIMUM EXPLOSION PRESSURE = 6.352042 BAR ABS AT 269 MILLISECONDS

AN OPTIMAL OVERDESIGN STRATEGY FOR FOULING

E. Kotjabasakis* and B. Linnhoff*

Fouling of heat transfer surface area is expensive. This paper introduces a concept aimed at reducing the cost of fouling. The concept is not to oversize the exchanger which fouls but to oversize other exchangers and to exploit interactions in the exchanger network. A numerical procedure is presented. The procedure takes full account of capital and operating costs, and constraints imposed through plant shut-down for cleaning, etc. It is based on the use of "Sensitivity Tables" as recently introduced elsewhere (1). A practical example is used to demonstrate the procedure and to compare the results with those obtained by more traditional methods.

INTRODUCTION

A recent survey (2) showed that the money lost due to fouling in oil refineries in the non-communist world is of the order of $4,500,000,000/annum (at 1979 prices). Considering that refineries represent only a minor proportion of the total fouling related expense in the economy we can safely conclude that the understanding and control of fouling costs is of major industrial importance. The incentive for the development of good practices is clearly present.

Fouling can be interpreted as a gradual change in plant operating conditions. As such it can be considered to pose a "flexibility" problem for plant design and operation.

A new approach to "design for flexibility" has recently been introduced in the literature(1). The approach is based on "Sensitivity Tables" and is structured in three stages:

Stage 1: Flexibility requirements are represented as DEVIATIONS from a defined base case.

Stage 2: The plant's PASSIVE RESPONSE to the deviations is established. Traditionally, this requires simulation.

Stage 3: DESIGN CHANGES are proposed to cancel out unwanted passive responses. Traditionally, this requires both the "invention" of suitable design changes and simulation.

*Centre for Process Integration, Chemical Engineering Department, UMIST, Manchester M60 1QD

"Sensitivity Tables" form an important part of the approach. They give an overview of the network interdependencies (or "sensitivities") involved. This helps the designer in two ways. First, he can establish PASSIVE RESPONSES without simulation. Second, he can see all possible DESIGN CHANGES at a glance, and can screen out the cheapest option quickly and effectively.

In this paper we demonstrate that the approach provides a powerful tool to combat the effects of fouling. Specifically, we demonstrate that the procedure:

*allows us to quickly determine the effects of fouling on the operation of an overall network;

*allows us to, just as quickly, determine how the effects can be compensated for most cheaply.

TYPICAL PROBLEM

Figure 1 shows the simplified flowsheet of a vacuum distillation unit. The relevant heat exchanger network is shown in Figure 2 using the grid representation proposed by Linnhoff(3). Data describing the current operation of the plant are given in Table 1a.

The plant is to be retrofitted to operate on a different duty for a single campaign once a year. The length of the campaign is approximately two months. A new feedstock is to be used. Flowrates within the plant need to be increased and some temperatures are changed. As mentioned in the introduction, Case B is described as a set of deviations from Case A. The data describing CASE B are given in Table 1b.

Case B operation has another complication. The temperature of stream 1 is increased significantly (see Table 1b) and this is known to result in exchanger fouling at temperatures in excess of 280 deg. Thus, exchanger no. 1 in Figure 2 will operate clean during Case A operation but will be subjected to fouling , to a significant extent, during Case B operation.

The design objective for the retrofit is to find the cheapest project, in terms of energy and capital costs (see Table 2), which will enable in future both Case A and Case B operation. All specified flows and temperatures must be maintained. Fouling and cleaning requirements (it usually takes two days to clean a heat exchanger) must be taken into account and plant downtime, which is very costly, must be avoided.

To assess what is "cheapest" we define the "Total Additional Cost", analogous to reference 1, as:

$$\left[\begin{array}{l}\text{Total}\\ \text{Additional}\\ \text{Cost}\end{array}\right] = \left[\begin{array}{l}\text{Annualised Cost}\\ \text{of Capital for}\\ \text{retrofit}\end{array}\right] + \left[\begin{array}{l}\text{Annual Cost of}\\ \text{Energy after}\\ \text{retrofit}\\ \text{Case (A+B)}\end{array}\right] - \left[\begin{array}{l}\text{Annual Cost of}\\ \text{Energy before}\\ \text{retrofit}\\ \text{Case A only}\end{array}\right]$$

We suggest that an engineer might tackle this problem by first simulating the performance of the existing plant for Case B. The results of such a simulation are shown in Figure 3. The target temperatures (TT) for streams 1,2, and 4 have been maintained by increased use of utilities. The target temperature of stream 3 cannot be maintained using existing equipment. Consequently, the engineer might specify a new heater for stream 3, see Figure 4.

Next, the engineer would need to specify overdesign on exchanger 1 to counteract the fouling problem. The extent of overdesign is not immediately clear. Plant operation must be considered: downtime should be avoided, even while an exchanger is cleaned. The capital cost of new area must be weighed against operating costs, based on non-constant operating conditions.

The rate of fouling is known. There is a reduction in the overall heat transfer coefficient of 1% per day. A quick simulation shows that the performance of exchanger no. 1 has deteriorated after less than forty days of Case B operation such that the load on the furnace exceeds its maximum firing capacity (9200kW). At that point, the plant can no longer operate. By the same token, it is clear that the plant cannot operate with exchanger 1 off-line for cleaning. It follows that we need either:

* a new furnace, or

* additional surface area in exchanger 1, installed in parallel to the existing exchanger.

A new furnace is too expensive. So the remaining question is what should be the size of the required new exchanger?

There is a trade-off between capital cost, operating cost and cleaning cost. To keep the example simple we will base our calculations on the minimum additional surface area required. This is easily determined. It is the smallest clean unit that will allow the plant to operate within the furnace firing limit (Figure 5). The result is 450 sq.m.

We have achieved an operable overall design for Case A and Case B subject to fouling. The Total Additional Cost for the retrofit is $ 50,000/annum.

SENSITIVITY TABLES

In a previous publication(1) we introduced a new approach to the cost effective provision of contingency in heat exchanger networks. It was based on simple "Sensitivity Tables" which enable the engineer to:

(i) evaluate the passive response of a network to deviations in feed temperatures, flowrates, heat transfer coefficents etc.

(ii) evaluate all possible design changes to eliminate unwanted passive responses.

A Brief Review

The heat balance equations for a single heat exchanger are:

$$Q = CP_H (T_1 - T_2)$$

$$Q = CP_C (T_4 - T_3)$$

where: T_1 = hot inlet temperature; T_2 = hot outlet temperature

T_3 = cold inlet temperature; T_4 = cold outlet temperature

and the design equation is:

$$Q = UA\ (\Delta T_{LM})$$

These equations can be transformed to give:

$$(1 - RB)T_2 + (B - 1)RT_3 + (R - 1)T_1 = 0$$

$$R(1 - RB)T_4 + (B - 1)RT_1 + (R - 1)\ BRT_3 = 0$$

where:

$$R = \frac{CP_C}{CP_H}$$

and:

$$B = \exp[(UA/CP_C)\ (R - 1)]$$

The equations are linear with respect to temperature but non-linear with respect to heat capacity flowrate and the heat exchanger parameters (U and A). If we know the exchanger parameters, the heat capacity flowrates, and any two of the four temperatures, we can solve the equations without iteration to find the other two temperatures. This is well known(4).

For networks of heat exchangers we can exploit the duplication of temperatures between exchangers and can describe the overall network by a system of simultaneous equations which again can be solved without iteration. Thus, if we know all exchanger parameters, all heat capacity flowrates, and network inlet temperatures only, we can calculate *all remaining temperatures* in a network. This holds true for any feasible network, with or without loops(1).

Non-linear effects such as non-linear enthalpy-temperature relationships, variations in local heat transfer coefficient, etc, are handled easily through the use of linearised sub-units(5).

The solution of the simultaneous equations requires only simple software(5). The result can be a simple but effective simulator which can be used with ease. However, an even more useful result exists.

We can solve the equations to obtain "unit network responses" for *any* input deviations and we can generate tables showing the correlation between inlet temperatures, flowrates, exchanger parameters, etc, deviations and their effects throughout the network(1). These "Sensitivity Tables" can be used for the rapid screening of sensitivities.

Three kinds of Sensitivity Tables are required for the full analysis of a system. They are the:

T(TS) Sensitivity Table

T(CP) Sensitivity Tables

and: T(UA) Sensitivity Tables

The tables are used in an apparently naive manner. Individual effects are simply added together with total disregard of non-linearities and interaction effects. This has the appearance of a non-linear problem being treated as linear. The appearance is misleading. Rather, the tables establish sensitivities within ranges and linearised sections. The errors generated are small. The results are, within reason, "accurate". The simplicity of the approach gives the user a near-exact birds-eye view of an otherwise intractably complex problem. Naturally the chosen design options should finally be confirmed through modelling and simulation. However, overall errors in Total Additional Cost are usually found to be less that 5% and optimal solutions identified quickly are confirmed as such even after protracted and "exhaustive search" using traditional simulation.

Sensitivity Tables for the Example Problem

We will now generate the relevant Sensitivity Tables for our example problem.

T(TS) Sensitivity Table

A single table (Table 3) is sufficient to provide all correlations between changes in all supply temperatures (TS) and the response in all other network temperatures (T). Given that the transformed base equations are linear with respect to temperatures, the table gives exact results for any combination of changes over any range of deviation.

The temperature notation used in Table 3 is explained in Figure 6. The correlation factors listed are simple multiplication terms. The use of the table is simple. For instance, we can see at a glance that if supply temperature TS1 changes by 10 deg then temperature T14 changes by 0.7 deg. If supply temperature TS2 changes by 10 deg then T14 changes by 2.27 deg., etc.

T(CP) Sensitivity Tables

The system of equations is non-linear with respect to heat capacity flowrates (CP). Therefore, we need one T(CP) Sensitivity Table for each stream if the effects of non-linearities are to be minimised. Each of these tables will cover a range of flow variations about the base case.

A typical T(CP) Sensitivity Table is shown in Table 4. It applies to stream 1 of our example problem. It lists temperature deviations as opposed to correlation factors. For instance, it tells us that if the heat capacity flowrate of stream 1 is increased by 30% then temperature T14 would increase by 3.53 deg, etc.

T(UA) Sensitivity Tables

The system of equations is also non-linear with respect to the exchanger parameters (UA). Thus, as in the case of flow, we need several tables. In this case we need one for each exchanger, covering a range of possible deviations in exchanger (UA) from the base case.

Using the T(UA) Sensitivity Tables we can consider operational factors such as fouling and the influence of flow upon overall heat transfer coefficient. We can also consider changes in design parameters such as area and overall heat transfer coefficient.

The T(UA) Sensitivity Tables for our problem are given in Tables 5a-d. Again, they contain temperature deviations and not correlation factors.

APPLYING SENSITIVITY TABLES

In this section, we will work with the Sensitivity Tables to solve our problem. First, we will address the new alternative duty (Case B), ignoring fouling. Second, we will address fouling.

Case B operation

We apply the tables in two stages. The determination of the passive network response is followed by the selection of a suitable design change.

Passive Response

The passive response is the final steady state condition of the existing network given the specified deviations from base case (see Table 1). It could be determined from simulation. Alternatively, we can use the Sensitivity Tables.

As mentioned earlier, traditional simulation indicated that all but the target temperature on stream 3 could be maintained through the use of additional utilities in existing exchangers (see Figure 3). The short fall on target temperature 3 was 7.35 deg. Let us see what the tables tell us.

Individual effects can be determined from the T(TS) and T(CP) Tables. For instance, for the increase in supply temperature 1 (40 deg) Table 3 tells us that the associated change in target temperature 3 (T14, see Figure 6) is:

$$\Delta T_{14}(TS1) = (0.069).(40) = 2.76 \text{ deg.}$$

The other effects are also quickly established. They are listed in Table 6. As stated previously, the overall effect is found by simple-minded addition. A shortfall of 7.2 deg is predicted. There is a difference between this result and that obtained from full simulation of less than 3%.

The Tables are both quick and simple to use and can be used repeatedly for other network temperatures to establish the full passive response. This may or may not be important, depending on the availability of a simulator. It should be borne in mind however, that the tables need only be computed once at the beginning of a project. Thereafter, the engineer is at liberty to test responses to the full without recourse to a computer.

Design Changes

In our example problem all four target temperatures need to be maintained. Three of these (TT1, TT2 and TT4) are easily maintained using existing coolers and heater. The remaining one (TT3, or T14 in Figure 6) should probably be maintained differently as no heater or cooler is situated conveniently.

The usual method of building flexibility into a heat exchanger network is by oversizing exchangers and the provision of bypasses. The T(UA) Sensitivity Tables give the necessary information for identifying and evaluating such options(1). They identify the effect of changes in exchanger parameters on every temperature in the network. Consequently, they allow us to easily discuss every possible design option - even the non-obvious ones. The use of the Tables is now in a reverse manner to that for determining passive response. The objective is to cancel undesirable elements of the passive network response. In other words, in our problem, we are looking for changes in the "effective area"(1) of any exchanger such that $\Delta T_{14} = +7.2$ deg.

Inspection of the T(UA) Sensitivity Tables shows us that there are four alternatives (in addition to the "obvious" solution involving a heater):

Heat exchanger 1	: Δ(UA1) = -82%	Bypass
Heat exchanger 2	: Δ(UA2) = -23%	Bypass
Heat exchanger 3	: Δ(UA3) = +66%	Additional Area
Heat exchanger 4	: Δ(UA4) = +40%	Additional Area

Using the Tables we see all these options at a glance. In contrast, the use of simulation would be cumbersome. We would have to run too many cases to be practical. Ultimately, we would need to make some intuitive choices and cross our fingers. It is this ability of the Tables to show *all options at a glance*, the "birds-eye view", that sets them apart from simulation. Simulation is a modelling tool. The Tables are also a design tool.

To evaluate the overall cost implications of the four options identified we need to know the overall passive response of the entire network for each of the options(1) when overlaid with initial passive response. Once more, this information is easily obtained from the Tables or from simulation. To enable a final assessment of accuracy to be made we stick, for the time being, to the use of Tables.

The "Total Additional Costs" of the four options is as follows:

Heat exchanger 1	: 190,000 $/year
Heat exchanger 2	: 73,000 $/year
Heat exchanger 3	: 123,000 $/year
Heat exchanger 4	: 71,000 $/year
Heater of Stream 3	: 85,000 $/year

In Figure 7 we show the Total Additional Cost of each of the options as a function of the Case B annual operating time. Using the Tables this information is quickly obtained. It can be seen that for Case B operating periods of two months per year or more the option involving heat exchanger 4 is cheapest. For shorter annual operating periods the option involving heat exchanger 2 is cheapest.

We would venture that the cheapest solution, i.e. overdesign of exchanger 4 (Figure 7), comes somewhat of a surprise and may not have been identified without the use of the Tables. The "obvious" solution (installation of a heater on stream 3) is 20% more expensive.

Finally we assess the errors associated with the use of the Tables for Total Additional Cost. Comparing it to the results of full simulation we found it to be around 3%.

Fouling

The traditional approach to fouling is to add area for contingency in the exchanger which is subjected to fouling. Is this the right approach? There is an interaction between exchanger design and fouling rate. It has been found in some notorious cases that an increase in exchanger size to cope with fouling has resulted in lower velocities etc., and therefore an increased rate of fouling! Secondly, we should recognise that the additional area is installed in a position where its effective use is impaired of necessity by fouling. Could we do better?

Why not ask the question "Can we cope with fouling by installing additional area in positions in the network other than where the fouling occurs?"

Sensitivity Tables help us again to implement this idea. They help us to evaluate the ways in which exchangers interact within the network. Fouling can be described as a reduction in UA of a particular exchanger. We can determine its effect from the Tables as a passive response and subsequently we can use the Tables to identify design changes elsewhere to overcome the effect.

Let us apply the Tables in this fashion to our example problem. As described above, there are two major operational concerns in our problem:

*the furnace firing capacity may be limiting

*we need to avoid plant downtime, even during cleaning.

From the information given in the T(UA) tables we can construct Figure 8, which shows the effect of changes in the effective area of exchangers 1,2 and 4 on the furnace inlet temperature. We see that we can compensate for the effects of fouling in three ways:

* additional area in the fouled unit (exchanger 1)

* additional area in exchanger 2

* additional area in exchanger 4.

The cheapest option appears to be an increase in the area of exchanger 4. *This option is significantly better than simply oversizing the fouling exchanger (exchanger 1)*. Next we must decide how much overdesign of exchanger 4 is required and how often exchanger 1 should be cleaned. The two factors are closely related.

First, we need to establish whether it is possible to remove exchanger 1 for cleaning whilst maintaining plant operation. The Tables tell us that this is now possible,, but only if the overdesign on exchanger 4 is large. This leaves us with two possible choices:

1) We can place new area in two locations: exchanger 4 for Case B operation and some fouling correction; and a new exchanger parallel to exchanger 1 to permit cleaning. This would imply an operating strategy analogous to that used earlier (Figure 5) but with lower capital cost implications.

2) We can place new area solely on exchanger 4. We would not remove exchanger 1 for cleaning until the production campaign (Case B) has finished. While exchanger 1 is cleaned (during Case A operation) we can keep the plant on line due to exchanger 4 being now larger.

This choice of options requires some detailed cost trade-off calculations. The first option was found to be more expensive than the second.

The chosen option is again evaluated with the Sensitivity Tables. For the purpose of comparison with the solution found earlier (Figure 5) we have based our calculations on the minimum additional surface area required on exchanger 4. The results are shown in Figure 9 in the terms of operating policy and furnace duty. The minimum overdesign on exchanger 4 is 330 sq.m.

This final solution as found using the Tables has a Total Additional Cost of $-42,000/year. It is $92,000/year cheaper than the "straight-forward" solution derived earlier (see Table 7).

DISCUSSION

The Sensitivity Tables

In the example presented the error associated with the use of Sensitivity Tables related to full simulation was found to be around 3%. General experience from practical application of the Tables indicates that the error in estimating Total Additional Cost is usually less than 10%, and often less than 5%.

The purist might be worried. Errors of this magnitude may affect comparisons. However, let us bear in mind that the screening of options is used in the present context for the early rejection of alternatives whose differences in Total Additional Cost typically exceeds 50%. The Tables ensure that all options are seen early, leading to the safe rejection of most. If necessary, more than one of the cheaper solutions can be fully evaluated using detailed simulation.

In essence the Tables present a "birds-eye" picture. The engineer gains a full understanding of available options and is able to evolve a practical, and perhaps non-obvious, solution with confidence and little effort.

Simulation does not present an alternative. It does not provide this "birds-eye" view. Too many repeated calculations would be required to build up an overall picture. The appropriate use of simulation is for final design validation, not for the invention part of design.

Compensating for Fouling through Network Interactions

To the authors' best awareness, this concept has not been previously suggested in the literature. There have been several papers addressing the behaviour of a given network to fouling (e.g. ref. 6). However, these papers only examine network response for given installed areas. When compensation is considered, the approach is invariably to install additional area in the exchangers which foul. The present paper suggests to install the additional area systematically and perhaps elsewhere in the network.

In the example presented we have seen that the traditional treatment of fouling (i.e. through concentrating on the unit in which fouling occurs) is not always the best solution. Fouling can be compensated for through additional area in "clean" units using network interactions. Why invest in extra heat transfer area that is going to foul?

This simple idea could have a far reaching economic impact.

Dealing with fouling by seeking to avoid it (e.g. chemical additives etc.) is an active area of research. However, it is a complex problem. Headway is being made in specific areas but a general solution is not in sight.

In this paper we present an alternative and a practical approach that works today:

> Once we have identified that fouling will occur let us compensate for it in places other than the unit which fouls. Let us then consider the overall sensitivities in the integrated process. Let us examine network interactions and check cheaper possible solutions. All this can be done by inspection and simulation. However Sensitivity Tables are nicely suited for the task.

CONCLUSIONS

Fouling can be considered to be a deviation in the operating condition of a plant. As such it can be considered to be a "flexibility" problem.

Sensitivity Tables exist for flexibility problems in design. In this paper, their use has been demonstrated for heat exchanger networks. They are quick and simple to apply. They give an overview of all options available to the designer. Sometimes, good options are non-obvious. The Tables are accurate enough for the easy recognition of such non-obvious but cheap solutions.

Full simulation is not suited for this task. The proper use of simulation is in final design validation. The use of repeated simulation for invention should be the last resort if all other means of invention have failed.

The cost of fouling in the process industries is high. The incentive to reduce these costs is real. One option that has not been recognised to date, as far as the authors are aware, is the use of exchanger interactions in design to compensate for the effects of fouling. In the immediate term this could provide the quickest route to a practical reduction in industrial fouling costs. The benefits are difficult to estimate. What is easy to say is that they will be vast.

ACKNOWLEDGEMENTS

Thanks are due to the Bodossakis Foundation, the Onassis Foundation and the Process Integration Research Consortium, who supported Mr. Kotjabasakis's research at UMIST. BASF, Union Carbide, and Dow helped this work greatly by providing meaningful industrial case studies.

Special thanks are due to Dr. G.T. Polley for his help and constructive criticism in preparing the manuscript.

REFERENCES

1. Kotjabasakis E. & Linnhoff B. "Sensitivity Tables for the Design of Flexible Processes." Chem.Eng.Res.Dev. 1986,64,197-211

2. Van Nostrand W. L., Leah J. R. & Havuska J. L. "Economic penalties associated with the fouling of Refinery Heat Transfer equipment". Paper presented at the International Conference on Fouling of Heat Transfer equipment, Rensselaer Polytechnic Institute, USA, 1979.

3. Linnhoff B., Ph.D. Thesis, University of Leeds, 1979.

4. Kays W. M. & London A. L., Compact Heat Exchanger, McGraw-Hill, 1958

5. Kotjabasakis E., Ph.D. Thesis, UMIST, 1987.

6. Fryer P. J., Paterson W. R. and Slater N. K. E., unpublished manuscript, University of Cambridge, 1985.

TABLE 1

Stream data describing current plant operation (CASE A) and the new alternative duty (CASE B).

CASE A (10 MONTHS/YEAR)

STREAM DATA

STREAM No.	TS	TT	CP (kW/°C)	h (kW/m^2°C)
1	285	218	81.49	0.35
	218	150	69.71	0.28
	150	120	66.97	0.24
2	272	151	49.17	0.50
	151	75	44.41	0.41
	75	40	43.29	0.40
3	55	115	55.87	0.38
4	40	102	76.45	0.33
	102	175	81.51	0.41
	175	236	89.51	0.47
	236	315	97.10	0.53

(a)

CASE B (2 MONTHS/YEAR)

STREAM No.	TS	TT	CP (kW/°C)	h (kW/m^2°C)
1	+40°	-	+30%	-
2	-20°	-	+10%	-
3	-	-	+20%	-
4	-	-	+20%	-

(b)

TABLE 2

Cost Data

Cost of Heat Transfer Area

*Process/process exchangers, heaters:

Installed cost ($): $26,000 + 330\ (A)^{0.95}$

A: area of exchanger in m^2

*Air coolers:

Installed cost ($): $2,000\ (A)^{0.76}$

A: area of cooler in m^2,

bare tube surface area based on:

$U = 400\ W/m^2k$

$MTD = \frac{1}{2}\ Ti - 12°C$

Ti = process stream inlet temperature

*Life time of equipment: 2 years

Cost of Utilities

*Fuel cost 2.6 $/GJ

Furnace efficiency: 0.75

*Steam: 54 psig, 140 °C

Cost of steam: 5.8 $/tonne = 12.25 $/MWhr

Heat transfer film coefficient: 6 $kW/m^2\ °C$

*Cost for air coolers operation = 0.8 $/MWhr

*Annual operating time: 8,500 hrs/year

Downtime

To be avoided

TABLE 3

The "T(TS) Sensitivity Table" gives the correlation between changes in stream supply temperatures (ΔTS_i) and the resulting changes of network temperatures (ΔT_i)

Example of use: If $\Delta TS_1 = +10°C$

then $\Delta T_8 = (0.445)\ (+10) = +4.45°C$

	ΔTS_1	ΔTS_2	ΔTS_3	ΔTS_4
ΔT_8	.445	.298	.000	.258
ΔT_9	.275	.184	.000	.541
ΔT_{10}	.211	.141	.000	.461
ΔT_{11}	.110	.362	.000	.528
ΔT_{12}	.023	.076	.790	.111
ΔT_{13}	.008	.028	.267	.040
ΔT_{14}	.069	.227	.372	.332
ΔT_{15}	.155	.104	.000	.741
ΔT_{16}	.088	.489	.000	.423
ΔT_{17}	.594	.219	.000	.188
ΔT_{18}	.375	.137	.0000	.119

TABLE 4

The "T(CP) Sensitivity Table" for stream 1 gives the correlation between changes in its heat capacity flowrate (ΔCP_1) and the resulting changes in network temperatures (ΔT_i)

Example of use: If ΔCP_1 = -10% then,

ΔT_8 = -6.11°C

ΔCP_1	-40%	-30%	-20%	-10%	0%	10%	20%	30%	40%
ΔT_8	-30.64	-21.35	-13.19	-6.11	.00	5.29	9.88	13.89	17.41
ΔT_9	-45.08	-32.01	-20.16	-9.51	.00	8.50	16.10	22.91	29.01
ΔT_{10}	-45.16	-32.64	-20.87	-9.98	.00	9.11	17.41	24.97	31.87
ΔT_{11}	-12.03	-8.40	-5.21	-2.43	.00	2.12	3.98	5.62	7.08
ΔT_{12}	-2.52	-1.76	-1.09	-.51	.00	.44	.84	1.18	1.48
ΔT_{13}	-.92	-.64	-.40	-.19	.00	.16	.30	.43	.54
ΔT_{14}	-7.56	-5.27	-3.27	-1.52	.00	1.33	2.50	3.53	4.45
ΔT_{15}	-16.90	-11.80	-7.32	-3.41	.00	2.98	5.60	7.90	9.55
ΔT_{16}	-9.64	-6.73	-4.18	-1.94	.00	1.70	3.19	4.51	5.68
ΔT_{17}	-17.31	-11.42	-6.77	-3.04	.00	2.51	4.60	6.37	7.87
ΔT_{18}	-10.92	-7.21	-4.27	-1.92	.00	1.58	2.90	4.02	4.98

TABLE 5

The "T(UA) Sensitivity Tables" give the correlation between changes in (UAi) at a given exchanger (i) and the resulting changes in network temperatures (ΔTi).

Examples of use: If ΔUA_1 = +20%
i.e. $(UA_1) = (1.2)\ (UA_1)$ base case

then, $\Delta T_8 = -5.8$°C

ΔUA_1	-80%	-60%	-40%	-20%	0%	+20%
ΔT_8	43.9	27.0	16.1	7.1	0.0	-5.8
ΔT_9	27.1	17.2	10.0	4.4	0.0	-3.6
ΔT_{10}	20.9	13.3	7.7	3.4	0.0	-2.7
ΔT_{11}	10.9	6.3	4.0	1.8	0.0	-1.4
ΔT_{12}	2.3	1.5	0.8	0.4	0.0	-0.3
ΔT_{13}	0.8	0.5	0.3	0.1	0.0	-0.1
ΔT_{14}	6.9	4.3	2.5	1.1	0.0	-0.9
ΔT_{15}	15.3	9.7	5.6	2.5	0.0	-2.0
ΔT_{16}	8.7	5.5	3.2	1.4	0.0	-1.1
ΔT_{17}	-31.2	-19.8	-11.5	-5.1	0.0	4.1
ΔT_{18}	-19.7	-12.5	-7.2	-3.2	0.0	2.6

(a)

ΔUA_2	-20%	-10%	0%	+10%	+20%	+30%
ΔT_8	-4.5	-2.16	.00	1.88	3.54	5.00
ΔT_9	-2.87	-1.33	.00	1.16	2.19	3.09
ΔT_{10}	2.21	-1.03	.00	.90	1.68	2.38
ΔT_{11}	9.97	4.63	.00	-4.04	-7.58	-10.71
ΔT_{12}	2.09	0.97	.00	-.85	-1.59	-2.25
ΔT_{13}	.76	.35	.00	-.31	-.58	-.82
ΔT_{14}	6.27	2.91	.00	-2.54	-4.76	-6.73
ΔT_{15}	-1.62	-.75	.00	.66	1.23	1.74
ΔT_{16}	-7.64	-3.54	.00	3.09	5.81	8.20
ΔT_{17}	-3.40	-1.58	.00	1.38	2.59	3.65
ΔT_{18}	-2.15	-1.00	.00	.87	1.63	2.31

(b)

ΔUA_3	-20%	0%	+20%	+40%	+60%	+80%
ΔT_8	.00	.00	.00	.00	.00	.00
ΔT_9	.00	.00	.00	.00	.00	.00
ΔT_{10}	.00	.00	.00	.00	.00	.00
ΔT_{11}	.00	.00	.00	.00	.00	.00
ΔT_{12}	4.91	.00	-3.63	-6.41	-8.6	-9.5
ΔT_{13}	1.79	.00	-1.32	-2.33	-3.1	-3.5
ΔT_{14}	-3.90	.00	2.89	5.09	6.8	7.6
ΔT_{15}	.00	.00	.00	.00	.00	.00
ΔT_{16}	.00	.00	.00	.00	.00	.00
ΔT_{17}	.00	.00	.00	.00	.00	.00
ΔT_{18}	.00	.00	.00	.00	.00	.00

(c)

ΔUA_4	-10%	0%	+10%	+20%	+30%	+40%
ΔT_8	-1.60	.00	1.51	2.95	4.31	5.60
ΔT_9	3.45	.00	-3.26	-6.35	-9.28	-12.07
ΔT_{10}	2.65	.00	-2.51	-4.89	-7.14	-9.28
ΔT_{11}	-3.28	.00	3.10	6.04	8.82	11.47
ΔT_{12}	-.69	.00	.65	1.27	1.85	2.40
ΔT_{13}	-.25	.00	.24	.46	.67	.87
ΔT_{14}	-2.06	.00	1.95	3.79	5.54	7.20
ΔT_{15}	-4.60	.00	4.35	8.48	12.39	16.11
ΔT_{16}	-2.63	.00	2.48	4.84	7.07	9.19
ΔT_{17}	-1.17	.00	1.11	2.16	3.15	4.10
ΔT_{18}	-.74	.00	.70	1.36	1.99	2.58

(d)

TABLE 6

The change in temperature T14 resulting from CASE B operation is found by the addition of the individual effects from each deviation.

Effect of change in TS_1:

Table 3, $\Delta T_1 = +40°C$

$\Delta T14 = (0.069)\ (+40) = +2.76°$

Effect of change in TS_2:

Table 3, $\Delta TS_2 = -20°C$

$\Delta T14 = (0.227)\ (-20) = -4.54°$

Effect of change in CP_1:

Table 4, $\Delta CP_1 = +30\%$

$\Delta T14 = +3.53°$

Effect of change in CP_2:

$\Delta CP_2 = +10\%$

$\Delta T14 = +7.2°$

Effect of change in CP_3:

$\Delta CP_3 = +20\%$

$\Delta T14 = -7.95°$

Effect of change in CP_4:

$\Delta CP_4 = +20\%$

$\Delta T14 = -8.2°$

The overall effect is:

$\Delta T14 = (+2.76) + (-4.54) + (+3.53) + (+7.2) + (-7.95) + (-8.2)$

$\Delta T14 = -7.2°C$

TABLE 7

Comparison of the two solutions

	Capital Investment ($/yr)	Energy Cost Difference ($/yr)	Total Additional Cost ($/yr)
SOLUTION 1	144,000	-94,000	50,000
SOLUTION 2	65,000	-107,000	-42,000
SOLUTION 2 as a percentage of SOLUTION 1	55%	14%	n.a.

SOLUTION 1: Design using a traditional approach (Figure 5)

SOLUTION 2: Design using Sensitivity Tables (Figure 9)

Capital Investment = [Annualised Cost of Capital for retrofit]

Energy cost Difference = [Annual Cost of Energy after retrofit Case (A+B)] - [Annual Cost of Energy before retrofit Case A only]

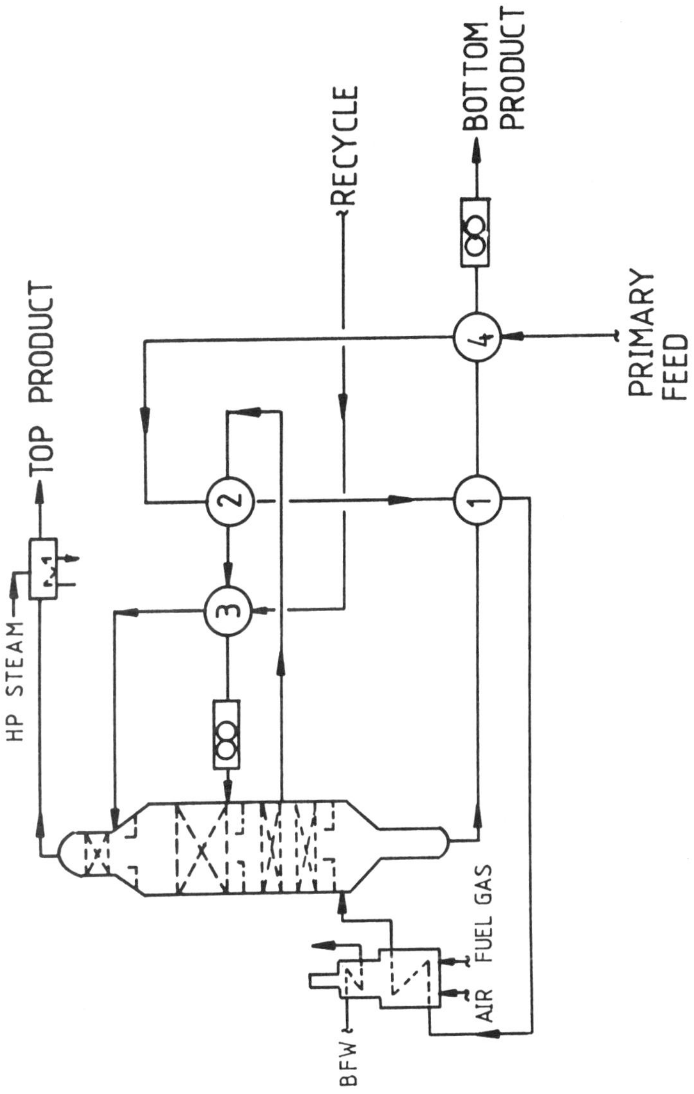

FIGURE 1

Simplified flowsheet for a Vacuum Distillation Unit.

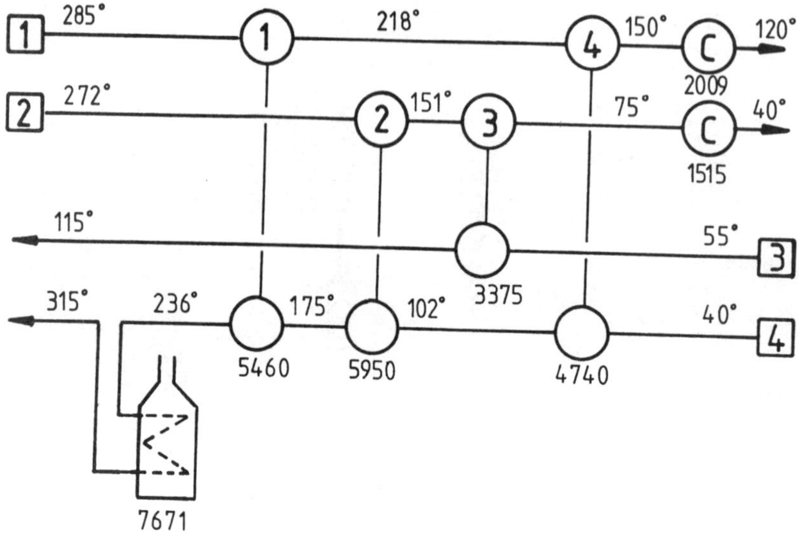

FIGURE 2

The existing network before retrofit (CASE A)

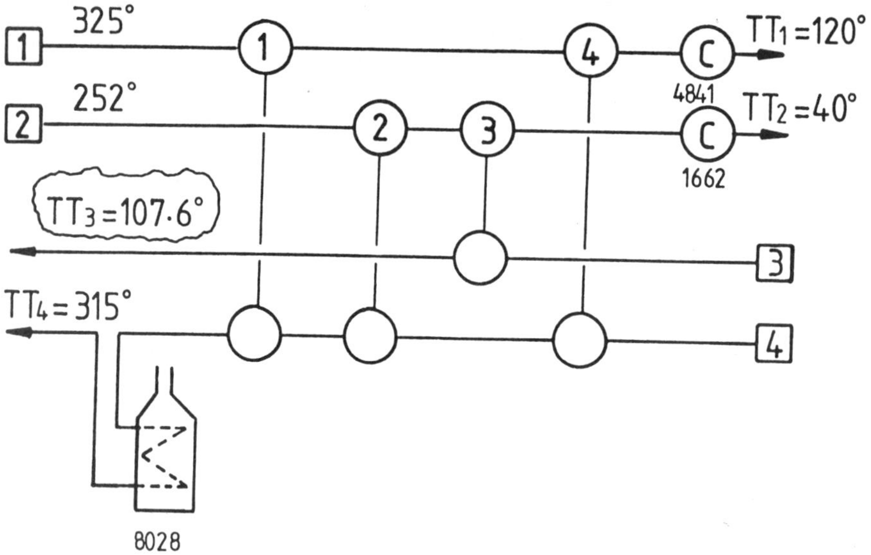

FIGURE 3

The "passive response". TT_1, TT_2 and TT_4 are maintained by utility exchangers. TT_3 is reduced by 7.4°C.

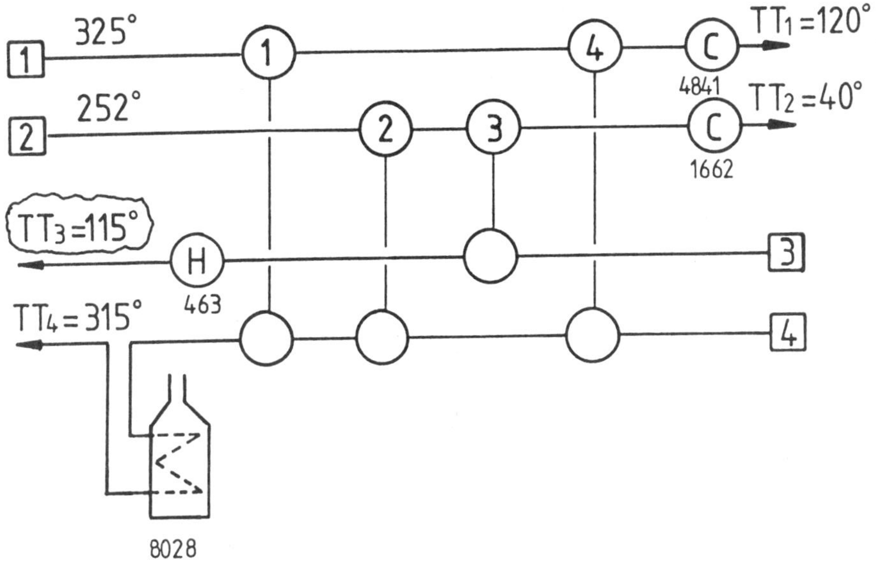

FIGURE 4

The "obvious" solution: install a steam heater on stream 3 for CASE B operation.

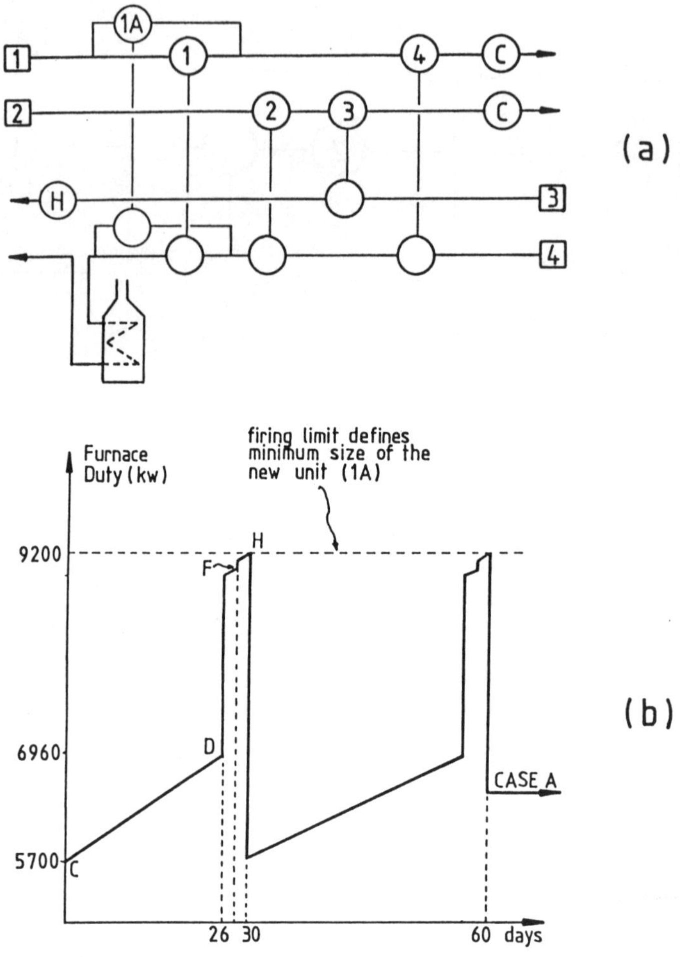

FIGURE 5

a) The final design using a traditional approach. In addition to the new heater on Stream 3 a new unit (1A) is required to compensate for fouling. Total Additional Cost: $50,000.

b) Operating policy during CASE B.
   C: Operation starts
   D: (1A) out for cleaning
   F: (1A) returned and (1) out for cleaning
   H: (1) returned

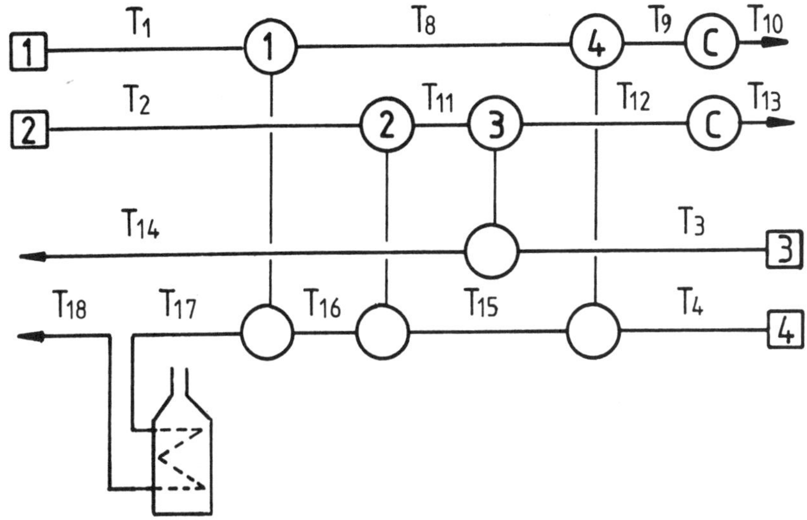

FIGURE 6

Temperature numbering system for the Sensitivity Tables.

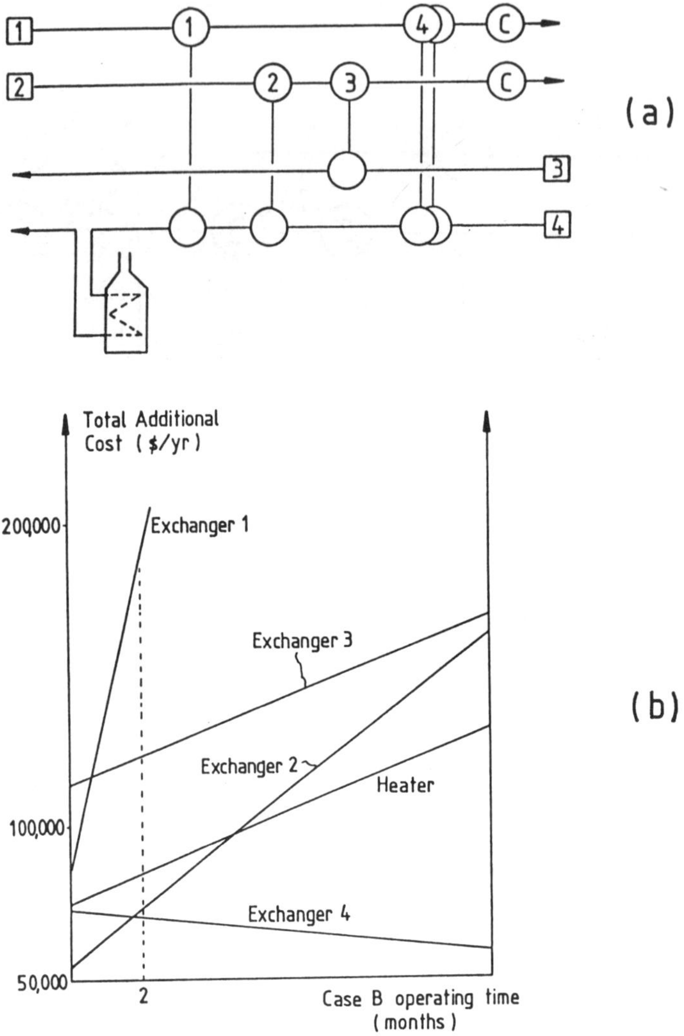

FIGURE 7

Design for CASE (A and B) without fouling

a) The cheapest alternative is to overdesign exchanger 4.
b) The effect of CASE B operating time on Total Additional Cost.

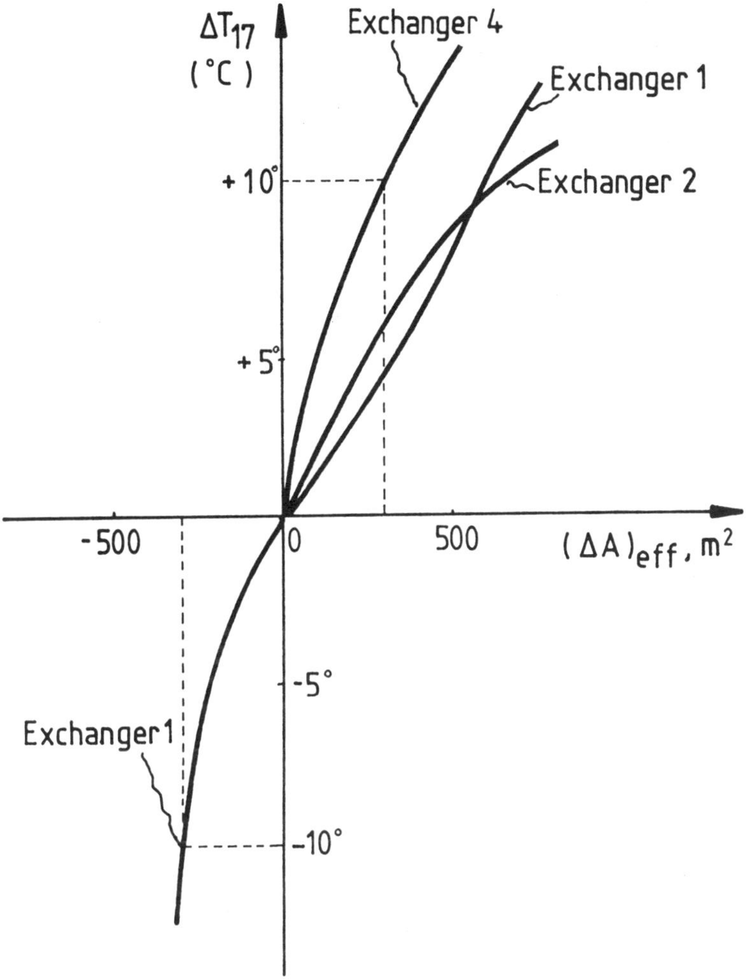

FIGURE 8

The effect of fouling in exchanger 1 on furnace inlet temperature (bottom left) and the effect of increase in effective area of exchangers 1, 2 and 4 on furnace inlet temperature (top left). The reduction of the effective area on exchanger 1 because of fouling (bottom left) can be offset by the smallest amount of additional area if this is installed on exchanger 4.

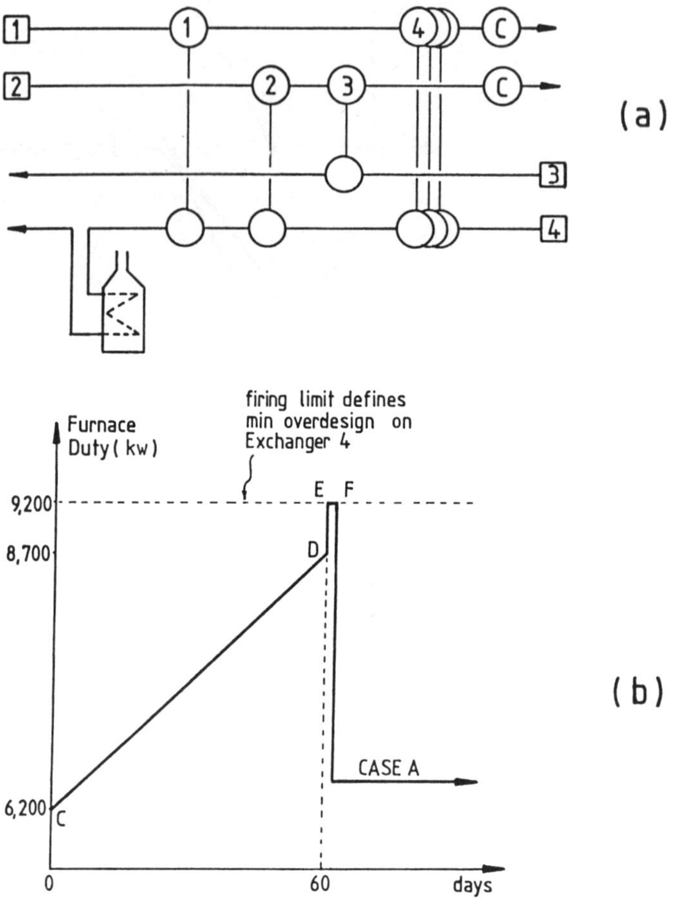

FIGURE 9

a) The final design as derived with the Sensitivity Tables. Overdesign exchanger 4 for CASE B operation as well as for fouling. Total Additional Cost: $-42,000/year.

b) Operating policy:
   - C: CASE B operation starts
   - D: End of CASE B operation. Exchanger 1 fouled by 60%, removal for cleaning.
   - E: CASE A operation. Exchanger 1 still out for cleaning.
   - F: Exchanger 1 returned.

"CONDITION BASED MAINTENANCE - THE OPTIMUM ROUTE TO ENHANCED PLANT AVAILABILITY"

P C Baguley*

This paper reviews the case for Condition Based Maintenance as an alternative strategy to the traditional modes of Preventative and/or Breakdown Maintenance. Monitoring techniques are examined and the steps required towards setting up a Condition Based Maintenance operation examined.

INTRODUCTION

One of the major English dictionaries describes the verb "to optimise" as follows:

> "To make the best of; take a favourable view"

This paper intends to, perhaps unusually, take a favourable view of the maintenance function and to identify ways in which that function can make a major and significant contribution to enhanced plant availability. The purpose of this will be to identify mechanisms and strategies which have and will enable "the best to be made", in terms of enhanced availability, of the plant and processes which that maintenance function is serving.

One author (1) has described this function as:-

> "a conglomerate of activities which strongly affects the surrounding functions which in turn affect the operation and performance of maintenance".

This suggests, rightly in the author's view, that maintenance cannot be viewed as a "stand alone" function and that what are required are "specific relationships with accounting, personnel, engineering and production management" (1). The nature of these relationships, and particularly the relationship with production management, is crucial to

* PA Management Consultants, Bowater House East, 68 Knightsbridge, London SW1X 7LJ

the effectiveness and value of the function. As with all relationships, a significant factor can be the mutuality or non-mutuality of the objectives or expectations. Various authors have attempted with a variety of success to identify, often proscriptively, the objectives of the maintenance function. Examples include:-

> "To ensure that, at minimum cost, production plant and equipment is available for productive use for the scheduled hours operating to agreed standards and at minimum waste."(2).

> "To control the condition of equipment"(3)

The contradictions and limitations of these typical and incompatible views will not be debated in this paper. It should suffice to comment that their diversity is symptomatic of an attitude which often limits the contribution which maintenance can make to the profitability of an organisation.

This paper shall start from the view that the objectives of the maintenance function lie in two complementary areas:-

- Optimal limitation of breakdown

- Optimal response to breakdown.

The implications, needs and demands of these objectives will be identified and examined.

EQUIPMENT FAILURE

The fact that all organic and inorganic entities and artifacts go through, on this planet at least, cycles of change and decay is a well established and integral feature of life. The level of integration of this fact into human culture is total, encompassing and influencing religion, philosophy, psychology, psychiatry, economics and marketing and many other areas of our lives. One numerate view of the failure patterns associated with this cyclical change is shown by the Bath Tub curve (fig 1). Whilst not presuming to be all encompassing this view of failure pattern does coincide with significant areas of experience and evidence for both plant, equipment and human beings. Venton (4) describes this curve as illustrating the form "which failure rate sometimes takes with complex equipment". The early life failure zone has been described by Buffham, Freshwater and Lees (5) to be of interest with chemical plant "in view of the importance of faster plant commissioning". Others (6) have listed those special mathematical forms which are often used in reliability modelling as probability density functions. Jardine (3) lists the following:-

- Hyper-exponential
- Negative exponential
- Normal
- Weibull

The use of the Weibull distribution is widespread (4 and 7) in reliability work and is noted as a distribution which is "an empirical distribution which appears to fit a large number of failure characteristics of equipment" (3). The use of these concepts and models for reliability measurement and hence failure prediction is an area which, for chemical plant has attracted limited interest. These authors also suggested a method of analysis of failure data for the Constant or Random failure zone of the Bath Tub Curve. Other authors have developed and described computer software for processing of reliability data. (8)

The analysis of this data is often complicated by two major difficulties:

- The presence of repair complicates the mathematics

- Systems cannot be regarded as independent.

The purpose and value of these activities should be to generate answers to the following questions:

- How reliable is the equipment?

- Most frequent causes of failure?

- Most costly items to repair?

- Significant downtime items?

- Variation of failures and repair costs with age?

Unfortunately the level of data collection and analysis, in most industrial concerns, is low and intermittent and usually aimed at day to day solution generation rather than problem solving at a causal level. The reasons for this are many and various but result in situations wherein the use of strategic concepts is very rare. This parlous state of affairs is not, however, due to lack of analytical tools. A recent survey (9) indicated that 60% of a sample of manufacturing companies "did not know how long or for what reasons their machines are out of action". This same source quoted many senior managers as seeing their maintenance departments as such a backwater that they cannot even be bothered to keep proper information on them. Whilst it can be argued, with some strength, that this is not a picture which is necessarily representative of the Chemical Industry, it is nevertheless true that most managers do not view maintenance functions as a source of benefit or profit enhancement.

MAINTENANCE POLICIES

The primary reason for the above is that most senior managers and their maintenance managers' attitudes to, and views of, maintenance, limit its role exclusively to that of a reactive crisis solving function.

Whilst there should be no doubt in anyone's mind that the prime objective of any maintenance function must be to ensure that that plant is available for operation whenever required by operation management, the manner in which that is implemented can make a fundamental difference to its effectiveness. It will also be evident that the plant or maintenance engineer is often constrained by a number of factors including budgets, plant design, maintenance personnel skill and numbers, etc. These internal maintenance factors are often exacerbated by increasing demands for plant availability which are forced upon operations personnel by an increasingly competitive and hostile economic environment.

Nevertheless, there do exist alternatives to a crisis solving mode of operation.

There are three basic maintenance policies as shown in Figure 2. The features and outline of these policies are as follows:

Preventive Maintenance

This policy involves the regular scheduled maintenance of plant and equipment at intervals defined by calendar time or hours run or throughput. It assumes that failure can be accurately predicted and occurs consistently, demands a high spares inventory and often results in component replacement very early in the life cycle. It is essentially a very fail-safe mode of maintenance which is, in practice, often not supported by failure data and which can itself be disturbing and damaging.

Corrective Maintenance

This policy involves only carrying out repair work when and after a failure has occurred. Its application is irregular and costly with a high potential for consequential damage in other plant areas by "Domino"

Condition Based Maintenance

BS 3811 defines this policy as preventive maintenance initiated as a result of knowledge of the condition of an item from routine or continuous monitoring. This maintenance policy utilises techniques (Figure 3) to monitor the condition of the plant,analyses the results and on the basis of that analysis, schedules the maintenance operation just before the impending failure.

All the above basic maintenance policies fit into the broad classification proposed by McCall (10) for maintenance policies involving random failures:

1) Preparedness models in which the state of the system is unknown until inspected

2) Preventative Maintenance models where the state of the system is known for certainty.

Evidence (9) indicates that in many industries Corrective or Breakdown maintenance is the norm despite its high consequential, labour and material costs. Equally, the use of Planned or Scheduled maintenance is expensive and ineffective due to the unpredictability of equipment failures. All of this evidence leads to the conclusion that Condition Based Maintenance is the best and most effective choice in maintenance policy.

To operate effectively, it requires a monitorable parameter that reflects the machine or plant condition such that detectable changes provide sufficient advance warning of failure or significant degradation. Its benefits are:

- Increased plant availability

- Reduced maintenance costs
- Better maintenance planning
- Improved safety through reduced failure rate
- Reduction in standing plant required
- Increased operational efficiency.

The aim of Condition Based Maintenance must always be to detect approaching failures so that they can be rectified at the next convenient non-production time. The cost benefit of employing this approach can be considerable with examples ranging from £2.9m over a two year period of monitoring turbine generators in CEGB SW Region (13) to $250,000/year saving resulting from application of simple vibration analysis in a paper mill (12). The costs of setting up and operating a condition based maintenance system include:

- Implementation: Initial survey
  Capital purchase and installation training
- Operation: Manpower for recording
  Measurement
  Analysis

The range of hardware costs is considerable from £1000-£1500/parameter for a hardwired computer based system to £200 for handheld portable sensor units. A major survey of Condition Monitoring (11) indicated 1% of plant capital value as a guide to setting up expenditure with equipment absorbing 40% and training/development 60% of this total. This survey also identified savings levels of the order of 1.2% of added value with the majority of that benefit originating from reduction of production losses. Our own experience in PA Management Consultants has indicated that these saving levels can be exceeded with levels as high as 1.5% of added value resulting from maintenance cost reductions alone. It has also been evident that the integration of machine condition sensing and data collection equipment into a process control computer system can generate considerable benefits when undertaken at the design stage of that control system.

CONDITION BASED MAINTENANCE

BS 3811 defines Condition Based Maintenance as:-

> "The preventive maintenance initiated as a result of knowledge of the condition of an item from routine or continuous monitoring"

In order to implement this maintenance policy, it is necessary to make decisions about the following factors:

- Which equipment is to be monitored?
- Which parameter is to be monitored?
- How is it to be monitored?
- How often is monitoring to be applied?

The choice of equipment to which monitoring should be applied should take into account the consequences of failure. The first step should be to identify plant which exhibit the following factors:

- Safety risks which arise from plant breakdown

- New unproven design

- Breakdown is expensive in terms of:

  o lost production
  o repair costs

- Operation based fault detection is difficult

- Need for advanced and accurate planning of maintenance activities

- Measuring equipment already installed.

An algorithm designed to facilitate the choice of maintenance policy and help identify the equipment to which Condition Based Maintenance should be applied, is shown in Figure 4.

It will be seen that Condition Based Maintenance is not, for most chemical plants, applicable to all equipment. It is, however, applicable to equipment which exhibits high cost for either repair work or loss of product.

For example, a multi unit, low risk or non-hazardous low product value, batch chemical plant with considerable spare capacity in terms of reactors, pumps, etc. will have an acceptable consequence of failure (low risk) and a low cost failure (ie: lost batch). Using Figure 6 this simple example leads to a breakdown maintenance policy. The introduction of flammable materials or high product value or high technology/high cost equipment all begin to shift the strategy towards Condition Based Maintenance. Plants which give rise to high productive capacity losses and high consequent cost of failure are usually in continuous operation in a single stream unit, minimum parallel or standby capacity and minimum storage capacity before or after. If you add to these factors others like dangerous process fluids, high temperatures, pressure or voltage operating conditions, high speed inertia components and very compact plant layouts, you begin to get a feel for equipment which responds well to Condition Based Maintenance.

Figure 3 identifies the major monitoring techniques currently in use. The proliferation of purpose built condition monitoring equipment often makes parameter and technique selection a difficult and complex process. The significant factors include:

- Machine condition/parameter correlation

- Instrument sensitivity

- Lead time to failure

- Frequency of monitoring

- Quantity and quality of data generated.

Vibration measurement is undertaken by two types of instrument, portable and fixed. The fixed type is used for critical areas of plant and equipment where continuous monitoring of vibration is essential if important symptoms are not to be missed. The portable type of equipment is hand held and is generally used to check the overall vibration level of the machine. More specialised equipment for defect analysis using Cepstrum or Kurtosis methods and analysis of shock pulse measurement can also be portable. The transducers used by both types of vibration measurement equipment can be attached to or held in contact with equipment to measure vibration (Seismic type) or be used to detect and diagnose a wide range of defects on shafts and rotors, journal and roller bearings, flexible couplings, gears, belt drives, machines with blades or impellers, foundations, pipework and structures. Figure 5 shows the relationship between some typical causes of vibration and the frequencies generated.

Temperature measurement and monitoring can be undertaken by a wide range of portable or fixed equipment covering the range from indicating paints and labels through to highly sophisticated optical devices. These are all well known and well established. Temperature levels can, for equipment as well as people, be used to detect the onset of trouble. Examples of equipment to which temperature measurement can be applied include bearings, friction devices, transformers, furnace insulation, electric cable insulation, cookers and heaters.

Assessing the appearance of plant and equipment by direct or assisted visual methods is often the simplest method of condition monitoring. The equipment used includes simple light probes, boroscopes, stroboscopes, TV cameras, dye penetrants and X-ray or gamma ray examination. It can be used to assess or detect surface wear or condition, levels of corrosion, surface cracking and blockages.

Performance monitoring represents the most direct form of condition monitoring. Providing accurate measurements are taken, any major deviation in performance can indicate the existence of a problem. Performance monitoring can also give a direct indication of machine efficiency and can be integrated in plant control systems. Generally, however, performance monitoring tends to a rather insensitive method of detecting incipient machine failure since serious machine component defects can become quite advanced before they produce a measurable effect on performance. To provide meaningful data, the equipment must be run at 'standard' conditions or be capable of being converted to 'standard' conditions. Examples would be delivery rates from pumps or conveyors, size distribution from a crushing mill, fuel consumption and electrical output of a diesel generator, temperature difference/flow relationship for a heat exchanger or pressure drop/flow relationship for a separator.

Wear debris monitoring involves monitoring machine lubricants for the presence of wear debris. This debris is generated by wear processess, for example, bearings, gears and pistons. The presence of debris in the lubricant can be detected by direct detection using electrical or optical methods; and debris collected using magnetic plugs, filters or centrifuges, or lubricant sampling and consequent laboratory analysis. The amount, size analysis, chemical composition and physical shape of the wear debris provides very powerful information about the nature of the incipient failure. Defects detectable by this form of monitoring include fatigue failure of plain, ball and roller bearings, adhesive wear on piston surfaces, surface wear of gears and oxidated wear of rubbing surfaces. Figure 6 indicates a range of typical contaminants and their source and likely cause.

With all these monitoring procedures and their results, it is essential that careful, accurate and consistent data recording takes place. This can be manual or microprocessor based but whatever method is used, it is essential that it effectively supports the analysis of the condition monitoring data. In order to do this, the data collation and analysis system must provide information on data trends and changes with time, absolute levels and comparison to established standards and sudden data level changes.

The steps to be followed in setting up a Condition Based Maintenance operation are as follows:-

Step 1

Initial investigation to establish feasibility and generate initial estimates of scope, cost and benefits

Step 2

Selection of critical plant and relevant parameter to be measured and technique/equipment to be used

Step 3

Selection/purchase of instrumentation and instruction of staff in taking routine measurements

Step 4

Record base line data

Step 5

Operation

- Data collection
- Data recording
- Data analysis
- Maintenance recommendations

Step 6

Review and modification of Maintenance programme.

The cost effectiveness of a Condition Based Maintenance programme has yet to be rigorously examined though recent authors have argued the case for improvement in management of Condition Monitoring schemes (14) and for the enhancement of these systems by introduction of AI (15) and Data collectors (16).

Generally, experience indicates that systematic and condition monitoring leads to improved equipment reliability and availability because it encourages immediate maintenance and thus closes the gap between defect onset and remedial action. This induces a change in approach moving from the maintenance only at fixed intervals, with random breakdowns and consequent repair, to a system involving much more actual remedial work. This system is more flexible and able to respond to work demands as they arise. However, condition monitoring techniques must be used with care and can often provide initially more questions than answers. The savings potential of Condition Based Maintenance can be considerable if it is applied to the righ process equipment using the correct monitoring equipment and backed up by experience, knowledge and effective recording and analysis systems.

CONCLUSIONS

It is evident that Condition Based Maintenance offers a viable and powerful option to the conventional options of Preventive or Breakdown maintenance. Subjective evidence indicates an increasing and growing usage and interest in this maintenance policy. However, there remains a need to establish, in a comprehensive and numerately accurate manner, the longer term implications and benefits of such a policy when consistently applied.

REFERENCES

1. PRIEL.V.Z. "Systematic Maintenance Organisation" London, McDonald and Sons, 1974.

2. CORDER.C.G. "Maintenance-techniques and outlook" British Productivity Council Seminar publication Section 6 pp73-77.

3. JARDINE.A.K.S. "Maintenance, Replacement and Reliability" London, Pitman, 1973.

4. VENTON.A.O.F. "How numerate is Terotechnology?" Inst.Mar.Eng.Trans., Vol 87, 1975, pp 202-208.

5. BUFFHAM.B.A. et al "Reliability Engineering - a rational technique" Inst.Chem.Eng.Symposium Series No 34, 1971.

6. SINGH.C. AND BILLINGTON.R. "System Reliability Modelling and Evaluation" London, Hutchinson, 1977.

7. LEONG.K.F. "Reliability of Complex Systems with Repair" M.Sc Dissertation, Univ.of Bradford, 1975.

8. HASTINGS.N.A.J. "RASP-Reliability analysis standard package" Bromsgrove, Albany Interactive Ltd., 1976.

9. THOMAS.D. "Maintenance - A sorry picture" Financial Times, 8th Dec.1985.

10. McCALL.J.J. "Maintenance policies for stochastically failing equipment" Man.Science, Vol 11, No.5, 1965, pp 493-524.

11. NEALE.M.et al "A Guide to Condition Monitoring" H.M.S.O., 1979.

12. ANON. "Machine Condition monitoring saves $250,000/yr" Pulp and Paper Canada, 85:6, 1984.

13. THOMAS.D. "Role of Condition Monitoring in Maintenance Management" 6th National Conference on Computers for Maintenance Managers, London, 1985.

14. HENRY.T.A. "Improving the Management of Condition Monitoring" CME, May 1986, pp 28-31.

15. MILNE.R. "Artificial Intelligence Applied to Condition Healthy Monitoring" CME, May 1986, pp 45-46.

16. NICHOLLS.C. "Condition Monitoring using Data Collectors" CME, May 1986, pp 50-51.

Figure 1

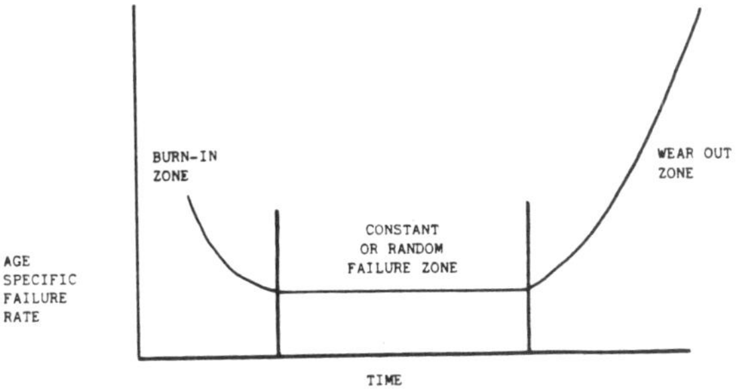

BATH TUB CURVE

Figure 2

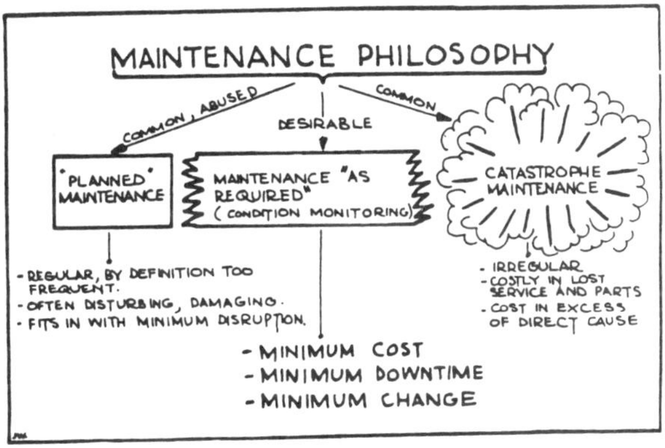

Figure 3

MONITORING TECHNIQUES

Vibration Monitoring

Wear Debris and Contaminant Monitoring

Performance Trend Monitoring

Thermal Detection Monitoring

Visual Appearance Monitoring

Noise Monitoring

Maintenance Strategy Algorithm

Figure 4

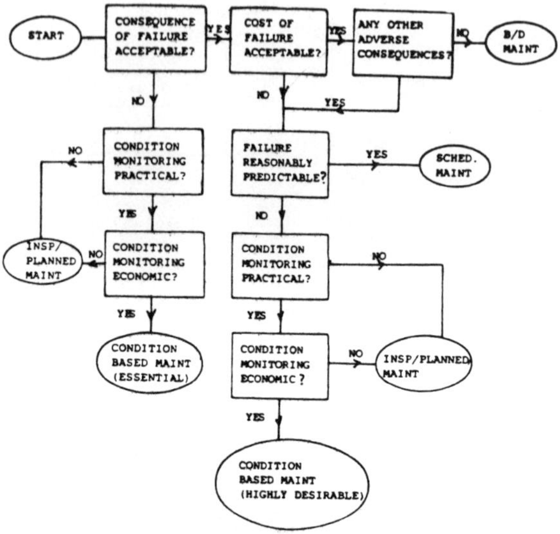

VIBRATION

Figure 5

Vibration Cause	Vibration Frequency	Features of Vibration
1. Out of balance	1 x RPM	Radial vibration, increases with out of balance
2. Misaligned coupling on bearing, bent shaft	1-2 x RPM	Large Axial component
3. Damaged Rolling bearings	very high 5-20 x RPM	
4. Rotor components loose on spindle	2 x RPM	This is critical with high speed machines
5. Gear mesh	No of teeth x RPM	Should show up on frequency analysis
6. Oil whirl	around 0.5 x RPM	A feature of oil film instability
7. Faulty belts	1-2 x RPM	

Figure 6

WEAR DEBRIS

Contaminant	Typical Source	Likely Cause
Iron	- Piston rings, liners - Wearing Mating surfaces - Rust contamination due to leaks	Wear, sticking rings, scuffing
Copper, Lead and tin	Bearing bushes and thrust rings	General wear or poor lubrication
Silicon and Aluminium	Mineral dust; excess silicon grease	Filter failure, inadequately cleaned system
Sodium	Seawater/Saline process fluids	Contamination of Lube oil or hyd. fluid with seawater

STATISTICAL PROCESS CONTROL IN THE PLANTATIONS INDUSTRY

Azman Firdaus Shafii*

This paper describes how Statistical Process Control (SPC) was introduced and applied in the palm oil milling industry. Examples include the use of SPC to minimise process and product variabilities, increase productivity gains and trouble shooting exercises. More recently, the concept of participative process management has been introduced, blending the tools of SPC and the broad-based ideas of Quality Control Circles (QCC).

INTRODUCTION

Recent literature (1) has indicated that despite the proliferation of computer-controlled plants, traditional Quality Control (QC) techniques are increasingly being used as a supplement to conventional process control and optimisation. An enhancement of such a technique, so successfully implemented in the palm oil industry that it has now become part of routine plant operations management, is Statistical Process Control (SPC). This paper describes how SPC was introduced and implemented in 10 palm oil mills as a means to minimising process and product variabilities, increased productivity gains, trouble-shooting and plant equipment (new or retrofit) assessment trials. More recently, the concept of participative process management has been introduced, blending the tools of SPC and the broad based ideas of Quality Control Circles (QCC).

THE PALM OIL MILLING PROCESS

The oil palm crop (*Elaeis guineensis*) and its associated palm products manufacture is the most important plantations industry in Malaysia. For the last five years, it has been the number two in the country's commodity exports earnings behind oil and gas. There are 272 mills and 36 refineries in operation today, processing about 8345 tonnes of Fresh Fruit Bunches (FFB) per hour. The output in 1985 totalled 4.1 million tonnes of palm oil and palm kernel oil.

*Engineering Services Scheme, PO Box 207, 42700 Banting, Selangor, Malaysia.

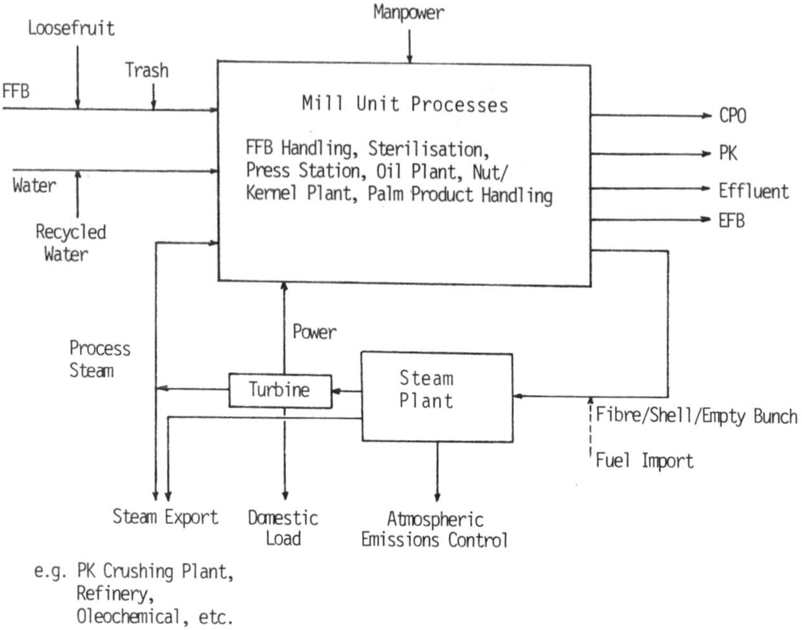

Figure 1 A Schematic Model of Milling

The first stage of palm product extraction is the palm oil mill. A schematic model of a mill is given in Figure 1 and a simplified flowsheet of the extraction process is shown in Figure 2.

Basically, in a palm oil mill , ripe FFB which are received from oil palm estates are steam-sterilised in time-controlled pressure cycles over 60 to 90 minutes, depending on FFB quality, to de-activate hydrolytic enzymes and to loosen the fruitlets for subsequent threshing in rotating drums. The detached fruitlets are then conveyed to steam-heated digesters and screw-pressed to separate oil from fibres and nuts. The highly variable crude oil liquor, typically 35-55 % oil, 20-55 % water and 3-9 % solids (fibres, cellular debris, dirt, etc.) is diluted, clarified or desludged, purified, vacuum-dried and stored as production oil or Crude Palm Oil (CPO). Typically, on a FFB basis, a 20 % CPO extraction rate is considered average. Similarly, the nuts are dried, cracked and the shells separated to obtain palm kernels (PK). Around 6 % PK extraction is usually possible .

In practice, real improvements may only be achieved if good control is exercised over raw material (FFB) quality, which is primarily a field variable, and on processing standards at the mill. It is widely accepted that CPO extraction rates and quality [in terms of Free Fatty Acid (FFA) level] are much more dependent on the former than the latter. This has important processing implications since the great variability in amount, nature and quality of FFB received at the mill can be thought of as a major disturbance entering the milling system. This disturbance, along with those due to processing discontinuities which prevent steady-state operations and the non-optimal process conditions due to the lack of physical and chemical understanding of unit processes, propagates down the system and the combined effects present formidable process and QC problems.

Thus, in the palm oil industry, productivity and quality begin in the field - a responsibility accepted both by the planters, or estate managers, and mill engineers alike. Good harvesting practices are therefore paramount in order to harvest just - ripe FFB (maximum oil yield per bunch) and to evacuate harvested fruits as fast as possible with minimum bruising (leads to rapid increase in FFA) and dirt contamination, to the mill for immediate processing. Good labour control is a pre-requisite to optimising harvesting rounds as well as to ensure that the minimum ripeness standard are fully abided to. This is necessary since FFB yield (per hectare) and oil yield (per tonne FFB) both vary with palm species, age, climatic and soil factors (Figure 3 and 4).

Most mills run on a 2-shift workforce except during peak cropping seasons when processing will be continuous. Thus, average extraction figures are obtained daily and typical values are shown in Figure 5 for 25 processing days in a month.

INTRODUCING STATISTICAL PROCESS CONTROL

In our experience, Statistical Process Control evolved over a considerable period and had its beginnings when the Engineering Services Scheme (Process Technology Division) was purely concerned with QC, standardisation of laboratory test methods for palm product quality parameters (FFA, oxidation indices, moisture, dirt, color, Iron and Copper content) and process R & D. Later it was found that the statistical tools developed could be effectively incorporated into routine process control functions at the company's 10 palm oil mills, if the relatively less academically qualified mill process supervisors and operators are sufficiently trained. To date, four annual training courses have been conducted. The course comprises lectures in mill unit processes, statistics, laboratory work and projects conducted at the participant's mill. The results have been highly successful.

For the very first time, the concept of a Mill Process Team was introduced in a highly conservative industry and the process personnel are delighted. Process personnel are now able to analyse process variable trends systematically, conduct Cause-Effect and Pareto studies, run Control Charts and Analysis of Variance (ANOVA) exercises — all at a specified level of statistical confidence. The gathering and interpretation of process data are becoming more meaningful and reliable thus, facilitating process and quality audits which are conducted by the company's 'visiting' engineers from the Engineering Services Scheme.

More recently, participative management concepts have been introduced, in particular, the broad-based ideas of the Quality Control Circles (2) (QCC). There was almost ready acceptance to the concept since mill personnel were already practising many of the quantitative tools of QCC. Mill process control is still largely manual. Proper consideration in the approach and philosophy of bringing together personnel of widely different backgrounds and turning them into an effective plant operations management team is therefore, vital.

ILLUSTRATIVE EXAMPLES

Three examples are described here to reflect on the success of implementing SPC in the company's 10 palm oil mills. These form only a tiny portion of the total documented experience to-date.

FFB Input Quality Control

To minimise FFA rise due to enzymatic hydrolysis, harvested FFB must be delivered to the mill as soon as possible but oil yield, in terms of CPO extracted per tonne of FFB, is also adversely affected by the harvesting of unripe (Black) or slightly under-ripe (Hard) FFB. For practical reasons, an oil palm harvester is only guided by a simple FFB minimum ripeness standard. A FFB, which contains hundreds of fruitlets, normally turns from black to a yellowish-red coloration as it ripens. Further, as it ripens, some of the fruitlets become detached from the stalk and may fall to the ground prior to harvesting. As a rule, the harvester is not permitted to harvest FFB if it is black (by visual observation, no detached fruitlets) or if it is hard (yellowish-red but with less than 10 loose fruitlets). Although hard bunches have almost fully developed oil cells, they give rise to performance difficulties during steam sterilisation and subsequently higher losses of fruitlets in the form of Unstripped Bunches (USB) after the threshing operation.

It is not suprising therefore that over the years, the issue of input FFB quality has been controversial and understandably sensitive to both supplying estates and mill management. To attempt an objective means of control in a problem involving subjective matters, it was decided that all incoming FFB be classified as simply being one of two types, namely a Black & Hard (B&H) bunch and a not B&H bunch.

Assuming statistical independence, the probability of a B&H is therefore 0.5 and the probability of getting B&H bunches when a given number of FFB are inspected may be assumed to follow a binomial distribution. Control charts can then be constructed based on regular FFB assessments at the mill reception yard. Such charts allow the detection of a minimum percentage shift (say, 5 %) in B&H bunches coming in and the performance curves are conveyed to the appropriate supplying estates to tighten daily harvesting control as and when required.

Figure 6 shows the dramatic changes in one mill. As B&H bunches are lowered, oil extraction rates have increased - an excellent example of productivity gain through simple statistical control of events.

Identifying the Causes of Unstripped Bunches

In another mill, when USB control charts were indicating worsening trends, the Process Team was requested to conduct an investigation into possible assignable causes. As the B&H count appeared to be under statistical control, there was a need to ascertain what were the high percentages of USB observed due to. A Cause-Effect diagram was worked out and eventually it was decided that the causes of USB may be quantified according to :

1. The nature of bunch (NB)
2. Inadequate sterilisation (IS)
3. Inadequate threshing (IT)

With reference to Figure 2, the above were defined as follows :

A USB due to the nature of bunch i.e. USB_{NB}, is one which after going through two sterilisation cycles and three threshing cycles, the bunch still remains as a USB. If the bunch is completely stripped after such number of passes, then the bunch is said to have undergone adequate sterilisation. On the other hand, a bunch which is completely stripped after two cycles of threshing is said to have undergone adequate threshing.

The following quantitative relations may then be derived :

$$\text{1st Pass USB} = USB_{IT} + USB_{NB}$$
$$\text{2nd Pass USB} = USB_{IS} + USB_{NB}$$
$$\text{3rd Pass USB} = USB_{NB}$$

Thus,

$$USB_{NB} = \text{3rd Pass USB}$$
$$USB_{IS} = \text{2nd Pass USB} - \text{3rd Pass USB}$$
$$USB_{IT} = \text{1st Pass USB} - \text{2nd Pass USB}$$

A 100 % count of USB was conducted in an experimental study and the mean values for the various USB categories were found to be, rounded to whole figures,

$$USB_{NB} = 4\ \%$$
$$USB_{IS} = 27\ \%$$
$$USB_{IT} = 69\ \%$$

It showed that basically the USB were not due to the nature of bunch, thus, reflecting the reliability of the B&H control system. Rather, the main cause happened to be one of poor threshing operation.

As a result, a similar study was subsequently conducted on the threshing operation to determine the effects of such factors as sterilised bunch throughput, drum rotational speed, height of lifter bars, etc. on USB. Consequently, a new set of thresher operating conditions was suggested to improve threshing performances.

Introduction of a Quality Control Circle

More recently, it was decided that the QCC concept, which has proven highly successful in the Japanese industries, be introduced into the plantations. One of the company's oldest mills was chosen to pioneer such a Circle.

It was proposed to the workforce that QCC is a participative management concept which aims at improving productivity (hence, reducing production costs) through people by developing both their aptitude and attitude. Thus, QCC serves as a mechanism for employees to contribute to the progress of the organisation.

QCC is basically voluntary participation and very quickly a pioneering circle was formed comprising ;

1	Facilitator	-	Mill Resident Engineer
1	Circle Leader	-	Process Supervisor
5	Circle Members	-	Station and Relief Operators

and the following values, expressed simply, were promoted :

Positive Attitudes	Positive Aptitudes
Respect for people	The need for training
Do not blame others	Data collection and analysis
The customer is king	Brainstorming
Speak with facts	Cause-Effect approaches
Enjoy our work	The 20 : 80 rule
	Applying Plan-Do-Check-Action cycles

At about the same time of circle formation, it was observed that the clarification station was indicating erratic dirt content in production oil as shown by Curve 1 in Figure 7. The in-house standard specification is 0.02 % dirt maximum. The QCC took the problem as its first task resulting in bringing it under good control as shown by Curve 2. As a result of this and other successful experiences, QCCs are now set up in all the company's 10 oil mills.

CONCLUSIONS

This paper has described several successful applications of Statistical Process Control in the plantations industry. Readily integrated with participative management concepts like the QCC, it is shown that the use of simple quantitative tools and a humanistic approach to work can contribute considerable benefits in the effort to optimise process plant operations and management.

ACKNOWLEDGEMENTS

The author wishes to thank Harrisons Malaysian Plantations Berhad for their permission to publish this paper.

REFERENCES

1. Quality Control Moves Upstream, Chem.Eng, p19-23, Apr 28 (1986).

2. Hajime, K., Invitation to QC - Its Challenges and Rewards, PHP Institute, Inc., Tokyo (1986).

3. Palm Oil Research Institute of Malaysia Bulletin - Special Issue, Nov (1984).

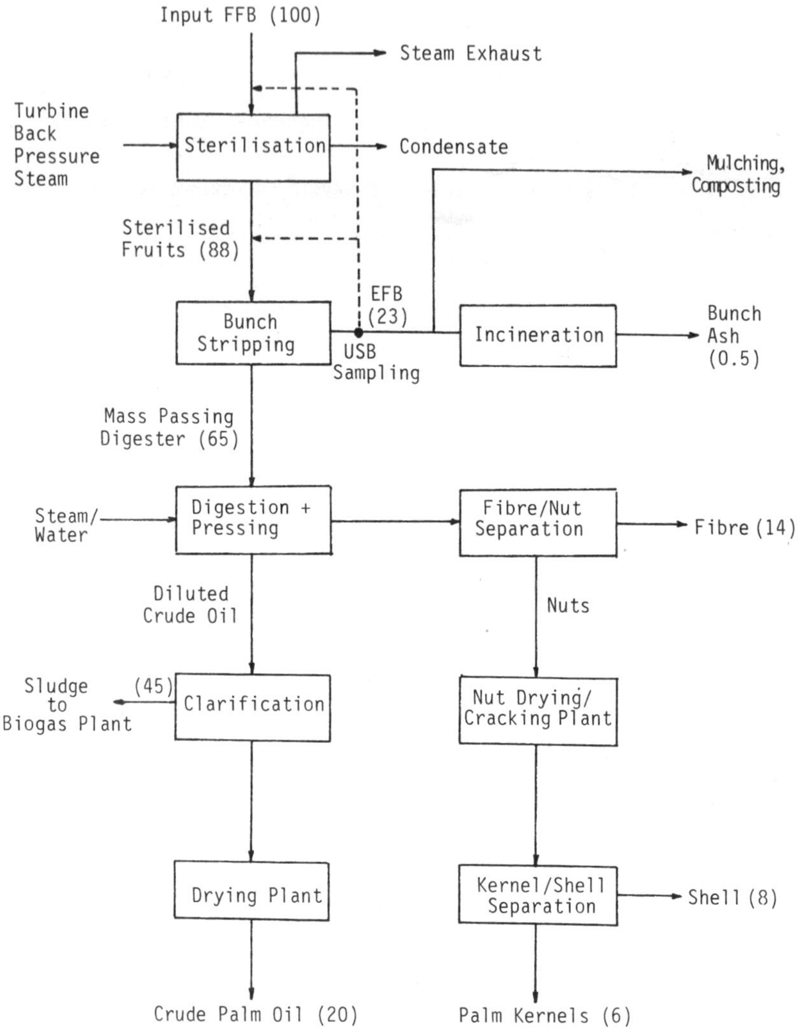

Figure 2 Simple Mass Flow Diagram for Crude Palm Oil and Palm Kernel Extraction - Basis : 100 tonnes FFB

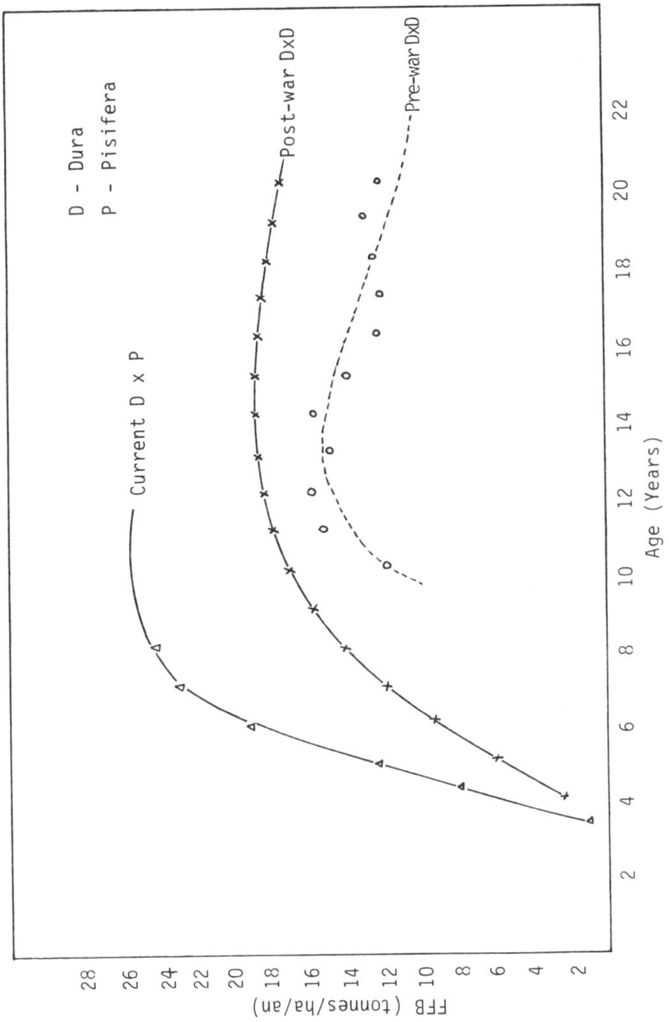

Figure 3 Yield Profile of Commercial Material from Different Periods (3)

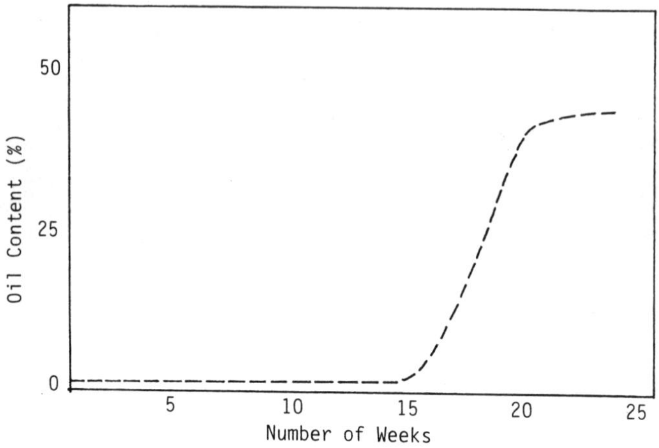

Figure 4 Development of Oil in Palm Fruit Mesocarp

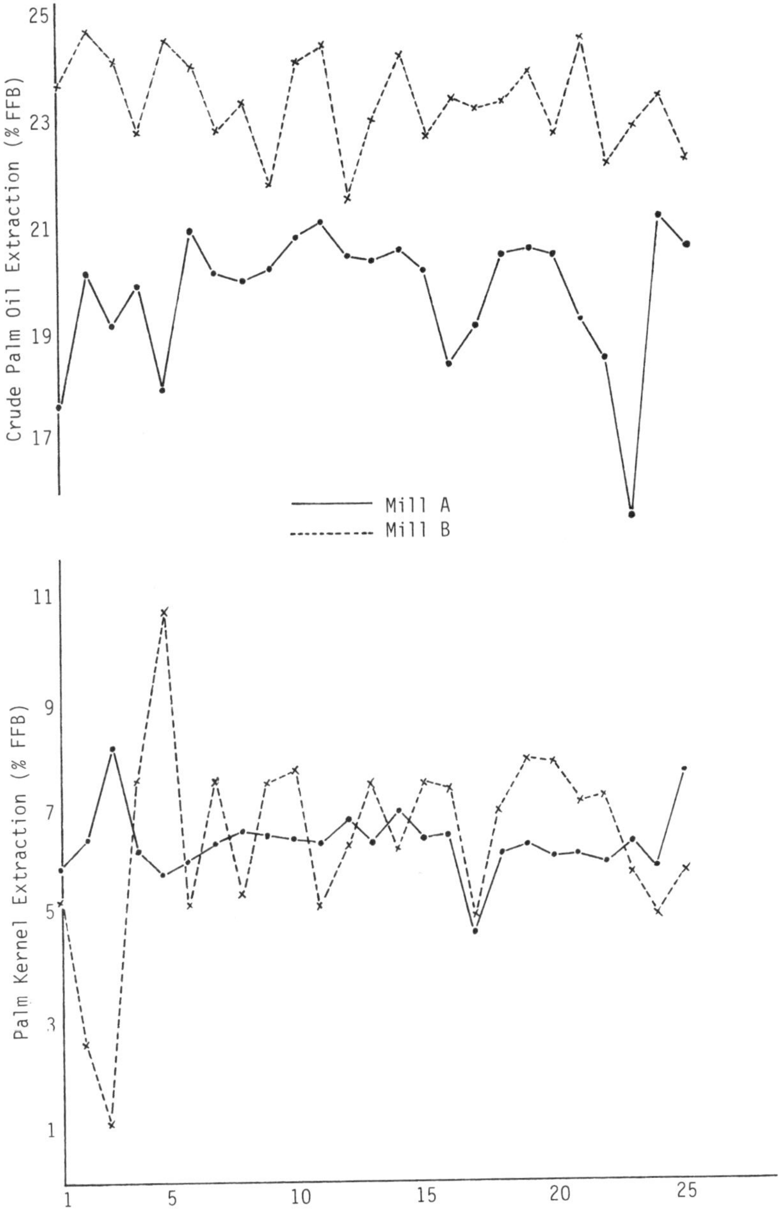

Figure 5 Daily Crude Palm Oil and Palm Kernel Extraction Rates

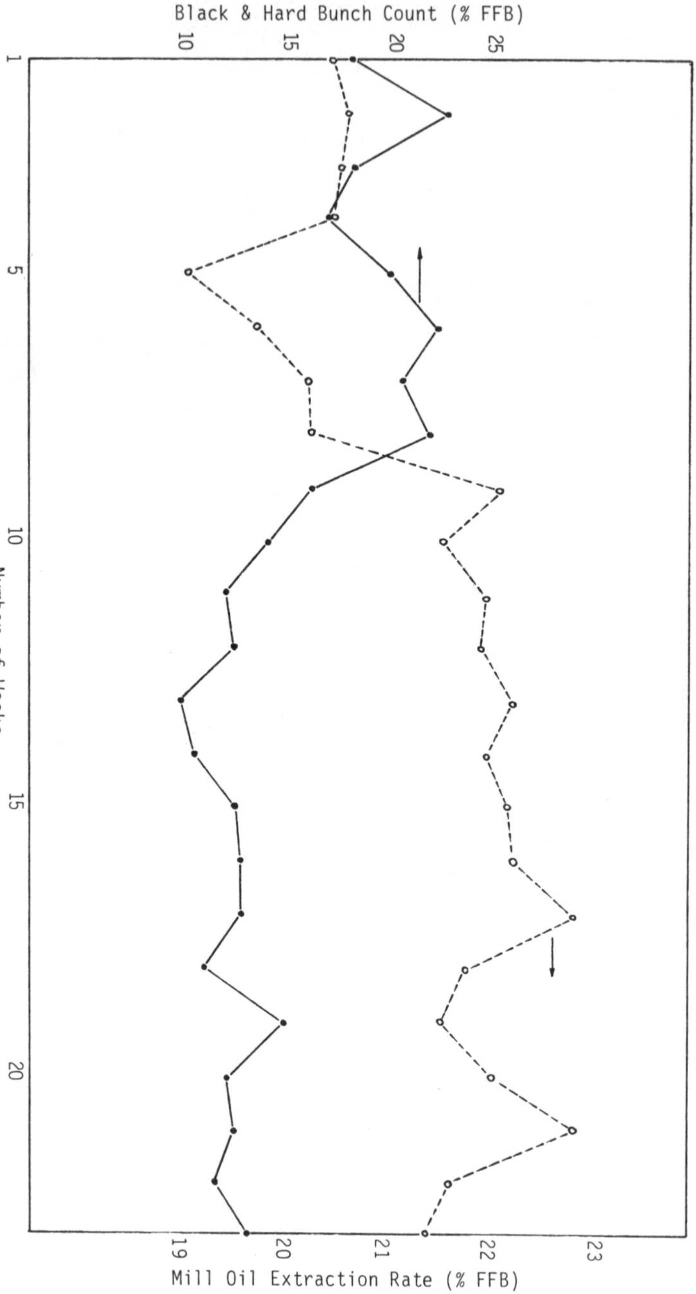

Figure 6 Effect of Input FFB Quality Control on Mill Oil Extraction Rates (Weekly Averages)

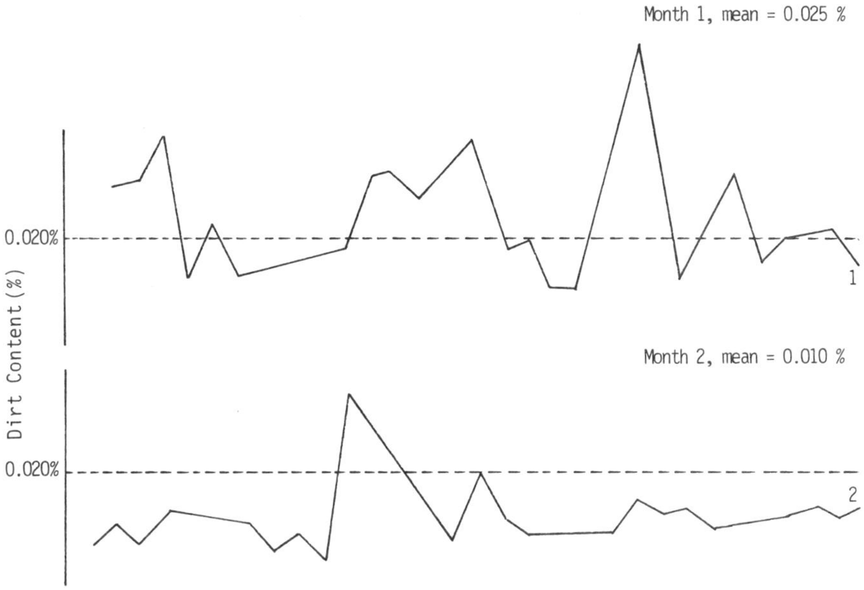

Figure 7 Daily Average Dirt Content in Production Oil

MAXIMISING WASTE UTILISATION IN PALM OIL MILLS

N.M.Sulaiman* and A.F.Shafii**

In 1985 industrial waste from the Malaysian palm oil milling industry is estimated to be 11.7 million tonnes of liquid effluent, 2.52 million tonnes of palm fibre, 1.44 million tonnes of palm shells and 4.14 million tonnes of Empty Fruit Bunches (EFB) or bunch stalks. Efficient waste utilisation technologies are therefore highly desirable to meet increasingly stringent environmental regulations and where possible, recycled for maximum re-utilisation. This paper summarises the efforts that have enabled plant management in 10 mills to approach a situation of zero-effluent discharge.

INTRODUCTION

The palm oil industry in Malaysia is a success story. Today it is labelled as the "sunshine" industry having moved from an activity involving only 46 mills producing 430,000 tonnes of Crude Palm Oil and no refining capacity in 1970, to the world's largest producer and exporter involving 272 mills and 36 operating refineries producing 3.6 million tonnes in 1985. Hourly, the industry processes 8345 tonnes of Fresh Fruit Bunches (FFB) harvested from 1.5 million hectares of oil palm estates. Since, on average, only 20% of FFB are extractable oil and 6% of FFB are recoverable palm kernels, the industry is continually faced with a massive waste disposal problem. In 1985, this is estimated to be 11.7 million tonnes of liquid effluent, 2.52 million tonnes of palm fibre, 1.44 million tonnes of palm shells and 4.14 million tonnes of Empty Fruit Bunches (EFB) or bunch stalks. Efficient waste utilisation technologies are therefore highly desirable to meet increasingly stringent environmental regulations and where possible, recycled for maximum re-utilisation. This paper summarises the efforts that have enabled plant management in 10 mills to approach a situation of zero-effluent discharge.

WASTE GENERATION CHARACTERISTICS IN PALM OIL MILLS

Most of the oil palm species (*Elaeis guineensis*) grown in Malaysia is of the Dura x Pisifera or DXP crosses. The DXP hybrid is called the Tenera and is characterised by thin shells and high oil in mesocarp content. The morphology of individual fruitlets is illustrated in Figure 1. Typically, up to a thousand fruitlets may be present on a FFB. The fruitlets are borne on spikelets emanating from a central peduncle. An average bunch weighs between

* Dept. of Chemical Engineering, University of Malaya, 59100, Kuala Lumpur Malaysia
** Harrisons Malaysian Plantations Bhd., P.O.Box 207, 42700 Banting, Selangor

20 and 30 kg but extremely large bunches, approaching 70 kg are not uncommon.

In a palm oil mill, ripe FFB which are received from oil palm estates are steam -sterilised in time-controlled pressure cycles over 60 to 90 minutes, depending on FFB quality, to deactivate hydrolytic enzymes and to loosen the fruitlets for subsequent threshing in rotating drums. The detached fruitlets are then conveyed to steam-heated digesters and screw-pressed to separate oil from fibres and nuts. The highly variable crude oil liquor, typically, 35 - 55% oil, 20 - 55 % water and 3 - 9 % solids (fibres, cellular debris, dirt, etc.) is diluted, clarified or desludged, purified, vacuum-dried and stored as production oil or Crude Palm Oil (CPO).

Inevitably, the milling process generates considerable wastes. On a FFB basis, these are summarised in Table 1.

Table 1 - Waste Generation in Palm Oil Milling

Class	Type	Availability (% FFB)	
Solid	Empty Fruit Bunch (EFB)	23	
	Palm Shell	8	
	Palm Fibre	14	
	EFB Incinerator Ash	0.5	
Liquid	Steriliser condensate	15	65%
	Sludge centrifuge waste	45	
	Hydrocyclone waste	5	
Atmospheric	EFB Incinerator Flue	-	
	Boiler stack emissions	-	

Over the years, the Department of Environment (DOE) Environmental Quality standards have become increasingly stringent. Table 2 gives the current discharge limits for palm oil mill effluents (POME) and atmospheric emissions from boiler stacks and EFB incinerators. The use of incinerators is strongly discouraged now and new mills (by definition, mills constructed after the introduction of the Clean Air Regulations of the Environmental Quality Act (EQA) 1978) are not permitted to incinerate EFB if they are situated near populated areas.

WASTE UTILISATION STRATEGIES

Solid Wastes

The three main solid wastes produced from the milling process are therefore empty fruit bunches (EFB), fibre and shell totalling 45 % of EFB processed. Biosolids which are by-products from the treatment of liquid effluent will be discussed under liquid wastes.

A study on the calorific values of fibre, shell and EFB was recently conducted at a 25 tonne/hr oil mill.(1) Data for average moisture and oil contents and average net calorific values for shell, fibres and EFB obtained are summarised in Table 3.

Table 2 - (a) DOE Palm Oil Mill Effluent Standard F
For Watercourse Discharge Effective From 1st. Jan 1984

Parameters	Limit of Discharge
BOD (3-day, 30°C) (mg/l)	100
Suspended solids (mg/l)	400
Oil and grease (mg/l)	50
Ammoniacal nitrogen (mg/l)	100
Total nitrogen (mg/l)	200
pH	5.0 - 9.0
Temperature (°C)	45

(b) Summary of Emission Limits and Standards Prescribed under Environmental Quality (Clean Air) Regulations 1978.

I. Dark Smoke From Fuel Burning Equipment

Type of Fuel	Equipment	Dark Smoke Limit
Solid Fuel	Existing and New	Ringelmann No.2

II. Emission Standards for Air Purities

Substance emitted	Standard
Solid particles	0.4 g/Nm^3

Table 3 - Average moisture and oil contents and average net calorific values for fibre, shell and EFB.

Parameter	Fibre	Shell	EFB
Moisture	37.6	16.4	63.0
Oil (% dry basis)	10.4	-	0.5
Net Calorific Value (kcal/kg)	2775	3982	1366

With the availability of fibre and shell averaging at 22% of FFB, the mixture is used as fuel for the boiler to raise steam for thermal as well as mill power generation purposes. Thus under normal conditions of rated capacity, the mill always produces surplus fibre/shell to meet all solid fuel requirements. The total fly ash resulting from the combustion of fibre/shell is produced at an average rate of 1.4 kg/min from a 20 tonnes FFB/h capacity mill and is blown back into the furnace for recombustion. Although EFB are available in excess they are not used as fuel due to the high moisture content. Further they are bulky, produce considerable ash and clinker thus requiring more furnace attention.

Utilisation of EFB is then done in either of two ways in the mill. Being rich

in plant nutrients, direct utilisation of EFB sprayed with a little raw or treated mill effluent as mulches has been an accepted estate practice. On average mulched EFB contains 0.37 % N, 0.04 % P, 0.91 % K and 0.08 % Mg. The use of mulched EFB in the field not only provides nutrients but has also improved soil physical and chemical properties such as increased soil organic content, improved soil structure, increased infiltration and aeration and improved soil water retention. Improved palm performance in terms of EFB yield and leaf nutrient status have also been recorded. Currently the rate of mulched EFB application is 36 tonnes/ha/ year or 250 kg/palm/year. In terms of fertiliser value, one tonne of EFB is equivalent to 8 kg urea, 2.9 kg CIRP, 18.3 kg muriate of potash and 4.7 kg. kieserite.
The alternative method of EFB utilisation involves slow burning of the EFB with large excess air in incinerators so as to maximise the yield of potash. This can then be used as a 3:2 substitute for muriate of potash fertiliser source. However this method of disposal significantly reduces its value as a source of plant nutrients and as organic matter for soil improvement. Both practices have contributed further cost savings as well as helping to solve the disposal problem.

Liquid Wastes

Untreated liquid discharges from palm milling activities are very high strength wastes, the biochemical oxygen demand (BOD) of which can range from 25000 to 35000 mgl^{-1}. Typical characteristics of liquid wastes found in Malaysian plantations industry are given in Table 4. The BOD and and suspended solids (SS) values for domestic sewage is included for comparison. As evident from the table, palm oil mill effluent (POME) stands out as the strongest organic waste with a BOD of about 100 times that of domestic sewage. It is not surprising to note that many well proven technologies including anaerobic/ aerobic ponding systems, have in the past failed to treat POME satisfactorily. Partly as a response to the enforcement of the EQA (1974), extensive and

Table 4 - Typical Characteristics of Liquid Waste in the Plantations Industry

Parameters	Palm Oil Mill Effluent	SMR* Rubber Wastes	Latex Concentrate Wastes	Coconut Water (liquid endosperm)	Domestic Wastes
pH	3.7	5.6	6.3	5.0	-
BOD	28910	460	2150	18700	200-400
COD	75040	1170	5810	39680	-
VFA	3350	-	980	830	-
Alkalinity	930	240	900	605	-
TKN	1190	90	430	191	-
AN	50	60	330	-	-
TS	63540	1180	3740	39530	-
SS	31460	250	370	720	100-300

(except for pH, all other parameters in mg/l)
* SMR - Standard Malaysian Rubber.

intensive in-house R & D programmes have been initiated in 1976 to treat the POME problem. Investigation on various treatment possibilities cover physical, chemical and biological methods. The physical treatment methods tested included electroflotation system, vacuum filtration and reverse osmosis. The chemical treatment mainly involved the use of chemicals to effect precipitation. However the physical and chemical methods were found to be incomplete in themselves and more importantly not economical. Extensive and intensive trials were then undertaken for various types of aerobic and anaerobic biological systems. A biological treatment method comprising a two-stage mesophilic anaerobic digestion followed by extended aeration has since proved to be successful. A typical flow scheme of the above effluent treatment plant (ETP) is shown in Figure 2.
The process can be divided into the following stages:

. an acidification phase in open ponds
. a methanogenic phase in digester tanks
. sedimentation of anaerobic liquor
. extended mechanical aeration in a lagoon
. sedimentation of aerobic liquor

The system operates on a batch system. At the start of each day anaerobic liquor combined with steriliser and clarifier wastes from the day's processing are pumped over a cooling tower into the acidification pond. The mixture is left overnight to react. Meanwhile acidified liquor from the previous day is fed to the mild steel unstirred anaerobic digester tanks where they remain for a minimum period of 21 days. The anaerobic liquor (BOD reduction exceeds 80%) is subsequently treated by extended aeration in a lagoon equipped with mechanical aerators. The minimum retention time of the aerobic phase is 20 days. The solids from the aerobic lagoon is separated by sedimentation in a 2 hr retention tank. The supernatant is then discharged or recycled to the mill as process water. In one mill, the latter has saved up to 57% of the cost in public water supply consumption.
A typical performance of the overall ETP in mill A in 1985 is given in Figure 3. Total compliance with standard F of the EQA is achieved for BOD, TN, pH and AN. The requirement for SS to be less than 400 mgl^{-1} is met 75% of the time.
The performance of the anaerobic digestion stage from the same mill in 1985 is shown in Figure 4, based on some of the more pertinent characterictics of the discharged anaerobic liquor. Reductions in BOD level exceed 94% and VFA has been reduced by 20% based on raw POME. The anaerobic liquor also contains between 20 - 50% of the original SS implying that at least half of the original SS remains in the tank.
This system has demonstrated that it is capable of treating POME satisfactorily. The system has a low energy demand (for pumping only), it is stable in operation (to shock loadings and inherent process variations) and supervision required is minimal.
The ETP also produces two potential sources of wastes that can be economically utilised as fertilisers. They are recovered intermittently as biosolids from the digester tanks and the aerobic sludges from the sedimentation tanks. Presently biosolids are supplied to the estates at the rate of about 0.1 tonne biosolids/tonne FFB from the digester tanks and 0.05 tonne aerobic sludge/tonne FFB from the sedimentation tank. The biosolids have proved to be rich sources of nutrients. Table 5 is a summary of nutrient composition of palm oil wastes from various stages of treatment. Nitrogen content has increased two fold and three fold respectively for the aerobic and anaerobic solids as compared to untreated POME. Yield data during 1983 - 1985 showed that effluent treated plots have cumulatively increased by 11% compared with plots receiving inorganic fertilisers.

Table 5 - Nutrient Composition of Palm Oil Wastes

Source of effluent	Mean Values (mg/l)				
	N	K	P	Mg	Ca
Untreated POME	1058	2280	177	667	448
Anaerobic Sludge	3552	2387	1180	1509	1249
Aerobic Sludge	2670	2378	461	1004	1190
Aerobic supernatant	52	2300	12	539	1451

Another by product obtained during the anaerobic digestion stage is biogas. Measurement showed that about 14 m^3 biogas/tonne FFB processed is produced comprising 65 - 69 % methane by volume. The present practice is to vent it off to the atmosphere. Economic analysis on the possibility of tapping the methane gas for mill power generation is presently unfavourable. The fact that palm oil mills by themselves are already fuel sufficient also reduces the attractiveness of this methane recovery.

Atmospheric Wastes

Atmospheric emissions from solid fuel boilers and incinerators in the palm oil industry must comply with the Department of Environment standards with regards to particulate and dark smoke emissions (DSE) as stipulated in the Clean Air Act. Thus particulate emissions and DSE should not exceed 0.4 gm/Nm^3 and an equivalent Ringelmann scale of No.2 respectively. This is a current problem in the palm oil industry where the DOE standards could not largely be achieved due to varied management standards and in particular old age boilers and boiler operation practices. In some installations the stipulated level of particulate emissions can be met by installing dust arrestors such as twin or multicyclones. In some mills the recovered debris is blown back into the furnace for further combustion thus becoming another energy source.
The DSE is a trickier problem to solve. Since 60 - 90% of the total number of particulates are found in the submicron sizes they are able to escape the cyclones to be emitted as dark smoke. This problem is further accentuated by a generally poor process control of boiler operation in most mills.
Incinerator emissions are mainly associated with high particulate levels since the plumes are usually white due to the high moisture content of EFB (see Table 3.) It is not advisable to use a dust arrestor as incinerator stacks are usually natural draft driven and thus would not bear the pressure drop across a dust arrestor if slow burning is to be maintained.
On the brighter side of things industrial management has, by and large, recognised the need to be environmentally conscious and has taken measures to ensure environmental problems are dealt with properly. As such, in new installations water tube boilers are preferred to fire-tube ones and evaluation of better control system and boiler practices are being evaluated. Current R & D activities with regards to air pollution management include modelling of atmospheric dispersions and stack height designs. To be truly successful these studies need to be integrated with studies on various aspects of boiler operation, combustion engineering and energy management as a whole.

CONCLUSIONS

The extraction of crude palm oil generates large amounts of strong organic liquid and solid wastes. Conventional waste treatment technology has not been fully successful in treating palm oil wastes. However, through intensive inhouse R & D activities, it has been shown that the palm oil wastes problem has not only successfully meet the stringent discharge regulations of the EQA but has developed simultaneous waste strategies that utilise by products. This strategy has contributed considerable cost savings in both the palm oil milling industry and the plantations.

ACKNOWLEDGEMENTS

The authors wish to thank Harrisons Malaysian Plantations Berhad for their permission to publish this paper.

REFERENCES

1. Shafii, A.F., K. Sivasothy, and N.B.H. Lim, Final Report of the Nordanal Oil Mill Energy Study, Engineering Services Scheme Report No.114, Jan 1986

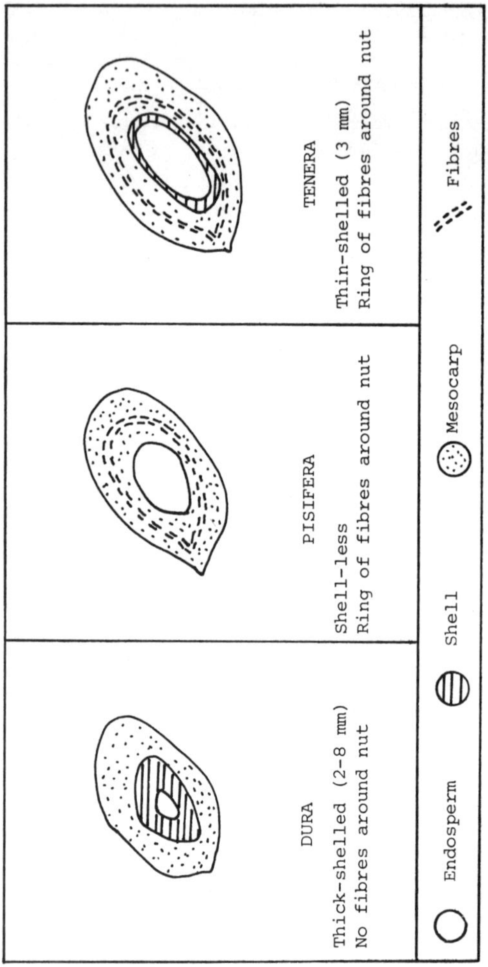

Figure 1 : Morphology of Palm Oil Fruitlet

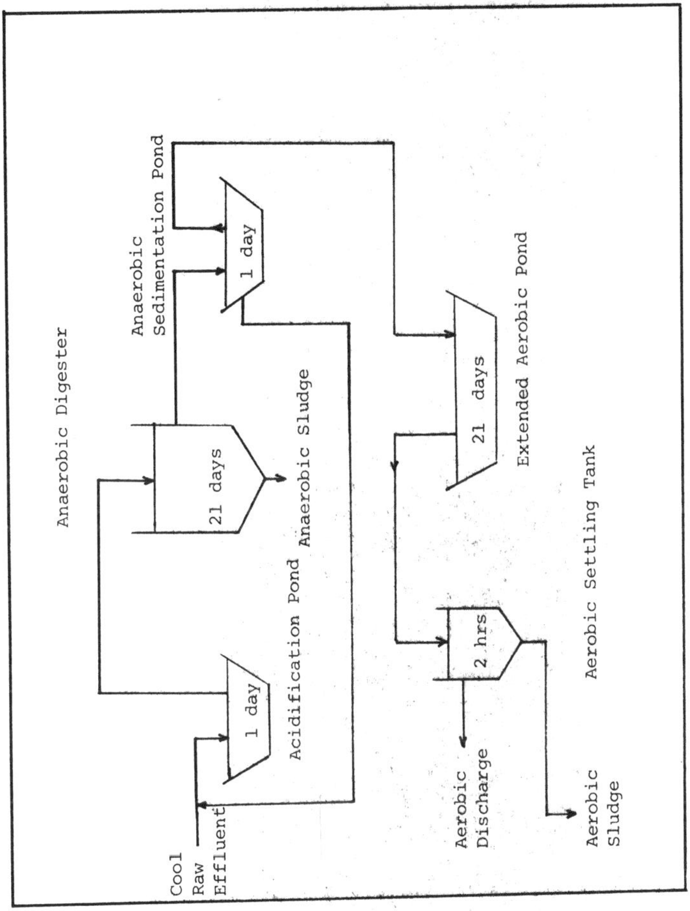

Figure 2 : Flow Scheme of Effluent Treatment Plant

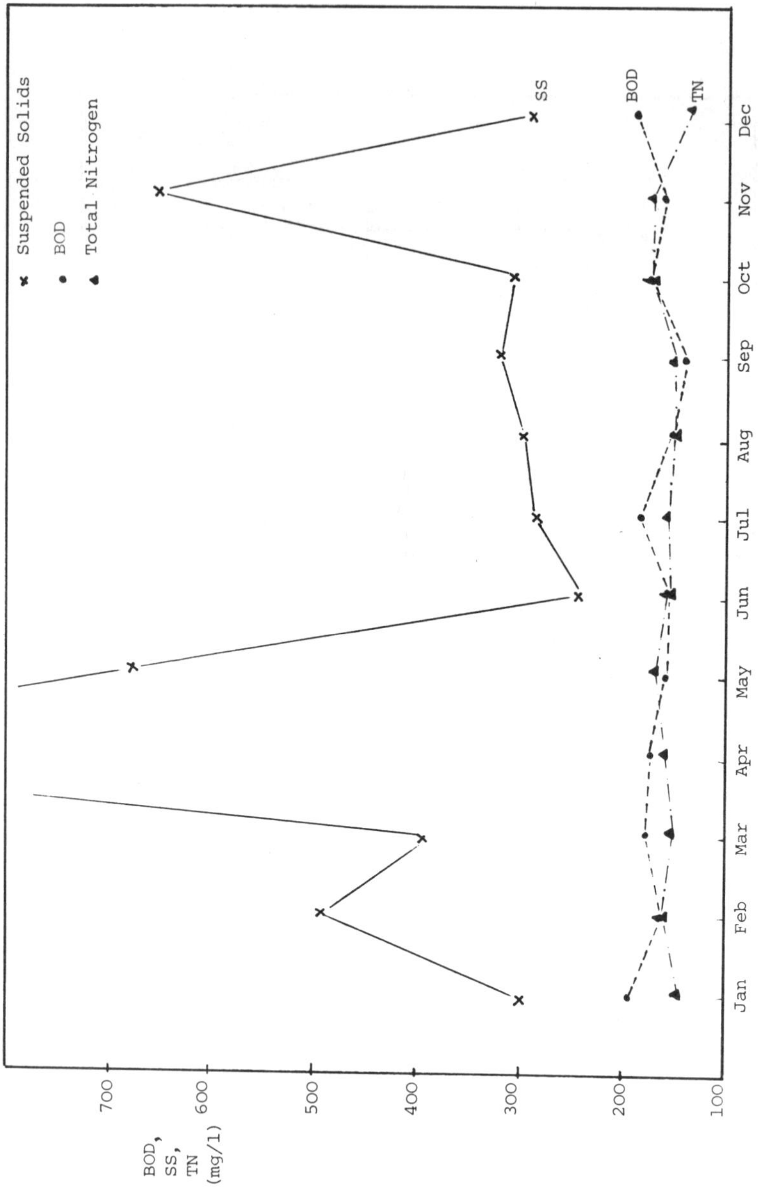

Figure 3 : Characteristics of Final Discharge (Aerobic Supernatant) from Mill A in 1985

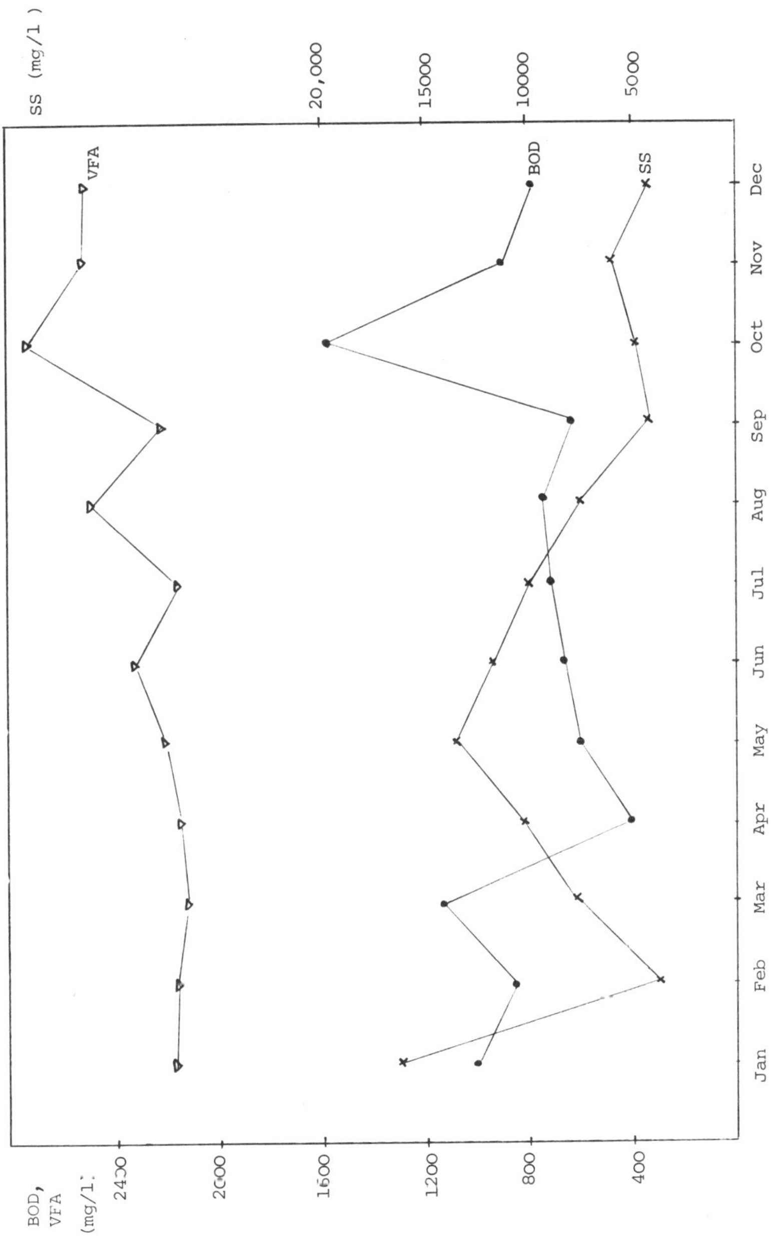

Figure 4 : Characteristics of Anaerobic Liquor from mill A in 1985

COMPUTER CONTROL OF AN OLEFINE PLANT USING AN ON-LINE OPTIMISER

P C Prichard*

ABSTRACT

ICI have used computer control on their last three Olefine Plants for on-line economic optimisation of the process. Plant performance data and commercial information are fed into the plant computer. This information is absorbed and a model with optimiser is used to generate the optimum set points for the main parameters on the plant. The optimum values are fed automatically every few minutes to a regulatory package which adjusts controller setpoints with attenuation for large step changes. The model is continually updated with new plant data. Constraints to operation are entered by the operator. Economic information is produced as a result of a model-based planning system which permits study of both business strategy and specific consequences on the plant itself. The operational aspects of the system are discussed.

INTRODUCTION

An olefine plant consists in its main elements of a set of furnaces arranged in parallel which process hydrocarbon feedstocks, followed by cooling, compression and separation systems to produce about 10 major products. The main economic factors are which feedstocks to use and at what conditions to process them. On each of their last three Olefine plants, ICI have used an on-line optimiser package to control the furnace conditions to give the maximum gross margin on the plant at all times. This paper describes the system and the major considerations in operating it automatically and continuously on a running plant.

* ICI Chemicals and Polymers Group, P O Box 54, Wilton Works Olefines South Offices, Middlesbrough, Cleveland TS6 8JA

The latest plant, No 6 Olefines, is owned 80% by ICI C & P group and 20% BPCL.

THE BASIC DESIGN OF AN OLEFINE PLANT
(see Diagram)

Hydrocarbon feedstocks, typically ethane, propane, butane, naphthas or gas oil, are fed into the furnaces, usually one feedstock per furnace but sometimes 2 or more types into the same furnace. They are mixed with steam ("dilution steam") and rapidly heated up to about 800 - 850°C where-upon they degrade ("crack") to a multitude of species. The prime ones are ethylene, propylene, methane and butadiene. After cracking the process streams are quenched to about 350-400°C to stop the reactions and the separation of the products begins. Fuel oil and some gasoline are separated first and the remaining gases are compressed and moved through a distillation train to produce streams of hydrogen, methane, ethylene, ethane, propylene, propane, mixed C4s, and some further gasoline; any C_2 and C_3 acetylenes are removed by hydrogenation; ethane and propane are generally recycled to the furnaces for cracking to extinction; energy is recovered when the process streams are quenched and this is used to carry out the separation. Such plants have to be large to be economic to operate, using around 2M tonnes p.a. of feedstock and producing about 0.5M Te pa of ethylene.

MAJOR PROBLEM AREAS

With the value of the products at around £300m per year and energy consumption of 500MW, there is a considerable incentive to maximise the cash flow even if improvements in percentage terms are small. A large number of techniques are used, many of them computerised. One of these is the on-line optimiser.

THE ROLE OF THE ON-LINE OPTIMISER

The purpose of the on-line optimiser is to maximise the cash flow (or gross margin) from the plant at all times. The particular function describing the cash flow is the value of the products less the cost of the feeds and energy. By knowledge of the performance of each furnace at the time and how it changes with conditions, the main control parameters on each furnace of feedrate, steam rate and exit temperature can be chosen; the optimiser ensures that these parameters do not exceed the limits set by the plant operators and that derived values for such items as product makes or machine power requirements do not exceed limits set for these as well.

CONSTRUCTION OF THE OPTIMISER

The optimiser consists of 2 major parts - the plant model and the optimisation routines. The model covers the whole of the plant in enough detail to ensure that the optimisation is correct within the limitations of accuracy of the measurements. The furnaces are modelled in great detail to ensure that the yield patterns and energy requirements are accurately assessed. A full energy balance is carried out over the whole plant. Critical pieces of equipment are modelled in detail. The whole model is a combination of theoretical and empirical relationships checked out against plant measurements and specific experiments. It is regularly and frequently updated to ensure that the model results match the current plant performance. The main elements updated are the furnace yield patterns and machine performance. Virtually all the updating is carried out automatically. A complex sampling and analysis system is used to measure the furnace effluent compositions. The optimiser handles 48 variables and 190 constraints.

OPERATION AND MAINTENANCE

The optimiser works continuously on its own computer. Plant information is gathered by a foreground control computer and sent to the optimiser every 3 minutes or for some of the slower moving items every twenty minutes. Many of the model coefficients are updated with this new information to ensure that the representation of the plant is the best possible. The constraints to operation are adjusted by the plant operators whenever required and these are also passed into the optimiser every 3 minutes. Economic data such as feed and product prices and the costs of services are entered when there are significant changes. The optimiser operates continuously on all this information giving optimised values every 3 minutes for the setpoints of the individual furnace feed rates, steam rates and cracking temperatures. These are not fed directly through to the plant control valves, but through a regulatory package to ensure that step changes in optimised setpoints only reach the plant over a period of time in a ramped fashion. Therefore any changes, especially those from the operators will be acted on within 6 minutes but a complete move of conditions may take up to 2 hours.

Maintenance on the system is heaviest on the furnace sampling and analytical parts. The continuous samplers are operating in an environment which causes fouling by tars and coke, and the analysers (GLC's) require regular attention to ensure proper sample conditioning and calibration. As with any computer software, the model and optimiser need virtually no day-to-day maintenance; such that is required is mainly directed towards making sure that faulty data is corrected at source. Modifications to the model and optimiser are carried out as soon as possible after any significant changes are made to the plant; as with changes to any control or real time programs these are tested vigorously before being commissioned.

DISCUSSION

Choice of Computer

The choice of which make of computer to use is relatively unimportant compared to the hardware layout required for an optimiser. There are a large number of machines in the mini-computer category which would operate satisfactorily.

For full automatic control, there is inevitably a control computer, and with many modern machines it would have enough space to hold an optimiser. However, the control programs have to run on-time and therefore it is difficult to arrange sufficient priority for the optimiser to ensure that it can run fast enough. Consequently the optimiser is run on a separate computer where the control and optimiser programs do not interfere with each other. They are connected by a conventional asynchronous link.

Plant Information

The plant information required for optimisation is collected by the control computer as a subset of the full data base held on that machine. Checking of each measurement is required to prevent out-of-range values corrupting the model. There is nothing special about the measurements except that flows may need compensating when meter conditions deviate from design. The operator is required to inform the computer of the status of each furnace and this is accomplished via a sequencing package which helps bring furnaces on-line from cold and take them off-line.

Constraint Setting

Furnace constraints are split into 2 types: those which are directly visible on the plant and those which are calculated. Examples of the first are furnace temperatures and feed rates and examples of the second are steam/feed ratio and firing heat requirement. There is another set of constraints, which relate to the whole plant and which the production from all of the furnaces must lie within; examples of these are the power capabilities of the main compressors, and hydraulic loadings on individual fractionating towers. The plant operators have to review all these constraints regularly making sure that they are properly matching the true limitations on the plant at that time and in practice they are making changes a few times on each shift. These changes help in a number of ways, but especially as a feedback mechanism to ensure that the plant is working at its true limit and also to give the operator interaction with the optimiser to the extent that he is confident that he controls the optimiser by controlling the range over which it can operate.

Economic Data

The prices actually used in the optimiser are realistic and up-to-date. They often cannot be derived easily because many of the products are used in other parts of the hydrocarbons processing complex. About 20 streams or components are given values. Consequently LP planning models are used to generate many of the prices, with those sold externally being chosen at market values. Because the range of operation is normally within the margin of price variation, marginal prices are used generally but the optimiser can handle up to 3 levels of price for any one product or feed. Prices need to be reviewed regularly and their effect checked in advance as far as possible to try and avoid surprises in actual operation compared to expectations.

Operation of the System

An optimiser of this type is obviously a sophisticated tool which produces results which are not necessarily what even an experienced operator would choose. However these results are being generated from the best set of data available and represent decisions which would require considerable human effort to produce by other means. It is vital to ensure that the operators know what the optimiser is likely to do and to prove to them by experience that it will always put out setpoints within the constraints they have set. Their role is to control the range of operation permitted and this is a marked deviation from their conventional task of choosing the setpoints themselves.

On a plant where conventional controllers have a fast response to changes in setpoints, it is important that the optimiser should also respond rapidly to changes made by the operators to constraints so that they can actually see something happening shortly after any change. Hence the 3 minute data transfer frequency has been chosen. This is probably a more important factor in choice of frequency than model updating, though the latter is also important to ensure that constraints to derived values are not violated or if so that corrective action is taken rapidly. The operator is usually unaware of this corrective action.

The difficulty most often encountered (but even then, rarely) is when the set of constraints is not self consistent, presenting the optimiser with an infeasible problem. This is flagged up with indications as to which constraints are in conflict. An example is when the sum of the minimum feed constraints on each furnace exceeds the maximum feed constraint for the whole plant; the operator has to make the choice over which constraints to alter and in the meanwhile the optimiser suspends operation.

The optimiser can be started or stopped via single commands to the control computer. This is particularly important in the event of an upset on the plant so that other control regimes can be implemented such as major rate reductions. A return to optimiser conditions is carried out via the regulatory ramping package since the plant conditions can be some distance away.

Sampling and Analysis

It is very important to ensure that the sampling and analytical system are working properly. When individual furnaces are being optimised against each other, they must all be being analysed at the time. There are many reasons why the sampling system for single furnace might fail; these are flagged up and action taken promptly to prevent optimisation on the wrong premises. Also all furnaces are analysed on each of the analysers to avoid systematic errors which might arise by splitting the furnaces between analysers.

In many cases, minor calibration errors are unimportant because the optimised solution is so firmly weighted to one extreme that major changes in calibration (or prices) would be required to cause a shift. However when the optimum is between maximum and minimum constraints and not hard up against either, the effect of calibration errors is significant in terms of chosen setpoints but less so in terms of profit because profit gradients under such conditions tend to be flat. Techniques to compare predicted yields against actual yields are used to give early warning of such errors.

General Considerations

The optimiser permits commercial decisions to be applied to plant operation directly. To do this effectively requires the close cooperation of all the disciplines involved - marketing, feedstock purchase, planning and operation. The optimiser is one of a hierarchy of models of the plant and business which collectively force the practitioners to think quantitatively about the problems of the business. The speed of data processing means that many of the decisions can be taken objectively.

The system has proved robust in practice but as with any decision-taking tool, has to be kept under review continuously to ensure that both structural changes to the plant and changes in external parameters are being correctly reflected by the optimiser. While it has proved difficult to explain the workings of such a sophisticated tool to the operators on the plant, they have gained enough knowledge of the principles and enough confidence in the operation that they let it choose the operating conditions while they concentrate on setting the constraints and pushing them back on the plant itself.

Economic Benefits

The optimiser provides benefits in 3 main areas which, bearing in mind the size of the plant, makes it a very profitable tool.

(i) Helping the plant to be run to its true limits.
(ii) Controlling the furnaces relative to each other so that the good ones are used harder than the bad ones.
(iii) Improving the fuel efficiency of the plant.

These are especially true in a production limited situation. In a market limited case, item (i) is obviously not valid, but is replaced by reducing the amount of feedstock required to achieve a given amount of product.

CONCLUSIONS

The use of this optimiser to control the plant directly is a good example of an application of high technology into the harsh environment of a big operating plant. It is easy to use and the operators have developed a high degree of confidence in it. It is one of several tools integrated together to provide effective business management of this and associated plants. It has proved very beneficial in a number of ways, not least financially.

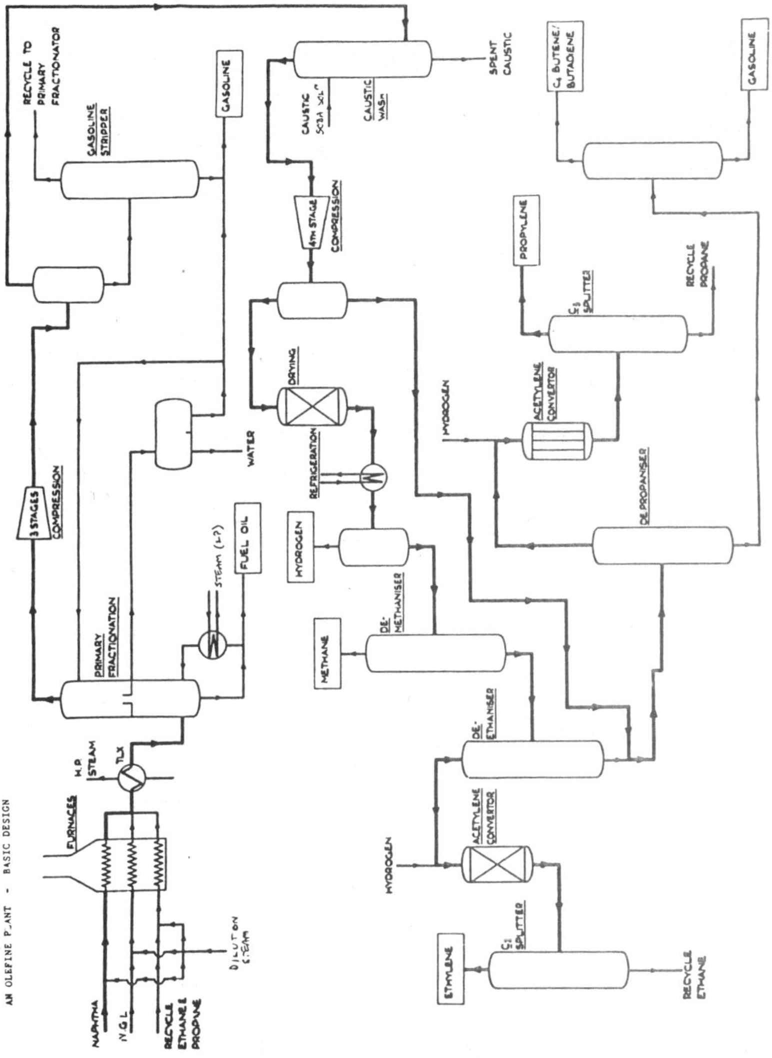

AN OLEFINE PLANT - BASIC DESIGN

ON-LINE OPTIMIZATION OF FLUIDIZED CATALYTIC CRACKERS

* Simon Wright, Peter Lawrence and Zak Friedman

This paper presents an on-line Fluidized Catalytic Cracking Unit (FCCU) optimization package using KBC's CATOP model together with an optimizer in a clearly structured computer control hierarchy. The paper discusses FCCU operating objectives and the techniques that are available today to meet these objectives. A case study is presented which assesses the merits of current practice computer control. The study is extended to identify constrained and unconstrained optima and to quantify the benefits that can be achieved through on-line optimization. It is concluded that KBC's FCCU optimization package is capable of identification of, and control at, the optimum operating conditions, and can capture credits of more than $3M/yr over and above current "state of the art" control.

INTRODUCTION

Catalytic cracking originated in the 1930's from the desire to extend gasoline supply, and production capacity was increased, through war-time co-operation, by development of fluidized catalytic crackers.

Heavy gas oils, which would otherwise be blended into fuel oil, are distilled in Vacuum Distillation Units (VDU) and fed via intermediate tankage to a stream of high temperature catalyst. The catalyst continuously circulates due to differential pressure between a reactor, fluidized by vaporized feed, and a regenerator which uses air both to fluidize and regenerate catalyst by combustion of laid down coke. Reactor effluent is desuperheated in a fractionating column, where heavy products are also withdrawn. Light overheads are compressed and further separated in a wet gas separation plant. FCCU products are either used for blending of gasoline, gas oil, diesel and fuel oil in pools of other refinery components or are further processed in petrochemical complexes. Figure 1 shows a simplified flow diagram.

Since the main purpose of the FCCU is to upgrade fuel oil, the most significant economic driving forces are fuel oil - gasoline price difference and fuel oil - gas oil price difference. The seasonality of gasoline - gas oil price difference and the volatility of fuel oil prices mean that both cat feed cost and value continually fluctuate. Typical margins can vary between $0/T and $30/T. As a result, and also due to substantial throughput flexibility, FCCU operation has a major impact on overall refinery profitability, and there is clearly scope to adjust operating throughputs and yields in response to the changing influences of the market. After safe operation, environmental control and product quality aims, the refiner's aim is to maximize FCCU profitability.

* KBC Process Automation Ltd., Chilworth Research Centre, Southampton, SO1 7NP

FCCU CONTROL

The goal of on-line FCCU optimization is achieved through a stagewise approach. A hierarchy of control is developed. Each level of the hierarchy can operate independently of the level above, and is a sensible step towards achieving some or all of the operating objectives.

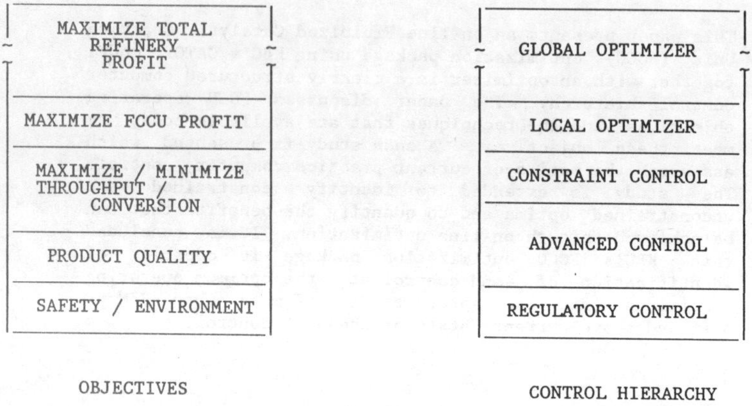

Regulatory Controls

Maintenance of a desired operation in response to outside disturbances, using sound regulatory control is a pre-requisite of all higher level control. This is achieved by feedback level, pressure, temperature and flow control. Some modern digital controllers include full validity checking of all inputs to test that signals are within reasonable limits, signal changes do not have high rate of change and that the signal is not frozen at one value. Regulatory control is the means by which safe FCCU operation is achieved and is the limit of automatic control for some refiners.

Advanced Controls

Enhancements to FCCU regulation include reactor heat and material balance controls, product quality controls and controls to minimize energy consumption.

Since coke, a product of catalytic cracking, is burned in the regenerator to supply most of the heat of reaction, the reactor heat and material balances are interdependent. For example, an increase in catalyst recirculation rate provides more heat to the reactor. The severity of the reaction is increased and hence conversion rate rises. More coke is produced which is burned off in the regenerator, reducing excess oxygen and generating more heat. The change in circulation rate affects both heat and carbon balance. Control of these balances is vital to stabilize the operation of the reactor - catalyst regeneration circuit, as well as offering potential energy consumption savings. In a unit that has limited air blower capacity, tight control of carbon balance using on-stream regenerator flue gas analysers or temperature indications within the bed of catalyst, ensures maximum coke production allowing maximum conversion. Heat balance can be maintained by adjusting the flow of hot regenerated catalyst to the reactor.

Advanced schemes to decouple the interactions of heat and carbon balance are in place in some refineries, together with feedforward compensation in the event of throughput changes.

Quality control can be achieved by inferring a quality from some other plant measurements such as temperature. Adjustments may be made as a result of laboratory testing of samples. On-stream analysers have recently improved their reliability and are now considered good quality control tools. Deadtimes resulting from the analysis method, sample point position and location of the analyser house can be compensated by dynamic modelling (Refer to Figure 2).

Energy consumption can be reduced throughout the process by tight carbon balance control and heat integration between feed circuit, fractionator and gas plant. If there is a preheat furnace, there is scope to maximize efficiency. Pressure minimization in light ends towers may be applicable, and efficient use of CO / waste heat boiler can also be ensured.

Safe operation can be maintained by advanced control schemes which can guard against non-stochiometric combustion, and advise when instruments fail. Compressor and air blower operation is protected by anti-surge controls.

Some refineries are at this stage of automatic control where FCCU regulation is ensured, and production quality control and profit maximimization have started to be achieved.

STEPS TOWARDS OPTIMIZATION

It has been recognized that with low fuel oil prices, high FCCU throughputs give high profit margins. When fuel oil prices are relatively high, as they were during the 1984/85 coal miners' strike or typically are during winter, the gasoline - gas oil price spread influences the trade-off between high throughput at low conversion and low throughput at higher conversion.

Current practice FCCU "optimization" is based around maximizing unit throughput at fixed reactor temperature or maximizing conversion at fixed throughput. Constraints commonly met include blower and wet gas compressor loads, maximum reactor or regenerator temperature, fractionator hydraulics and product draw-offs. Automatic constraint control strategies have been successfully applied to FCCU's and are widely reported (1,2,3). For both strategies, process gains and dynamics of each potential constraint must be obtained from plant tests to identify the "active" constraint. Once the unit has been driven to the limit, a skilled operator can adjust other variables to relieve the constraint. By this approach it is possible to meet two or more constraints simultaneously. The more sophisticated constraint control packages will do this automatically based on a pre-determined hierarchy. There are also strategies like pressure balancing for reactor and regenerator for non-constrained degrees of freedom.

Economic studies using historical data or plant simulations, such as KBC's CATOP, can examine alternative strategies and serve to signal a switch in manipulated variable or a re-ordering of the hierarchy. These are typically performed daily or weekly and can result in a change of strategy three or four times a year.

With a limited FCCU operating envelope and few degrees of freedom, a well-supported dynamic constraint controller probably captures most tangible benefits. However, with wider operating ranges that successive plant de-bottlenecks offer and the more degrees of freedom that are made available with recycle operations, automated catalyst addition facilities and residue processing, a constrained FCCU may no longer be an optimized FCCU.

IN SEARCH OF THE "TRUE" OPTIMUM

The industry's current practice has a number of draw-backs. Firstly, little or no account is taken of feed quality which may vary considerably as a result of crude distillation unit (CDU) feed changes, blocked operations that provide FCCU feed as by-products or variable import qualities. Secondly, "optimization" is attempted around the reactor / regenerator circuit with little attention paid to optimal product cuts associated with reactor and regenerator operating conditions and constraints. Thirdly, unconstrained optima will not be identified.

Model-based control must be used to tackle these areas where the cost of not reacting correctly is potentially very high. The FCCU model, CATOP, developed by KBC, is used by refiners to monitor unit performance, to predict product rates, qualities and conditions, and to optimize unit operation within fixed constraints. A sketch of the program structure is shown in Figure 3. Off-line, CATOP has access to three files, namely a crude assay database, a mechanical data file and an input file. Feedstock properties can be measured by the laboratory and entered in the input file, or can be accessed automatically from the crude assay database. The mechanical data file contains information which is relevant to a specific unit such as flowsheeting data, hardware specifications and unit factors for catalyst type and for yields, and product qualities. The input data file contains information which varies from run to run such as feed quality data, unit operating conditions and economic data for pricing feeds, products, utilities and operating costs. CATOP then characterizes the feed either from input properties or from a TBP cut extracted from the crude assay database, and based on the observed operating conditions the reactor effluent yields are predicted. These are then processed through the main fractionator and gas plant models to derive the yields and fractionated product qualities. For recycle operations, the HCO and / or slurry recycle is further processed through the reactor model until the model converges on recycle qualities.

Optimization of the FCCU operation in most cases means finding the set of conditions which maximizes overall unit profitability. The CATOP optimizer rates FCCU yield performance for different sets of operating conditions and uses sequential linear programming techniques to identify which operating conditions give the most attractive reactor effluent yield pattern and fractionator cut points within unit constraints.

WHAT'S THE INCENTIVE?

A typical 50,000 B/SD FCCU has been studied using KBC's CATOP program to identify the incentives to install a model-based optimizer above a computer constraint strategy.

The economic basis for the study used a fuel oil - gasoline price difference of \$70/Te and a gas oil - gasoline price difference of \$10/Te. These are typical winter product prices differentials and have not been affected greatly by recent massive crude and product price fluctuations.

The first part of the study examined current practice constraint control strategies, which adjust conversion to fill both blower and compressor with remaining degrees of freedom fixed, for validity over a wide range of economics. In addition to the base case operation, economics were varied considerably to simulate gas oil pool cloud giveaway, fuel oil pool viscosity giveaway, a fuel blending density constraint and refinery gas balance problems. The study concluded that when throughput, reactor pressure and key fractionator parameters are set at their optimal values, current practice computer control would optimize the unit adequately. Table 1 illustrates how the incentives are derived. The strategy is stable to the large changes in economics explored, although the incentive varies between $3M/yr and $1M/yr as pool blending considerations and potential refinery gas balance problems continually alter fractionated product marginal values. As a result, continuous implementation of these strategies is estimated at $1.5M/yr. Approximately $1M/yr is obtained through regulatory and advanced control, $0.5M/yr through implementing on-line constraint control strategies and since we estimate that the choice between which strategy to adopt is not clear 30% of the time, there is a further $0.25M/yr associated with this sort of strategy which can be captured by model based control.

The study then concentrated on the degrees of freedom which are not addressed by current computer control practice.

Since air blower and compressor loadings play such a major role in FCCU optimization, variations in reactor pressure capable of tilting the balance between the two machines are also important. The study concluded that a 5% change in pressure can affect the profitability by as much as $1M/yr, and that there is an optimal pressure setting which varies with operating conditions. Figure 4 illustrates the loss of profitability, and variation of other process parameters, as reactor pressure is changed at maximum conversion.

A second important degree of freedom is throughput. While it is clear that both the blower and compressor should be fully loaded, there may be an optimal unconstrained throughput. Indeed, for the case examined here an optimal throughput was identified. Operation 10% above or below the optimum results in a loss of $1.5M/yr, even though both blower and compressor are fully loaded. Considering that throughput flexibility is typically of the order 5 to 10%, it is reasonable to claim that model-based control would capture between $0.5M/yr and $1.0M/yr, and more if feed mix is also a variable.

Quality control using laboratory analysis, and on-line analysers, is hindered by long process dead-time, relatively low availability of analysers and need of extensive technical support. Model-based control can practically eliminate all those difficulties since control can be done based on the model prediction which is always available and does not have a time delay. The analyser is then used for calibrating the model. This simplifies use of analysers. For example, the model prediction of pour point may be used for control, whereas a cloud-point analyser is used to calibrate the model. Quality control is generally worth $1M/yr, with standard practice analyser feed back control capturing about 70% and model-based control capturing about 30% of this benefit.

Intermediate product values are usually based on the quality which is the limiting specification of the pool into which they are blended. Hence changes in product properties and splits can change unit profitability significantly. The optimizer can calculate product values based on predicted qualities and, with knowledge of the constraints, assess the optimal fractionator cut points. This study showed that the incentive to optimize LCO/HCO cut was \$1.7M/yr, by increasing LCO cloud point from $-4.5^{o}C$ to $-0.8^{o}C$. For many operations, such a quality change may be acceptable since it simply results in the addition of a marginal blending component to compensate, which was taken into account in the pricing mechanism. The benefit captured by model-based control is assessed at between \$0.3M/yr and \$1.0M/yr, reflecting the fact that FCCU economics are driven by blending considerations.

There may also be scope to optimize fractionator pumparounds, balancing improved separation against heat recovery, though at current fuel prices, heat recovery is not too valuable.

Having established that there is usually an optimal set of FCCU operating conditions that varies with external influences such as local blending considerations and global economics, and that the incentives to operate at the optimum point can exceed \$3.0M/yr, the next section looks at capturing these benefits using KBC's ON-LINE CATOP model-based control package.

CLOSING THE LOOP - ON-LINE MODEL-BASED FCCU OPTIMIZATION

CATOP users have tried their best to optimize operating parameters off line and let plant operators know what they are. This is a good inexpensive approach but due to a number of problems inherent in off-line optimization techniques, they have only been able to capture about half of the \$3M/yr worth of incentives. These problems can shortly be described as follows:

Firstly, frequent feed quality changes due to poor knowledge of crude characteristics, variable imported feedstocks and untimely laboratory analysis mean that sound feed characterization is not continuously possible.

Secondly, "constants" are not always constant. Catalyst activity and selectivity varies and even constraints may change with ambient conditions. In other words, the Mechanical Data Base needs continuous updating of some parameters.

Finally, and most serious, is that the conditions output by the optimizer may not be valid some time after they have been set by the operator as, due to some unforeseen circumstance, one of the operating points changes. For example, if a control valve sticks, cooling water temperature is higher than anticipated, or, for his own reasons, the operator fails to accomplish one of the objectives, then the remaining objectives are meaningless in the sense of optimization.

KBC's ON-LINE CATOP model-based FCCU optimization package is designed to resolve these issues, in the form of a hierarchical control strategy that is readily constructed above existing control schemes. Figure 5 shows the hierarchy building from regulatory control, advanced control and constraint control to model-based optimization.

The model-based optimization module uses dynamic correction of all data which may already be part of existing advanced control strategies. It typically consists of a first order representation of process dynamics plus dead time estimates and is used to construct continuous "steady state" conditions for running KBC's CATOP model.

The purpose of the data reconciliation module is to determine whether the set of measured data is consistent within itself. It checks the data against a number of physical laws which must be obeyed, for example, mass balances and energy balances. Failure of data reconciliation at this point quickly pin points instrument problems and incentives to perform maintenance.

Input data can now be passed to the steady state CATOP model where a prediction of operating conditions is made. These are recycled to the data reconciliation module where any deviations between actual and predicted operating conditions are put down to changes in feed quality and / or catalyst activity. Again, unacceptable feed characterization or catalyst activity implies an inconsistent set of input data. In this way, the model recalibrates itself continuously. No other mechanical constants are updated on line since CATOP is generally calibrated on the basis of a rigorous test-run where steady operating conditions are maintained for periods of four hours or more and feeds and products are thoroughly analysed. The quality of data used for such a calibration test is better than that that is being controlled normally.

The sequence Dynamic Correction - Data Reconciliation - Model Update - CATOP Steady State Model is low on computer resources and so can run quite frequently, testing plant data. Only when a significant change in operating conditions is detected need the optimizer be called. Since the optimizer is relatively high on computer resource, the frequency at which it is called can be tuned to the magnitude of change in operating conditions.

There are two sorts of constraint that the optimizer needs to be able to identify. "Explicit" constraints are mechanical limits such as maximum metal temperatures, maximum vessel pressures and minimum exchanger throughputs. "Implicit" constraints either cannot be read directly on the plant, such as flooding, or cannot be directly input into the CATOP model, such as control valve position. Some independent model which may be as simple as a linear gain that translates change in valve position into change in flow, or as complex as implying compressor feed rate limit from its maximum speed limit, is used to translate "implicit" constraints into "explicit" constraints that the optimizer must honour. The steady state model can also be used to forecast constraint values to avoid overshooting limits.

Subject to all existing and anticipated constraints, the optimizer generates a set of optimal operating conditions that maximizes the objective function. When these conditions are at their constrained values it is necessary to determine how to operate at that constraint. It is highly unlikely that, when the model assumes a constraint will be met, in practice the constraint is met exactly. If the constraint is undershot, it is likely that profit is not being maximized. This is because it has been determined that in most cases operating at a constraint with all other degrees of freedom fixed is optimal. The model can identify which is the best variable to manipulate to meet the constraint. Similarly, when backing away from the constraint the correct manipulated variable can be identified. The cause of not meeting a constraint may be inaccuracies in CATOP's steady state predictions, in which case the data reconciliation module will soon compensate, or errors in translating "implicit" constraints to "explicit" in which case these models will self-calibrate.

If the optimum operating conditions are unconstrained then the optimizer will effectively disable the constraint controls. They will still be active to back away from unexpected constraints, however.

In summary, the optimizer outputs setpoints to those degrees of freedom not currently addressed by computer control, and strategies to the constraint control packages that are in use today. Should a variable change unexpectedly or the operator fix certain variables, for safety reasons for example, then the optimizer will determine the optimum set of conditions for the remaining variables. Also, disabling the optimizer only removes one level of control since lower levels in the hierarchy will still operate, and sucessful implementation of this control strategy can capture incentives exceeding $1.5M/yr.

SUMMARY AND CONCLUSIONS

This paper illustrates how a sophisticated computer control hierarchy can be developed to optimize FCCU operation and achieve benefits of over $3M/yr. It shows that KBC's on-line FCCU optimization package, using CATOP, can identify optimum operating conditions and operate continuously at these conditions subject to moving constraints.

The optimization package uses data validation, dynamic correction and data reconciliation to update the CATOP FCCU model. An optimizer uses the model output to define optimum operating conditions. When a constrained optimum is identified the package instructs constraint controllers how to behave at the constraint.

The package is part of the continued development of overall refinery optimization. KBC has a refinery flow sheeting system, PETROFINE capable of modelling many other refinery processes. It is both flexible and accurate in the way it models processes and is used for decision support for planning, scheduling and plant control. It is already being used off line by several U.K. refiners for overall refinery optimization. The FCCU optimization described here could ultimately become a lower level of an extending hierarchy of control that balances one process unit with another.

REFERENCES

1. Improve FCCU Profitability with Better Control J. C. Brice and K. V. Krikorian, KBC Process Consultants, Weybridge, Surrey; Hydrocarbon Processing, May 1983.

2. FCC Process Variable Questioned, Panel discussion on FCC process variables, Oil and Gas Journal, 23rd March 1981.

3. FCCU Advanced Control, Advanced Process Control Handbook, Hydrocarbon Processing, February 1986.

TABLE 1

	BASE	MAX CONVERSION
FRESH FEED (T/H)	320.8	320.8
FUEL GAS (T/H)	16.2	15.7
LPG (T/H)	46.0	45.6
STABILIZED GASOLINE (T/H)	114.6	117.0
LCO (T/H)	102.1	100.5
HCO (T/H)	4.7	4.6
SLURRY (T/H)	21.7	20.8
COKE (T/H)	15.5	16.6
REACTOR TEMP (^{o} C)	532.0	526.0
REACTOR PRESS (KG/CM^2 G)	3.7	3.7
AIR RATE (KM^3/H)	180.0 *	180.0 *
COMPRESSOR FEED (T/H)	87.0 *	87.0 *
REGEN O_2 (MOL%)	2.4	1.0 *
CONVERSION (LV%)	74.4	75.2
$\triangle$ PROFIT (M$/yr)	<------ +$3.1M ------->	

* Constraint value.

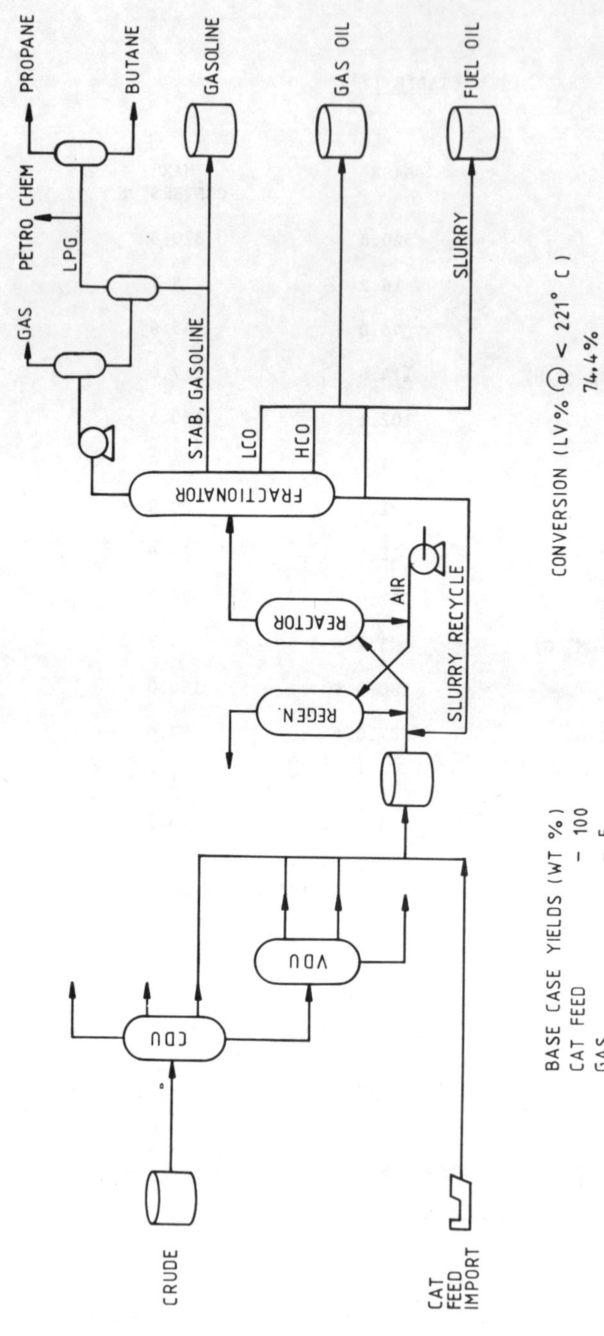

BASE CASE YIELDS (WT %)	
CAT FEED	– 100
GAS	– 5
LPG	– 14
STABALISED GASOLINE	– 36
LIGHT CYCLE OIL	– 32
HEAVY CYCLE OIL	– 1
SLURRY	– 7
COKE	– 5

SIMPLIFIED FLOW DIAGRAM

FIGURE 1

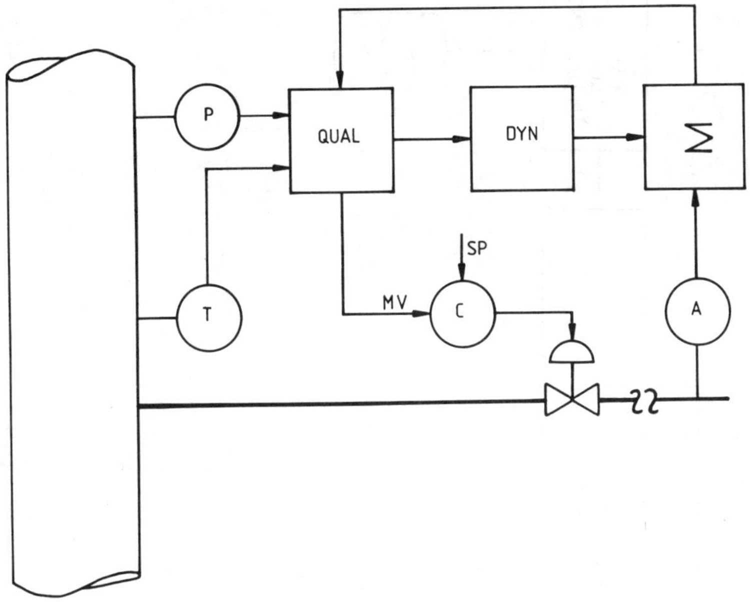

DYNAMIC MODELLING ELIMINATES DEAD TIME FIGURE 2

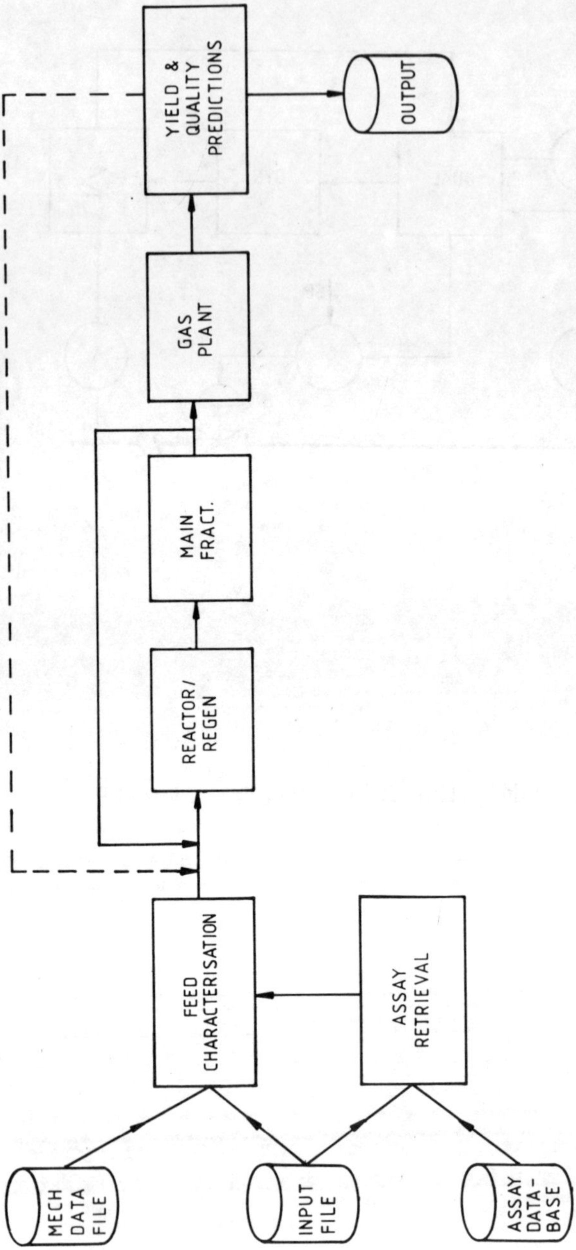

CATOP PROGRAM STRUCTURE

FIGURE 3

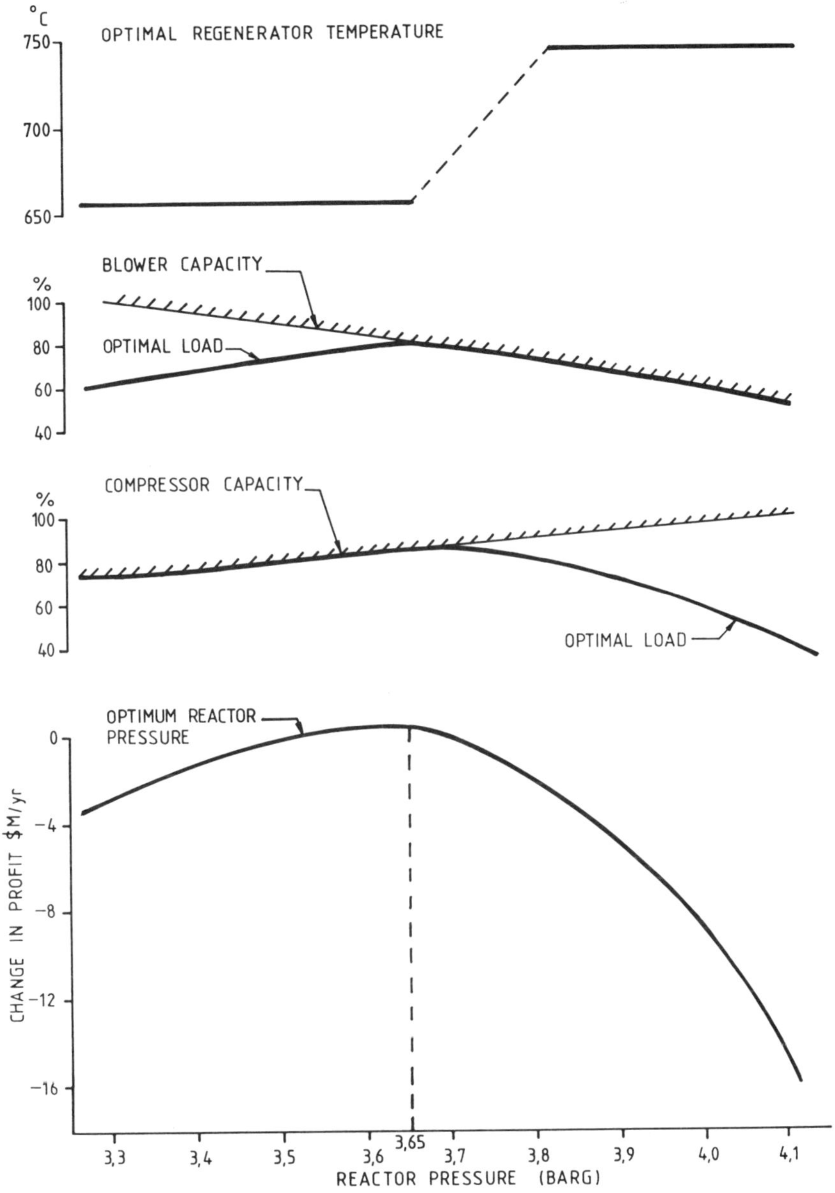

REACTOR PRESSURE INFLUENCES OPTIMUM OPERATING CONDITIONS FIGURE 4

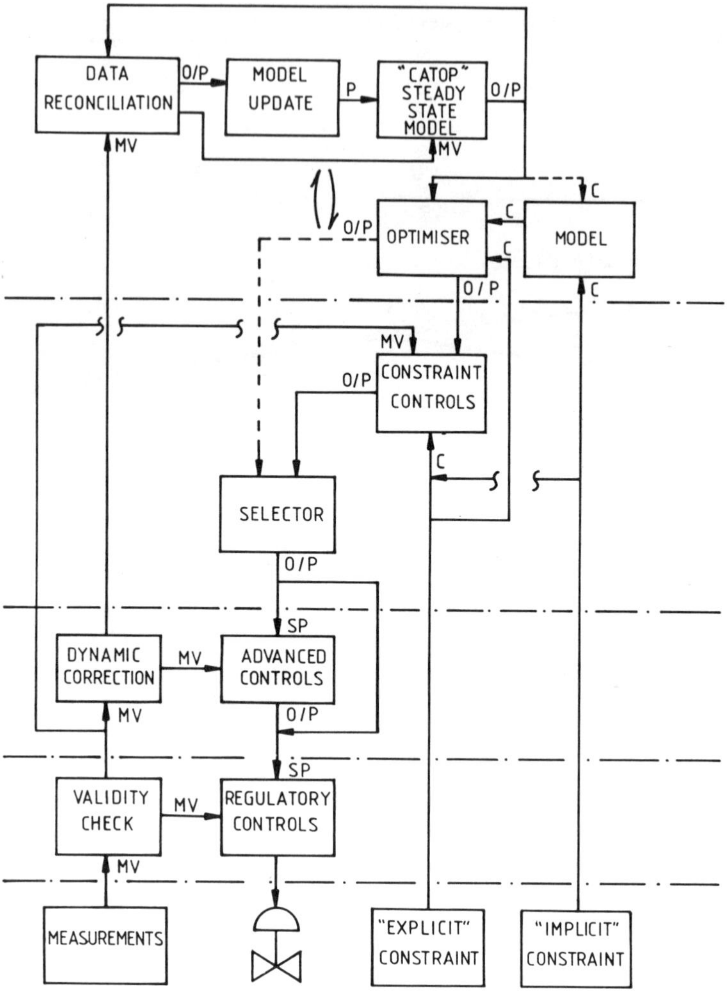

CONTROL HIERARCHY FIGURE 5

STRATEGIES FOR DESALINATION OPTMISATION

J.R. Flower* and J. Pardoe*

This paper considers the application of simultaneous optimisation techniques to Multi-Stage Flash (MSF) desalination processes. MSF units are highly integrated, countercurrent processes, and so are suitable for efficient solution by simultaneous optimisation methods. Exploiting the particular characteristics of the process leads to a more efficient solutions when using these types of methods.

INTRODUCTION

The simultaneous approach to process simulation and optimisation has been widely discussed in the chemical engineering literature as an improvement on the sequential modular approach to flowsheeting. Though this approach is well suited to an interconnected process such as desalination, little published work has exploited these ideas for this process area. A recent paper by Helal et. al. (1), has shown that despite the reluctance to use this approach, an equation-oriented solution procedure which exploits the special features of a multi-stage flash distillation process can work efficiently for simulation. However, no published study has used a simultaneous approach to the steady state optimisation of desalination plants.

THERMAL DESALINATION PROCESSES

Thermal desalination, or distillation, involves the boiling or evaporation of sea water to form water vapour which is condensed to yield a salt free stream. The principle types of process are Multi-Stage Flash distillation (MSF) and Multiple Effect evaporation (ME). The 'multi' term means that a cascading effect is used, where multiple re-use of the energy content of the heating steam is achieved.

In MSF distillation, sea water is heated under pressure to an elevated temperature between 90-125 C without boiling. The hot sea water or brine is then passed through a series of stages at successively lower pressures, a portion of the flashing stream being evaporated at each stage, (Figure 1). The controlling parameters on cost and output include: the temperature drop in

* Department of Chemical Engineering, Leeds University, Leeds LS2 9JT, U.K.

each stage; the stage heat transfer coefficients; and the overall flash range, i.e. the temperature difference between the hottest brine temperature and the reject brine temperature.

With multiple effect evaporation, the steam generated in one effect condenses in the next effect, evaporating more liquid. ME evaporation is less popular than MSF distillation, hence the emphasis in this paper is on the latter. The general process models have similarities and the principles applied here to MSF plant can easily be applied to ME evaporation.

Features of Desalination

Desalination, whether using MSF units or ME evaporators is a highly integrated, energy intensive process. The cost of feed is very low, and initial processing costs are minimal. The major costs of the process are the capital costs and the energy costs when the plant is in operation. The units are of relatively simple design, but of large size. Additional process costs are for cooling water and pumping, but these are small in comparison with the capital costs and energy costs. Pumping requirements are for large volumes of fluids, rather than obtaining high pressures.

DESALINATION PROCESS OPTIMISATION

Stage-wise Methods

In the late sixties and early seventies some work was carried out in the field of process optimisation of desalination processes. In the absence of simultaneous or integrated optimisation methods, the stagewise nature of desalination plants favoured stage-by-stage solution methods, given the constraints on computing facilities at the time. The two methods discussed at this time were dynamic programming and the discrete maximum principle.

Dynamic Programming Approach

Coleman (2), and Itahara and Stiel (3), have applied dynamic programming to MSF distillation plants. Dynamic programming requires a stagewise structure for the problem, finding the optimal solution for each stage regardless of the input variables, or rather for every possible value of input variables.

Discrete Maximum Principle Approach

Beamer and Wilde (4), have applied the discrete maximum principle to desalination processes. The stagewise nature of a MSF unit makes it suitable for this solution method.

MSF plants have countercurrent flow so to apply these stagewise optimisation procedures either the input conditions to the cold stage must be fixed, or an iteration procedure must be placed over the dynamic prgramming calculation. The latter approach would be very laborious, and the former approach is used. Several key variables have to be fixed; the cooling water flow, rejection temperature, product temperature, ratio of recycle flow to product flow, and ratio of salinity of recycle stream to sea water, (feed). This ensures that the model can be solved. This is probably the major limitation of the method, in that potential optimisation variables have to be fixed a priori.

The case study approach

The recent published work on simultaneous methods do not seem to have been taken up by the desalination industry. Recent papers appear to use the case study approach or attempt to set up simple functions which incorporate all the major design variables, e.g. Giannuzzi et. al. (5).

Discussion

These methods work reasonable well given their limitations. However their rigid form means that some variables must be fixed which conceptually may be difficult, especially the brine reject temperature. This, in effect, specifys the temperature difference for heat transfer at the final, coldest, end, which might be better left to the optimisation procedure.

By forcing some key variables to be fixed, so that each stage can be solved independently, ensures, at least, that the problem is solvable, and that any intermediate answer has some value or use. Many optimisable variables within an integrated plant like a MSF one, interact strongly which does cause problems. By fixing some of these variables, these interaction can be reduced and a solution is almost assured.

SIMULTANEOUS OPTIMISATION

The countercurrent stream topology of a MSF process makes it unsuitable for the traditional sequential-modular approach, because of the large number of iterations required. These kinds of process models are much more easily solved, and the choice and combinations of specifications available is greater, using the explicit equation-oriented approach, Shacham et. al. (6). The setting up of the process model for use with an optimiser has been described elsewhere, e.g. Hutchinson et. al. (7).

Nonlinear Programming Methods

The general nonlinear programming (NLP) problem can be written as

$$\text{minimise} \quad f(\mathbf{x}) \tag{1}$$

$$\text{subject to } l_i \le g_i(\mathbf{x}) \le u_i \qquad i=1,m$$

$$l_{m+i} \le x_i \le u_{m+i} \qquad i=1,n$$

where $\mathbf{x}$ is an n-component vector with components xi, and represents the stream variables, etc. l_i and u_i are the specified lower and upper bounds (which are the same for equalities). $\mathbf{g}$, the constraint functions, can be linear or nonlinear. Inequalities can be converted to equalities by introducing slack variables to $\mathbf{x}$.

There are several algorithms available to solve this type of problem, (8). Successive quadratic programming (SQP), as its name implies, solves a sequence of quadratic (sub-)programs (QPs), i.e. quadratic objective, linear constraints. These subproblems are approximations about the current point. Variations of the SQP algorithm have been applied to several types of flowsheeting systems, e.g. Hutchison et. al. (7). An alternative algorithm used in process optimisation is MINOS (9), which successively solves a subproblem in the independent variable subspace.

STRATEGIES FOR DESALINATION OPTIMISATION

MSF Process Model

Equation oriented form of MSF process model. For each stage, n=1,N;
Boiling point rise

$$0 = T_{Bno} - T_{Dno} - fn\ (T_{Dno}, X_{no}) \tag{2}$$

Flashing brine energy balance

$$0 = F_{Bni}.H_{Bni} - F_{Bno}.H_{Bno} - (D_{no}-D_{ni})\ H_{Dno} \tag{3}$$

Overall enthalpy balance

$$0 = F_{Bni}.H_{Bni} - F_{Bno}.H_{Bno} + F_{Cn}(H_{Cni}-H_{Cno}) + D_{ni}.H_{Dni} - D_{no}.H_{Dno} \tag{4}$$

Heat transfer equation

$$0 = q - \frac{U.A.(T_{Cno}-T_{Cni})}{\ln\left[\frac{(T_{Dno}-T_{Cni})}{(T_{Dno}-T_{Cno})}\right]} \tag{5}$$

where,

$$q = F_C(H_{Cno} - H_{Cni}) \tag{6}$$

or,

$$q = (D_{no} - D_{ni}).\ L_{Dno} \tag{7}$$

Distillate mass balance

$$0 = B_{ni} - B_{no} - D_{no} \tag{8}$$

The model is set up to exploit the special features of a desalination process. This has only two main components, water and salts, though more sophisticated studies may distinguish the different salts present. Salt is only present in the brine streams, so only in these streams do salt variables exist. The salt flow only changes if the brine streams are split or mixed, and a new salt variable is only created in these types of operations.

Types of Objective

There are a variety of objectives in the optimisation of a process, and it is likely that for a flexible design, the design parameters would give good, if not optimal, results to several objectives, implying a robust design. The most obvious objectives at the design stage are total overall profit or total overall cost. If the product flowrate is specified, total overall cost is the most useful. Here total overall cost is taken to mean total operating costs and total annualised capital costs.

Capital Costs

Capital costs can be divided into two parts; the cost associated with the size of each unit and that associated with the number of units. The costs associated with the size of a unit a function of the process variables, for example, the larger the flow through a unit, the larger its size. These relationships are not necessarily smooth but can often be approximated as such, over short ranges. These costs can usually be optimised using the NLP techniques under discussion. The costs associated with the number of units in the process cannot be optimised using NLP because it is difficult, even impossible to smooth these cost functions. In this study only fixed flowsheets will be considered.

Objective Functions for Desalination Plants

For an initial design study, the objective could be reduced to considering only the cost of heat transfer areas, steam, and feed processing, (2). Other costs, particularly those of the MSF unit shell are relatively fixed for a given size of plant.

A typical simplified form of the design objective might be;

$$\text{Minimise } c_1F^a + c_2A_H^{\ b} + \sum_1^N c_nA_n^{\ c} + c_4S^d \tag{9}$$

(subject to process heat and mass balances, and process constraints)

where; c_1, c_2, c_3 & c_4 are cost factors; F is the brine feed rate; A_H is the heat transfer area in the steam heater; A_n is the condensing heat transfer area in stage n; and S is the steam flow in heater.

c_1 is the unit cost of sea water, including factors its pretreatment, pumping, etc. c_2 is the product of the unit cost of the heat trasfer area and the capital recovery fraction. c_3 is the product of the unit cost of the condensing heat transfer area and the capital recovery fraction. c_4 is the unit cost of steam.

PARTICULAR PROBLEMS WITH THE MSF PROCESS

Complexity of the objective

The final form of the objective in equation (9), has quite a simple form. A simplified total cost objective could be linear. However, this disguises quite a complex problem and simple linear programming techniques are not appropriate. The relationships between heat transfer area, for both MSF units and steam heaters, and process temperatures are nonlinear though well defined if logarithmic mean temperature differences (LMTD) are assumed. Qualitatively, the optimisation could be said to centre around the brine heater-feed temperature, which is dependent on the operation of the flash unit, give that the brine heater exit temperature is closely bounded by process conditions. The higher this feed temperature is, the smaller the heater; the lower it is, the smaller the MSF unit.

The inclusion of the explicit area term (for either MSF or heater), means that one of the constraints is the nonlinear heat transfer equation (5), with either (6) or (7).

Most optimisation or extreme seeking algorithm are more stable if nonlinearities are in the objective function rather than the constraints. Combining (5) and (6) and rearranging gives;

$$A = \frac{F_1 (H_{1o} - H_{1i})}{U (T_{1o} - T_{1i})} . \ln\left[\frac{(T_v - T_{1i})}{(T_v - T_{1o})}\right] \quad (10)$$

$$\simeq \frac{F_1 . C_{pl}}{U} . \ln\left[\frac{(T_v - T_{1i})}{(T_v - T_{1o})}\right] \quad (11)$$

which can be substituted into the objective, equation (9), for both A_H and A_n.

Scaling considerations

Few optimisation or extreme seeking problems can be solved effectively as they stand. Often they require some form of scaling, the importance of this has been highlighted by Biegler & Hughes (10). The more complex version of the objective can be analysed to improve the scaling of the objective, and of the variables.

To simplifiy the analysis sightly, some elements of (11) can be assumed to be constant, particularly the overall heat transfer coefficient and the average specific heat. Their value depends on the units of flow and energy used.

The derivatives of (11) with respect to the temperatures are;

$$\frac{\delta A}{\delta T_v} = \frac{(T_{1i} - T_{1o}) F_1 . C_{pl}}{(T_v \quad T_{1i})(T^v - T_{1o}) U} \quad (12)$$

$$\frac{\delta A}{\delta T_{1i}} = \frac{- F_1 . C_{pl}}{(T_v - T_{1i}) U} \quad (13)$$

$$\frac{\delta A}{\delta T_{1o}} = \frac{F_1 . C_{pl}}{(T_{vap} - T_{1o}) U} \quad (14)$$

Points to note about these derivatives for the MSF unit are that;

(a) The flow of 'coolant' liquid through each flash stage and the heater will be the same, (see Figure 1), and it is usually of a similar magnitude to the feed rate.

(b) Each coolant temperature usually appears twice in the objective, as an output from one unit and an input to the next.

(c) The technology and economics of MSF process are such that very small or very large temperature changes across a stage are unlikely.

The overall result is that there are large differences between the different partial derivatives of the objective function, and a wide range of variable values. This results in a rather poor performance of NLP techniques, which will be illustrated later.

Given the importance of well scaled variables on NLP algorithms, several scaling strategies will be discussed. The flow and temperature variables have different characteristics and will be treated separately.

Flow Scaling

The unscaled flow variables tend to be very large, typically ranging between 4000 tonne/hr for the feed rate, to 20 tonne/hr for the smallest distillate flowrate. The derivatives with respect to the flowrates also vary, those associated with the coolant flows are very small, while the value of the steam derivative is large, due to its relatively high cost. Any scaling procedure for the flow variables has to take account of these points.

There are two possible approaches. The ratios between flowrates and heat transfer areas are dependent only on the temperature profile of the plant. Thus the process can be optimised for a unit feed or product, with flowrates and heat transfer areas being suitably transformed afterwards. However, this will tend to produce some very small variables, and the transformed problem will require scaling itself.

Alternatively, the units for flowrate and energy can be changed to give reasonable values, e.g. from kg to tonne, kJ to MJ. Changing both types of units means that the physical properties are unaffected. Only the overall heat transfer coefficients need be altered, and heat transfer area terms are unaffected. From equations (10)-(14), it can be seen that change the flow and energy units will leave the temperature gradients unaffected as both flow and U are changed, but the coolant flow derivative will increase when the value of U decreases and vice-versa.

The size of the cost factors must also be taken into consideration in any alteration of the units of flow rates. The high cost of steam relative to other costs means that as the units are changed, the value of the cost factor of steam will corresponding increase. Using the scale factors likely to be encountered, this affect does not seem to be detrimental to finding the solution.

Temperature Scaling

Temperatures in desalination plants are usually between 380 K and 400 K. If temperature variables were transformed so that they were between the recommended limits of 1 and 10, (Gill et. al. (8)), the size of the derivatives with respect to the temperatures would increase. Also, the MSF unit is inherently ill-conditioned, in that very small changes of stream temperatures produce much larger changes in stream flowrates. Transforming temperature variables so they are smaller, will increase this ill-conditioning in the constraints.

As the maximum range of the temperture variables is fixed quite tightly by the nature of the desalination process, a second scaling method should be considered. The variables can be transformed, (8);

$$x = d.y + c \qquad (15)$$

where x is the original variable; y, the transformed variable; and, c and d, constants. Given the constraints on the temperatures the worst estimates of d and c would be about 300.0 and 10.0 respectively, producing transformed variables approximately in the range 1 to 10. Experience suggests though, that given a better estimate of c, such as one based on the the initial estimate of the variable, d can be reduced.

EXAMPLE SOLUTION

The benificial affects of the analysis of the problem can be illustrated by a small example comprising a brine heater and the first two stages, (the hotest stages) of a MSF unit, Figure 2 and Table 2. The problem was optimised using a SQP algorithm, (11). To ensure that this artificial problem gave similar results to one of a larger size, the exit brine temperature had to be constrained, hence the rather unusual reject conditions.

To illustrate how the calculation proceeds, Figure 3 shows the value of the objective at each major iteration of the solution procedure, for each type of problem scaling. Two sets of costs were used, as set out in Table 1. The graphs illustrate that scaling the problem, either just the flow variables or all the variables, improves the rate of convergence to the true optimum.

The simple 2-stage flash with heater problem illustrated some of the details of the strategy, however, the variable interactions are naturally much less than in a full-sized plant. A typical plant, as illustrated in Figure 1. also shows the benifits of scaling the original problem. Unlike previous, stagewise optimisation methods, (2)-(4), the simultaneous approach allows the exit conditions to be left free. Particularly, the cooling water flow, and the reject temperatures may be simply constrained instead of fixed. In this case the recycle ratio has been set to simplify the problem, but this may be left as a free variable (12). Other constraints are; steam temperature, 400 K; feed temperature, 300 K; maximum brine temperature, 393 K; and maximum reject temperature, 310 K.

The graphs of Figure 4 illustate clearly the more rapid solution of the scaled problems over the unscaled problem. The greater difference in convergence between the problem with only scaled flow variables and the completely scaled problem is is not great, however, due to a poor starting point. The optimal solutions are presented in Table 3.

MINOS

Recent work (12) has shown that the problem formulation and scaling suggested here for desalination processes being optimised be the SQP approach is also valid for the MINOS algorithm.

CONCLUSION

There are considerable advantages for desalination optimisation of simultaneous methods, if they can be used efficiently. These methods can be applied to desalination, though some of the features of of the process can make this difficult. Scaling methods which exploit characteristics of the process variables enhance the performance of nonlinear programming methods on desalination design problems.

REFERENCES

(1) Helal, A.M. et. al., Comput. Chem. Eng., **10**(4), 327-342, (1986).
(2) Coleman, A.K., Desalination, **9**, 315-331, (1971).
(3) Itahara, S. & Stiel, L., Desalination, **4**, 248-257, (1968).
(4) Beamer, J.H. & Wilde, D.J., Desalination, **9**, 259-275, (1971).
(5) Giannuzzi, L.F. et. al., Desalination, **39**, 437-446, (1981).
(6) Shacham et. al., Comput. Chem. Eng., **6**(2), 79-95, (1982)
(7) Hutchison, H.P. et. al., Comput. Chem. Eng., **10**(1), 19-29, (1986).
(8) Gill, P.E. et. al, "Process Optimisation", Academic Press, Florida,(1981).
(9) Murtagh, B. A. & Saunders, M.A., Math. Progam. St., **16**, 84-117, (1982).
(10) Biegler, L.T. & Hughes, R.R., AIChE J., **28**(6), 994-1002, Nov., (1982).
(11) Gill, P.E., et. al., Report SOL 83-12, Dept. Operations Research, Stanford Universiy, (1983).
(12) Pardoe, J., PhD., forthcomming.
(13) Aly, G., Desalination, **45**, 119-131, (1983).

SYMBOLS

A Heat transfer area
B Water flow in brine streams
C Coolant stream
D Distillate stream
F Total stream flow, or Feed
H Specific enthalpy
L Latent heat of vapourisation
N Total number of stages
S Steam flow
T Temperature
U Overall heat transfer coefficient
X Salt concentration

c_i Cost i
q Heat transfered

Subscripts

B Brine stream
C Coolant stream
D Distillate stream
H Heater
i Input to a unit/stage
l Liquid phase
n Stage number
o Output from a unit/stage
v Vapour phase

Table 1: Typical Deslination Costs

Variables	Actual Data	Cost/hr	Actual Data	Cost/hr
Steam	10$/t	10$/t	0.772$/t	0.772$/t
Area	$25200.A^{.778}\$$	$0.0874.A^{.778}\$$	3.153 $/m	3.98×10^{-4} $/m
Total Feed Cost	neglig.	neglig.	3.6×10^{-6} $/t	3.6×10^{-6} $/t
Plant Life	10 years		30 years	
Interest	15%		5%	
Availability	8000 hrs		330 days	
Reference	Aly (13)		Coleman (2)	

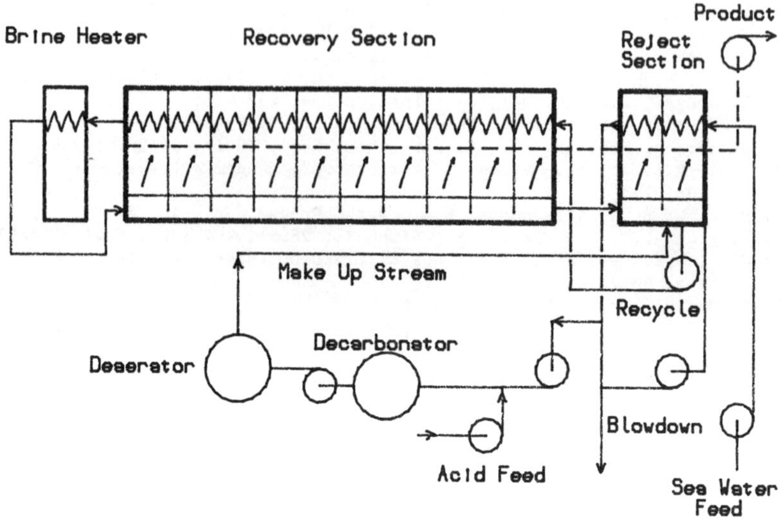

Figure 1. MSF Desalination Plant, With Recycle

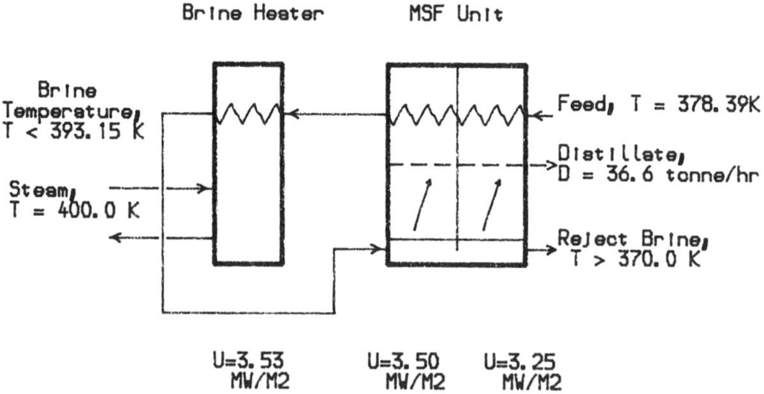

Figure 2. Example Problem

Table 2: Two Stage Flash Example Solution

Variables	Starting point	Scaled values	Optimum
Temperature K			
Feed/coolant input	380.0	1.0	378.39
Coolant interstage	381.0	1.0	381.48
Coolant ex MSF	384.0	1.0	384.59
Brine ex heater	393.0	1.0	393.15
Steam	400.0	1.0	400.00
Distillate interstage	390.0	1.0	389.16
Brine interstage	390.0	1.0	390.23
Distiallate ex MSF	387.0	1.0	386.10
Brine ex MSF/reject	388.0	1.0	370.00
Flows tonne/hr			
Feed - water	4000.0	40.00	3056.0
Feed - salt	200.0	2.00	169.3
Steam	70.0	0.70	51.76
Distillate interstage	20.0	0.20	18.46
Brine interstage	4000.0	40.00	3037.4
Distillate ex MSF	40.0	0.40	36.61
Brine ex MSF/reject	4000.0	40.00	3019.3

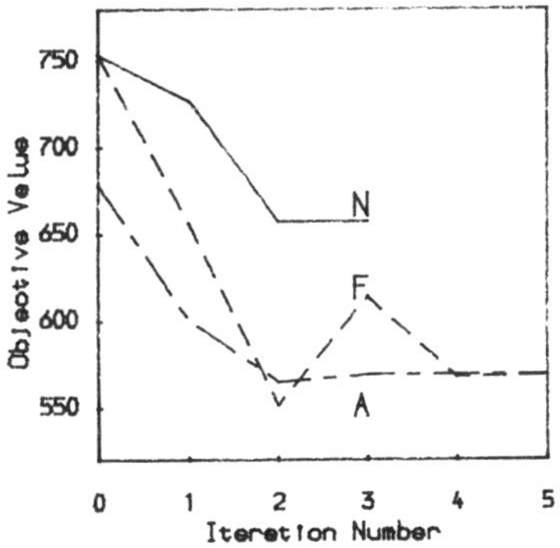

a. Costs From Aly (13)

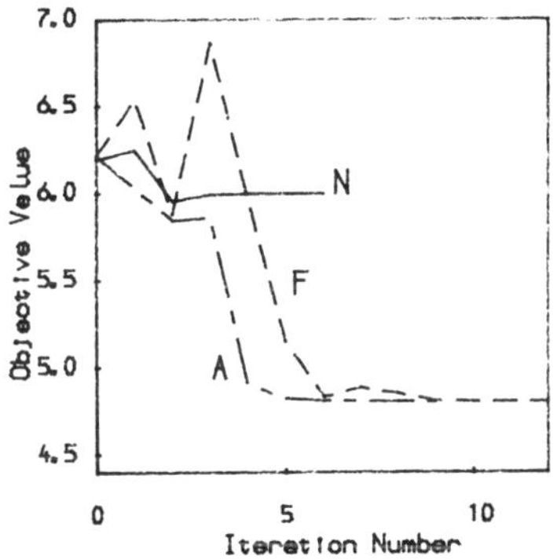

b. Costs From Coleman (2)

N=No Scaling, F=Flows Scaled, A=All Variables Scaled

Figure 3: Objective vs Iteration Number For The Two Stage Flash Example

Table 3: Optimal Solution of Ten Stage Example

Process Parameters	Starting Point	Aly (13) Optimum	Coleman (2) Optimum
Product tonne/hr	300.	500	500
Steam tonne/hr	64.7	27.3	26.7
Heat transfer Area m2			
Heater	1100	615	605
Recovery stage 1	450	515	819
2	450	737	834
3	450	742	837
4	450	747	842
5	450	751	848
6	450	757	854
7	450	761	860
8	450	766	865
9	450	770	871
10	450	775	877
Rejection stage 1	750	4865	3970
2	750	4157	3482

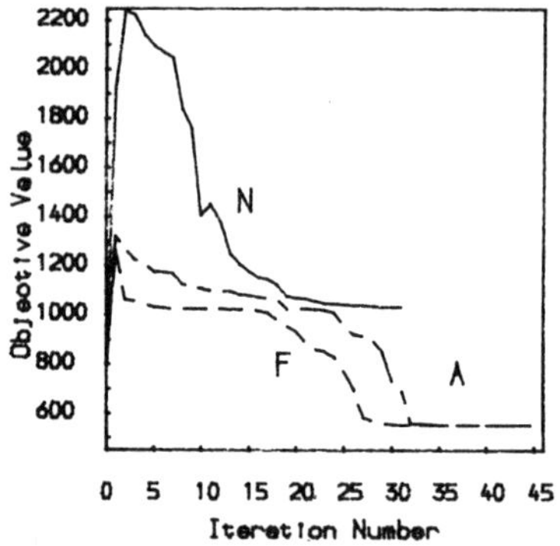

a. Costs From Aly (13)

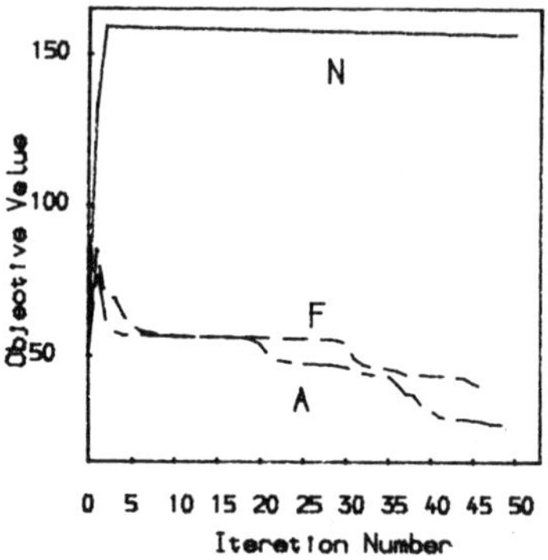

b. Costs From Coleman (2)

N=No Scaling, F=Flows Scaled, A=All Variables Scaled

Figure 4: Objective vs Iteration Number
For The Ten Stage Flash Problem

OPTIMAL UNIT SELECTION IN A COMBINED HEAT AND POWER STATION

D Foster*

The immediate strategic problem of power station operation concerns the adjustment of the continuous operating variables to meet the energy demands at minimum cost. Superimposed upon this is the background task of identifying that selection of equipment which can reduce this cost even further. The paper describes a dual system which addresses both of these problems and which is operational at ICI's Wilton power station on Teesside. Some difficulties experienced with equipment selection are discussed together with the ways that they have been overcome in practice.

INTRODUCTION

The scheduling of equipment within a power station to meet the varying energy demands of its customers is an established operational problem. An industrial power station whose context is shown in Figure 1 can pose a more difficult problem than most because several different types of exported energy may be involved and usually this leads to an increase in complexity within the station itself. Apart from this the option of electrical transfer to or from the National Grid under some agreed, but time-dependent tariff creates another major strategic dimension to the options that are available to reduce operating costs.

In its simplest form the objective of a scheduler can be viewed as an attempt to get the best steady-state performance out of the on-line equipment without recognition of future events. In mathematical terms this reduces to an application of non-linear constrained optimisation and is exactly the problem solved by the HAMBLE system at the ICI power station at Wilton on Teesside. Figure 2 depicts a schematic of the plant with each strategic option marked by a suitable character. In total the non-linear optimiser manipulates about 40 degrees of freedom to satisfy the demand equality constraints at minimum cost within the region defined by over one hundred inequality constraints. The programme runs automatically every 5 minutes and takes about 1 minute of CPU time on a VAX 11/780.

*** Engineering Department, Imperial Chemical Industries, Billingham, Cleveland**

Although the benefits of HAMBLE by real-time optimisation have been very worthwhile, amounting to a reduction in fuel and electrical import costs of about 1%, even greater savings have been achieved by using the package in off-line form to improve the choice of on-line equipment. A typical study may involve the definition and investigation of many dozens of sub-optimisations selected by the "expert" as likely contenders for the best configuration. Even so, important combinations can be missed and the search may be limited by time and clouded by conceptions of what is best. Hence the need for automatic optimal unit selection, the main subject of this paper. Formally this may be viewed as a mixed integer non-linear programming problem involving 21 discrete binary variables and 40 continuous variables.

At this time the problem addressed relates to a simple time period of known duration. A steady-state formulation is supposed to apply over this period but an initial switching cost is included as part of the objective function. Thus the optimisation problem is to achieve a best compromise between the accumulated steady-state benefits of a configuration over the duration and the cost of setting it up. This work is the forerunner of an attempt to solve a multi-time period version (Figure 3) of the same problem which will truly play off the cost today of having surplus equipment on-line against the benefits of having it available tomorrow. Using a forecast of the expected demands on the station together with a knowledge of the import/export tariff for each time-period the aim would be to compute the strategy which would meet these demands at the least overall cost. This cost could include the switching costs associated with a change in the on-line configuration. Expressed as a mixed non-linear programme this problem is many times larger then the one described in this paper but it is hoped that the accumulated experience will identify the steps which must be taken to produce useful solutions of this problem in a realistic time.

HAMBLE - ON-LINE OPTIMISATION

The on-line optimisation problem

The daily problem faced by operating personnel is to decide how best to meet the site energy demands at minimum cost within any internal restrictions on operation presented by the power station itself. Should the net in-house generation match the site electrical demand exactly or should the National Grid import/export facility be used? What is the best loading on available boilers and turbines remembering that efficency is a non-linear function of load and inlet temperature? Is it worth dumping steam to increase power generation? These are some of the questions which must be addressed to provide an effective solution. Figure 4 summarises the objective function, degrees of freedom and constraints in the system which together constitute the optimisation problem presented to the process operators.

The mathematical model

The mathematical model (Figure 5) which forms the basis for the optimisation is a detailed heat, work and mass balance of the entire station. Given a set of values for the degrees of freedom of the system it will predict a complete energy flowsheet of the station including steam exports, in-house electrical generation and overall operating cost. Within the HAMBLE model each boiler, turbine and heat-exchanger has its own non-linear representation which has been matched to its physical counterpart. The unit models are based upon the underlying thermodynamic mechanisms and are linked together in an "equation-based" manner to describe the network in considerable detail. As an example Figure 6 illustrates what has been done to model a feed-water heating train to account for the interaction between the feed-water heating side and the secondary turbines. In total, the overall mathematical description consists of about 100 non-linear simultaneous equations which can be solved in about 30 seconds on a VAX 11/780 using a reasonable starting point.

Optimisation Algorithm

In the context of the HAMBLE optimisation these equations are posed as a set of equality constraints (g) to augment the normal set of constraints (h) as shown below. The internal model variables (x) now augment the decision variable set (u) to provide a very much larger optimisation problem. However, this equation-based approach does eliminate the numerical noise associated with the traditional approach of solving the model equations as an inner loop and avoids the problem of convergence failure for some values of the degrees of freedom. The mathematical statement of the problem in this form is

$$\min_{x,u} f(x,u) \qquad \text{(cost)}$$

$$\text{subject to} \quad \begin{aligned} gi(x,u) &= 0 \quad & i &= 1,..n \quad & &\text{(model equations)} \\ hj(x,u) &= 0 \quad & j &= 1,..3 \quad & &\text{(energy demands)} \\ hj(x,u) &> 0 \quad & j &= 4,..m \quad & &\text{(general constraints)} \end{aligned}$$

This can be solved directly by MJD Powell's 1977 Variable-Metric algorithm (ref 1) but a modification of this appropriate to systems with many equality constraints provides much faster execution times. In effect this modification applies a projection step to reduce the size of the problem before the solution of each major QP iteration within the Powell algorithm and takes into account the structure of the Jacobian matrix dgi/dxj i e - the structure of the model equations (Figure 7). With this approach it is possible to reap the formulation and numerical benefits of equation based modelling and at the same time achieve very fast execution times. Note that in the on-line system numerical methods have been used for providing optimiser gradient information. A faster run-time could have been achieved by using exact derivatives but the effort was not considered worthwhile in view of the already adequate performance for the purposes of an on-line advisory system.

The on-line system

At Wilton, because the demands on the power station change frequently and because they change by significant amounts, there is justification for optimising performance on a regular basis. Thus HAMBLE has been developed into an on-line system in which the optimiser runs continually. The system is shown in Fig 8. Through instrument links with the plant the current operating point can be assembled by the computer almost immediately. The HAMBLE model can then be used without the optimiser to predict the current energy demands. With knowledge of the demands only constraint limit data needs to be input before the optimiser can be run and the results compared directly with plant performance. Another feature of the system available to the process operators is the "what-if" version of HAMBLE which allows hypothetical scenarios of operation to be set up, run and analysed very easily. The main use of what-if has been to improve the choice of on- line equipment needed to meet a given regime of energy demands. It is this function that is performed automatically by the "unit-select" algorithm described next.

AUTOMATIC UNIT SELECTION

The discrete optimisation problem

An examination of the discrete options available in the power station under study indicate that the range of flowsheet possibilities can be described by 21 binary decision variables, shown in Figure 9. Most of these correspond to on/off states for equipment but some indicate alternative sources for feed-water heating steam from the secondary turbines. It is thus possible to view the process of optimal unit selection as finding that set of values (0 or 1) for these binary variables which cause the objective function of the continuous optimisation to achieve its least possible value overall. Clearly enumeration and evaluation of all the possibilities is not practical or sensible and a more efficient approach to manipulating the 0-1 variables is called for. In reality, however, the choice of combinations is not as extensive as it may seem. An "expert" plant operator trying to solve the problem manually by repeated runs of HAMBLE knows this fact and just does not try certain combinations. Special sets of decisions are not allowed for logistic reasons, rules of thumb tell him whether he has enough boilers on line to generate approximately the required steam and the fuel cost and availability may dictate particular strategies on fuel. Even so the number of remaining configurations is too large to evaluate exhaustively and a methodical search technique is needed to reduce the number of alternatives that need to be tested. Branch-and-bound algorithms are one such approach to intelligent enumeration and can provide an efficient method for solving combinational problems of the type considered here.

Branch-and-bound

The theory and implementation of branch and bound algorithms is well established so only a brief description relevant to the current use is needed. In essence the application of a branch-and-bound algorithm to a binary decision problem involves (implicitly) constructing a tree which, if completed, would generate every possible combination of the 0-1 variables at its lowest level. Fig 10 illustrates the tree for four 0-1 variables. Each level (depth) of the tree is characterised by an equivalent number of the 0-1 variables having been fixed at 0 or 1. If the fixed combination of

0-1 variables which define a node are allowable then the node is said to be feasible. The "value" of a feasible node is the result of a calculation which provides a lower bound on the objective function which can be achieved by searching deeper from that node. The two branches from a viable node to its lower left and right "sons" correspond to fixing some chosen variable at 0 (left) or 1 (right). As the tree is searched by the algorithm the "best solution so far" is recorded and used with the node feasibility and value to eliminate large sections of the tree from the search. In the worst case branch-and-bound algorithms can enumerate all combinations of the binary decision variables but the visitation of a moderate number of nodes is the more likely outcome. Clearly a good lower bound calculation is necessary to make the search efficient. One feature which can improve the search dramatically involves branching from the node only if the value of the node betters the best solution so far by at least a certain small amount. This avoids the algorithm identifying configurations which are similar in cost. Another simple feature which can improve the search involves using a priority scheme at each node to select the branching variable.

The lower bound calculation

To provide the lower bound calculation continuous analogues of the 0-1 decision variables can be embedded in the existing HAMBLE optimisation in such a way that they provide a continuously differentiable path between the 0 state and the 1 state. The global minimum of this extended problem with the additional independent variables will clearly provide a lower bound on the remaining free options. Unfortunately if the embedding is done without thought the extended problem can be difficult to solve and be non-convex even though the original smaller problem was convex. Although it is possible to do something about non-convexity introduced in this way it is worth taking the trouble to try to build-up in the 0-1 variables in a way which avoids this difficulty from the start. Even so a larger and more complex optimisation problem than a simple HAMBLE run is involved and it is worth spending effort on refining this calculation so that an unprofitable or infeasible node is identified as quickly as possible. This has been done in three ways. Firstly, a hierarchy of preprocessing detects likely infeasibility before even attempting to solve a non-linear programme. Secondly, the non-linear optimisation is based on a simplified version of the HAMBLE model which is derived automatically on request from the on-line system. Thirdly, the progress of the non-linear optimisation itself is monitored intermittently and only allowed to complete if it seems there is some chance of improving the best solution so far.

Preprocessing

Every configuration generated by the branch-and-bound procedure undergoes three levels of preprocessing. Each of these levels involves the use of linear programming to solve a linear simplification of the node calculation but to a different degree of refinement. The first level simply tests and applies the logical constraints that must be satisfied by the 0-1 decision variables if the configuration is to be allowed. No objective function is required because the LP simply detects feasibility. If as a consequence of applying the constraints some decisions are "forced" then this is used to simplify the problem even further. A restriction that falls into this category might require at least one of two specific machines to be on-line at the same time, for instance. Another might be a set of constraints which brings several identical boilers on in a specific order to avoid evaluating what are effectively duplicate configurations. The next stage, the "coarse linear balance" is equivalent to the rules of thumb that might be applied by an experienced operator. The rows of this LP provide very rough balances on the steam mains to ensure that the available equipment can meet the demands within certain loose tolerances. Individual items of equipment are not considered and as before the LP only needs to detect feasibility. The final stage involves generating an LP version of the full Non-Linear Constrained Programming (NLCP) problem with each item of equipment represented individually. The models used are not point linearisations but pregenerated least-squares fits over the full range. This LP can be used to detect infeasibility at a more refined level or be provided with an objective function for optimisation. In the optimisation node the result can be used as a starting point for the NLCP calculation or, alternatively, the NLCP calculation can be skipped and the results of the LP used for the node calculation. This gives the opportunity to compare the results of using a linear and non-linear model for solving the unit selection problem.

The simplified model

Although the original HAMBLE optimisation provides an adequate response for on-line optimisation it has proved a too time-consuming calculation for the extended problem of the lower bound calculation. Investigation showed that it was possible to eliminate the pressure variables from the HAMBLE model without much loss and fit general quadratics in the rate and enthalpy variables to describe operation of the key units. The opportunity was taken to eliminate as much internal iteration as possible and introduce extra variables and constraints to decouple the problem into sections which interacted only in a limited way. This enabled optimiser gradient information to be provided algebraically rather than numerically with a consequential improvement in computational efficiency. The simplification provided a ten-fold decrease in run time for the usual on-line optimisation calculation.

The non-linear optimisation

One very noticeable feature of the problem which defines the lower bound calculation is the abundance of general linear constraints. These stem from the way in which the 0-1 decision variable are incorporated into the formulation. Some node calculations can involve up to 60 linear constraints and it is clearly worth taking advantage of this if it is possible. NPSOL (ref 2) is a sequential quadratic programming routine that provides this facility and gives very worthwhile improvements in computational efficiency over the routine used for the on-line calculation. Further, it provides the opportunity to temporarily terminate the optimisation after a chosen number of iterations, investigate current solution status, and continue the calculation with minimal overheads, if required. Thus the non-linear optimisation has been broken down into several sections with intermediate tests on progress to determine whether or not it is likely that the "best node so far" can be improved upon. If any of the tests fail the optimisation exits prematurely and no branching from the node is done. Figure 11 depicts the various stages of processing involved in the lower bound calculation. Another improvement in the performance of NPSOL was engineered by ensuring that redundant constraints were not present at each optimal solution. As equipment is driven off-line by the non-linear optimiser the associated constraints set which applies in the on-line state degenerates into a linearly dependent set. A step which improved the performance of the algorithm was to apply small status-dependent perturbations to chosen constraints that avoided the constraint becoming active in the off-line state but applied the full constraint in the on-line state.

OPERATING EXPERIENCE

Starting configurations

Originally the branch-and-bound search was started from cold with no estimate of what the optimal configuration might be. Under these circumstances it could take maybe twenty node evaluations before the first "integer" solution was discovered and the search tree began to be truncated Typically a complete run would involve forty node visits but often the solution was non-optimal as a result of a non-global cost being found at some nodes. A major improvement in search efficiency was achieved by providing an expert guess as the starting point and simply regarding the unit select as a refinement tool. Sometimes the expert guess proved right and the algorithms exited after evaluating the root node and maybe its two sons but sometimes there was something to go for and new solutions were revealed. After the modifications discussed in the next paragraph the need for a good starting configuration became more urgent because the number of nodes visited often exceeded one hundred.

Local optima

Although for the most part the underlying optimisation problem of HAMBLE is convex there have been difficulties with non-convexity in the lower bound calculation. Clearly if a node calculation does not produce a global minimum the result could be that a fruitful branch from the node is not pursued. Figure 12 illustrates the results of the lower bound calculation for a range of values of a chosen 0-1 variable. 12.1 is an ideal situation where the best value of the decision variable (z) is chosen naturally by the algorithm. 12.2 could cause a consequential branch on the variable and 12.3 could cause the lower bound calculation to find a non-global minimum. Where this was a problem (and it was in two specific areas) terms of the form $Kz(z-1)$ were added to the objective function to attempt to produce the situation 12.1. At the worst, if K was over-estimated the result would be occasional branching on the variable Z but at least the search tree would not be truncated prematurely. Another approach which helped was to modify the tree search so that nodes were only rejected if they were WORSE than the best solution so far by more than some specified amount. Clearly this generated a larger search but helped if the lower bound calculation identified local minima that were near the global minimum.

Example of a unit select calculation

Figure 13 shows some output from the calculation. Node 0 corresponds to the 'expert' starting point and node 1 the root node with all 0-1 variables (the columns) allowed to vary over the range (0,1). The column entries are the results of the node calculation for each 0-1 variable. A blank means "fixed at 0" and ON means "fixed at 1". "=*" means a free variable has been driven to its lower bound, "*=" means a variable has been driven to its upper bound and '**' means the final value for the 0-1 variable lies inside the range. The branching variables at each stage are selected dynamically from the '**' set according to the embedded priority scheme.

The "SOL" column indicates the integer solutions found with their associated and gradually improving cost. The last line is a re-evaluation of the optimal network discovered earlier.

Conclusions

In practice the solution of a mixed integer non-linear programming problem by using a branch-and-bound technique in conjunction with solving an extended version of the continuous non-linear optimisation is not as straightforward as the literature might indicate. However, the main difficulty seems to be not in reducing the number of non-linear optimisations performed by this approach but in ensuring that the way in which the extended problem is formulated avoids the inclusion of local minima which prematurely truncate the tree search. This can happen even if the original non-linear problem is convex. Since other approaches such as solving a sequence of related mixed integer linear programmes use the same extended problem formulation they too can suffer from the same failing and are likely to provide non-optimal solutions on real problems. In the author's opinion the experts proved surprisingly good at identifying optimal networks and it was only through sustained effort in analysing non-optimal solutions that this was matched eventually by the automatic system.

References

1. "A Fast Algorithm for Nonlinearly Constrained Optimisation Calculations", Powell, M J D

2. SOL Report - 1986, "Users guide to SOL/NPSOL", (Version 4.0), Gill, Murray, Saunders and Wright.
   Department of Operation's Research, Stanford University

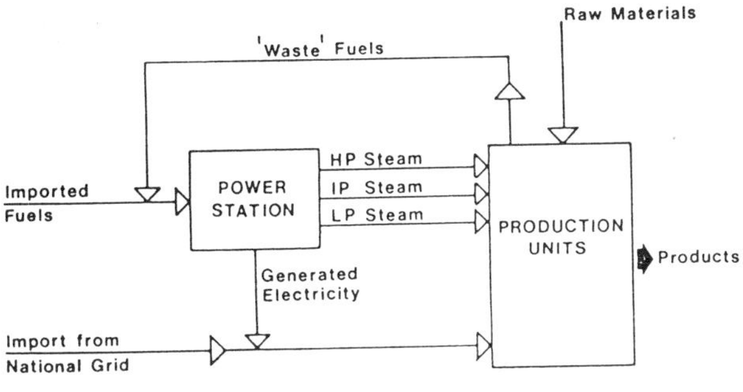

Fig.1 A CHP Station in the Industrial Context

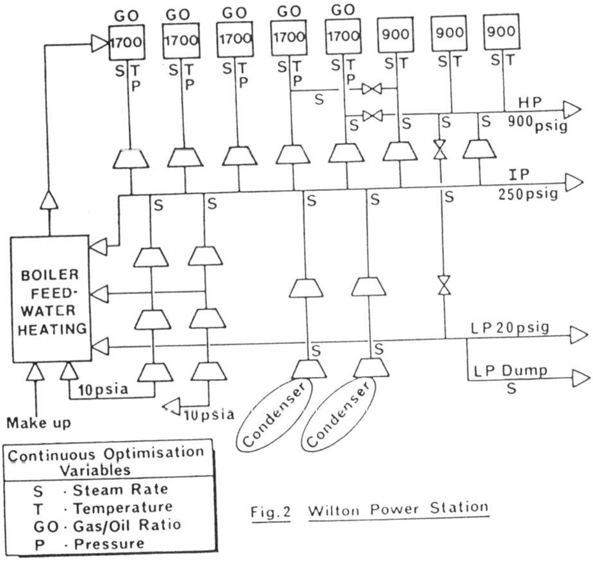

Fig.2 Wilton Power Station

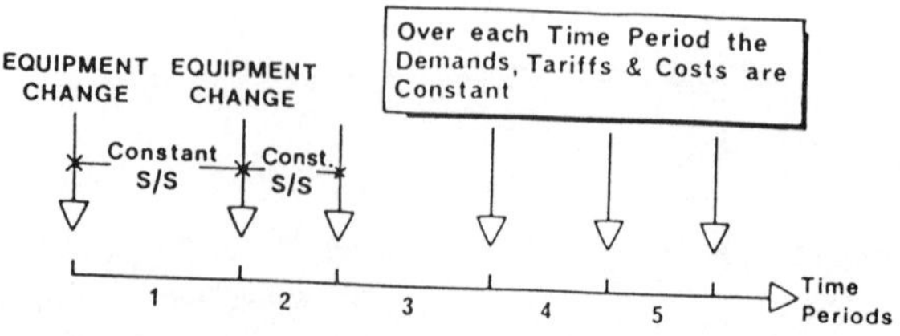

Fig.3 The Multi-Time Period Problem

OBJECTIVE	: MINIMISE OVERALL OPERATING COST	
DEGREES OF FREEDOM	: BOILER DELIVERY RATES	8
	BOILER DELIVERY TEMPERATURES	8
	BOILER DELIVERY PRESSURES	5
	BOILER FUEL OIL TOPPING-UP	5
	PRIMARY TURBINE MW	2
	SECONDARY TURBINE MW	4
	CONDENSER RATES	2
	LP STEAM DUMP RATE	1
	DESUPERHEATER RATES	4
CONSTRAINTS	: STEAM DEMANDS (EQUALITY)	3
	MIN & MAX ON ALL DEGREES OF FREEDOM	78
	MIN STOP-VALVE PRESSURE DROPS	15
	MAX PRIMARY EXHAUST TEMPERATURES	8
	MIN LP PASS-OUT TEMPERATURES	4

Fig.4 The Power Station Optimisation Problem

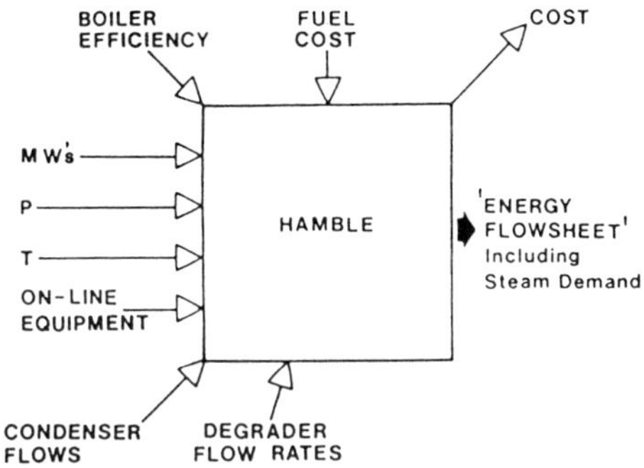

Fig.5 Heat Work and Mass Balance Steady State Simulation (HAMBLE)

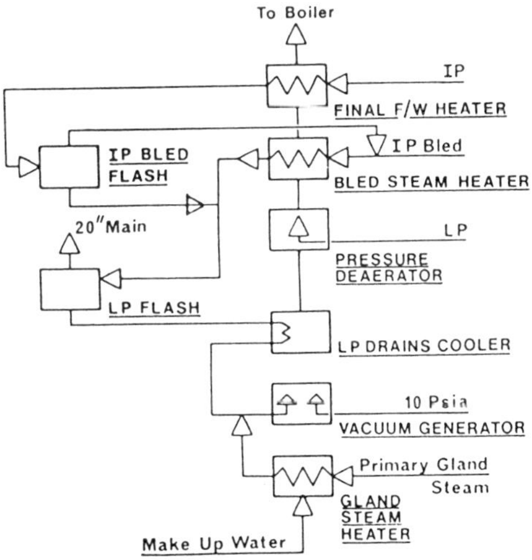

Fig.6 1700 Boiler Feed Heating Train

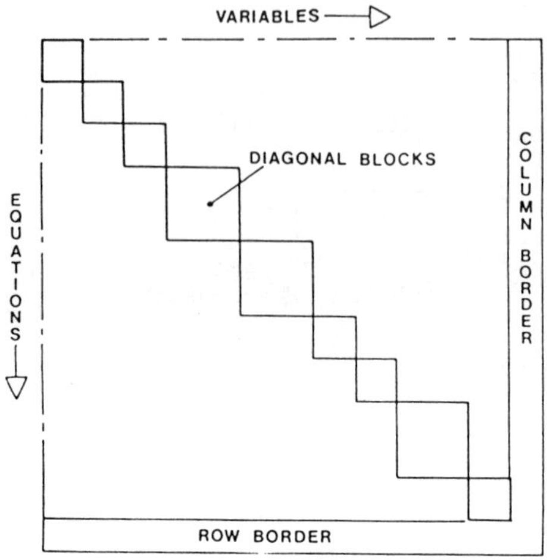

Fig.7 Structure of Hamble Equations (G)

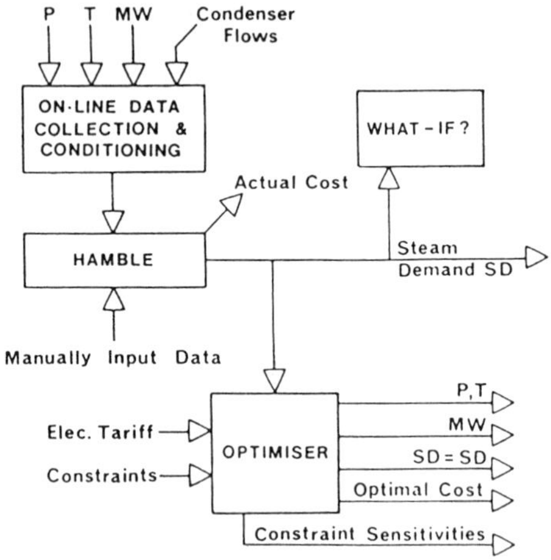

Fig.8 ON-LINE System

ON/OFF	– All Boilers	– 8
OIL OPTION ON/OFF	– 1700 * Boilers	– 5
ON/OFF	– 900 * Primaries	– 2
ON/OFF	– Secondaries	– 4
IP BLED LINKAGE	– Secs 3 & 4	– 2

Fig. 9 The Discrete Optimisation Variables

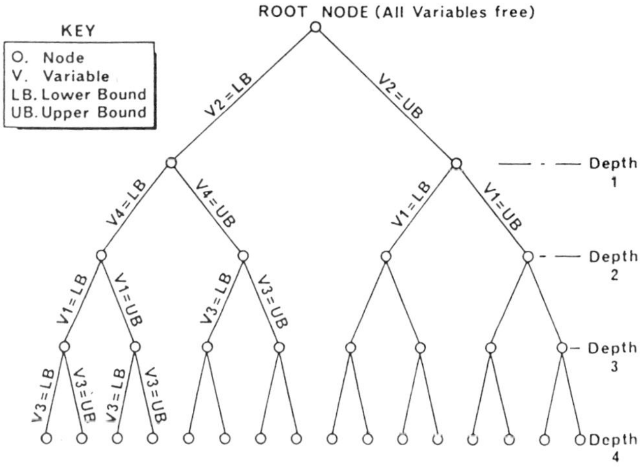

Fig. 10 A Branch-and-Bound Tree for 4 Variables

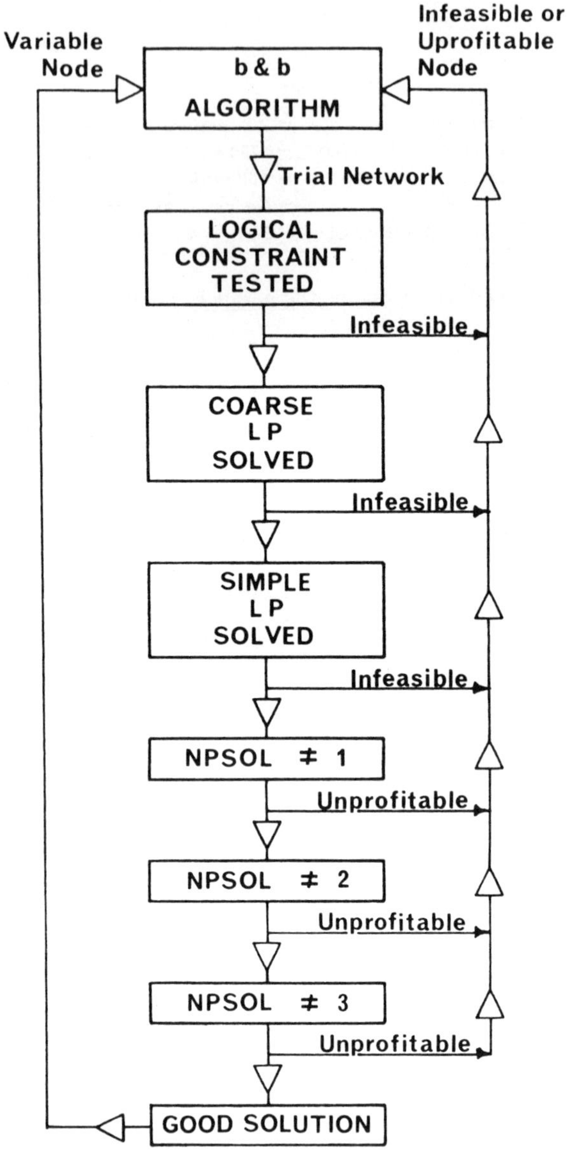

Fig.11 Progressing Hierarchy for Lower Bound Calculation

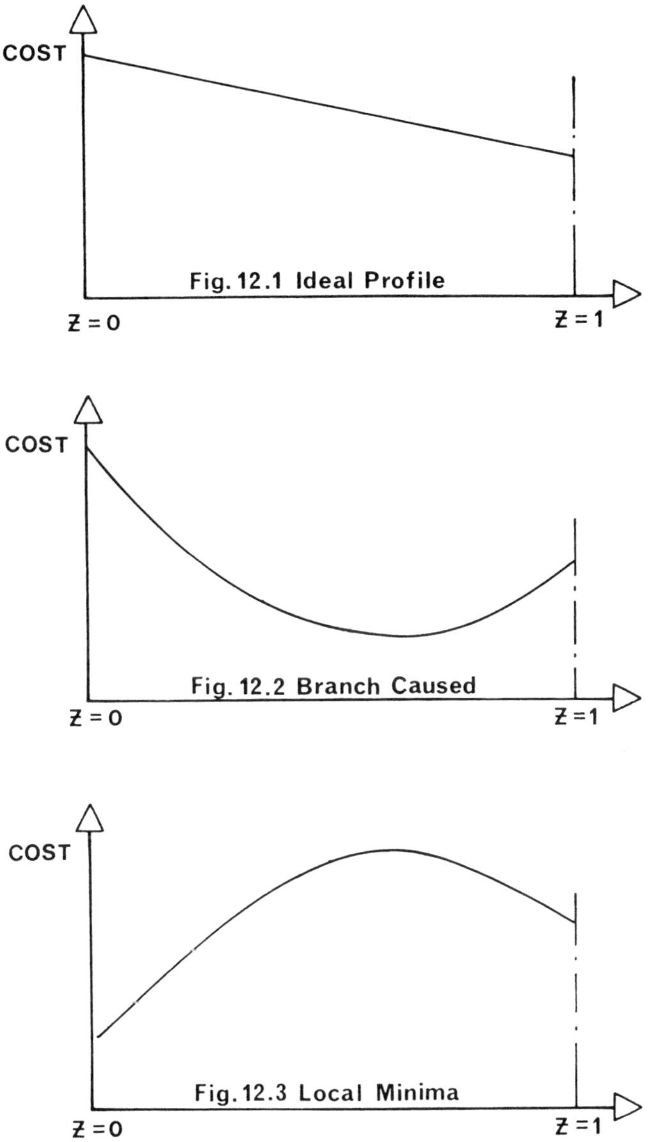

Fig. 12.1 Ideal Profile

Fig. 12.2 Branch Caused

Fig. 12.3 Local Minima

Fig. 12 Dependence of Cost on 0-1 Variable

BRANCH & BOUND TREE NODES

NODE	B1	B2	B3	B5	B6	B7	B8	B9	O1	O2	O3	O5	O6	O7	O8	O9	P1	P2	S1	S2	S3	S4	G1	C2	NG	SOL	FEAS	COST
0			ON			ON	ON	ON			ON			ON	ON	ON	ON		ON	ON		ON	ON	ON		I	1	0.518277E+04
1	=*	=*	**	*=		*=	*=	*=	=*	=*	*=	=*		=*	*=	*=	=*	=*	**	**	**	*=	**	**			1	0.456730E+04
2	=*	=*		*=		*=	*=	*=	=*	=*	=*	=*		=*	*=	*=	=*	=*	**	**	**	*=	**	**			1	0.462017E+04
3	=*	=*		*=		*=	*=	*=	=*	=*	=*	=*		=*	*=	*=	=*	=*	**	**		*=	*=	*=			1	0.467605E+04
4	=*	=*		<*		*=	*=	*=	=*	=*	=*	=*		=*	*=	*=	=*	=*	**			*=	*=	*=			1	0.517614E+04
5	=*	=*				*=	*=	*=	=*	=*	=*	=*		=*	*=	*=	=*	=*	**			*=	*=	*=			2	
6	=*	=*		ON		*=	*=	*=	=*	=*	=*	=*		=*	*=	*=	=*	=*	**			*=	*=	*=			1	0.529598E+04
7	=*	=*		*=		*=	*=	*=	=*	=*	=*	=*		=*	*=	*=	=*	=*	**	ON		*=	*=	*=			1	0.468542E+04
8	=*	=*		**		*=	*=	*=	=*	=*	=*	=*		=*	*=	*=	=*	=*		ON		*=	*=	*=			1	0.502954E+04
9	=*	=*				*=	*=	*=	=*	=*	=*	=*		=*	*=	*=	=*	=*		ON		*=	*=	*=			2	
10	=*	=*		ON		**	*=	*=	=*	=*	=*	=*		=*	*=	*=	=*	=*		ON		*=	*=	*=			1	0.505478E+04
11	=*	=*		ON			*=	*=	=*	=*	=*	=*		=*	*=	*=	=*	=*		ON		*=	*=	*=			2	
12	=*	=*		ON		ON	**	*=	=*	=*	=*	=*		=*	*=	*=	=*	=*		ON		*=	*=	*=			1	0.506129E+04
13	=*	=*		ON		ON		*=	=*	=*	=*	=*		=*	*=	*=	=*	=*		ON		*=	*=	*=			2	
14	=*	=*		ON		ON	ON	*=	=*	=*	=*	*=		=*	**	=*	=*	=*		ON		*=	*=	*=			1	0.507471E+04
15	=*	=*		ON		ON	ON	*=	=*	=*	=*	*=		**		**	=*	=*		ON		*=	*=	*=			1	0.507761E+04
16	=*	=*		ON		ON	ON	*=	=*	=*	=*	*=				**	=*	=*		ON		*=	*=	*=			1	0.507765E+04
17	=*	=*		ON		ON	ON	*=	=*	=*	=*	*=					=*	=*		ON		*=	*=	*=		I	1	0.521839E+04
18	=*	=*		ON		ON	ON	*=	=*	=*	=*	*=				ON	=*	=*		ON		*=	*=	*=		I	1	0.508679E+04
19	=*	=*		ON		ON	ON	*=	=*	=*	=*	**		ON		**	=*	=*		ON		*=	*=	*=			1	0.509006E+04
20	=*	=*		ON		ON	ON	*=	=*	=*	=*	**		=*	ON	=*	=*	=*		ON		*=	*=	*=			1	0.507583E+04
21	=*	=*		ON		ON	ON	*=	=*	=*	=*			**	ON	=*	=*	=*		ON		*=	*=	*=			1	0.507780E+04
22	=*	=*		ON		ON	ON	*=	=*	=*	=*				ON	**	=*	=*		ON		*=	*=	*=			1	0.507794E+04
23	=*	=*		ON		ON	ON	*=	=*	=*	=*				ON		=*	=*		ON		*=	*=	*=		I	1	0.520681E+04
24	=*	=*		ON		ON	ON	*=	=*	=*	=*				ON	ON	=*	=*		ON		*=	*=	*=		I	1	0.509444E+04
25	=*	=*		ON		ON	ON	*=	=*	=*	=*			ON	ON	=*	=*	=*		ON		*=	*=	*=		I	1	0.508598E+04
26	=*	=*		ON		ON	ON	*=	=*	=*	=*	ON		=*	ON	=*	=*	=*		ON		*=	*=	*=		I	1	0.507583E+04
27	=*	=*		<*		*=	*=	*=	=*	=*	=*	=*		=*	*=	*=	=*	=*	ON	ON		*=	*=	*=			1	0.473468E+04
28	=*	=*				*=	*=	*=	=*	=*	=*	=*		=*	*=	*=	=*	=*	ON	ON		*=	*=	*=			2	
29	=*	=*		ON		*=	*=	*=	=*	=*	=*	=*		=*	*=	**	=*	=*	ON	ON		*=	*=	*=			1	0.473636E+04
30	=*	=*		ON		*=	*=	*=	=*	=*	=*	**		=*	*=		=*	=*	ON	ON		*=	*=	*=			1	0.483362E+04
31	=*	=*		ON		*=	*=	*=	=*	=*	=*			**	*=		=*	=*	ON	ON		*=	*=	*=			1	0.494585E+04
32	=*	=*		ON		**	*=	*=	=*	=*	=*				*=		=*	=*	ON	ON		*=	*=	*=			1	0.511792E+04
33	=*	=*		ON		*=	*=	*=	=*	=*	=*			ON	*=		=*	=*	ON	ON		*=	*=	*=		I	1	0.498507E+04
34	=*	=*		ON		*=	*=	*=	=*	=*	=*	ON		=*	*=		=*	=*	ON	ON		*=	*=	*=		I	1	0.494538E+04
35	=*	=*		ON		**	*=	*=	=*	=*	=*	=*		=*	*=	ON	=*	=*	ON	ON		*=	*=	*=			1	0.473660E+04
36	=*	=*		ON			*=	*=	=*	=*	=*	=*		=*	*=	ON	=*	=*	ON	ON		*=	*=	*=			2	
37	=*	=*		ON		ON	**	*=	=*	=*	=*	=*		=*	*=	ON	=*	=*	ON	ON		*=	*=	*=			1	0.473725E+04
38	=*	=*		ON		ON		*=	=*	=*	=*	=*		=*	*=	ON	=*	=*	ON	ON		*=	*=	*=			2	
39	=*	=*		ON		ON	ON	*=	=*	=*	=*	=*		=*	**	ON	=*	=*	ON	ON		*=	*=	*=			1	0.473915E+04
40	=*	=*		ON		ON	ON	*=	=*	=*	=*	**		=*		ON	=*	=*	ON	ON		*=	*=	*=			1	0.483088E+04
41	=*	=*		ON		ON	ON	*=	=*	=*	=*			**		ON	=*	=*	ON	ON		*=	*=	*=			1	0.494225E+04
42	=*	=*		ON		ON	ON	*=	=*	=*	=*	ON		=*		ON	=*	=*	ON	ON		*=	*=	*=		I	1	0.494304E+04
43	=*	=*		ON		ON	ON	*=	=*	=*	=*	*=		=*	ON	ON	=*	=*	ON	ON		*=	*=	*=		I	1	0.473931E+04
44	=*	=*		*=		*=	*=	*=	=*	=*	=*	=*		=*	*=	*=	=*	=*	**	**	ON	**	**	**			1	0.473166E+04
45	=*	=*		*=		*=	*=	**	=*	=*	=*	=*		=*	*=	*=	=*	=*	**	*=	ON		=*	=*			1	0.490591E+04
46	=*	=*		*=		*=	*=	*=	=*	=*	=*	=*		=*	*=	*=	=*	=*	**	**	ON	ON	**	**			1	0.475961E+04
47	=*	=*	ON	*=		*=	*=	*=	=*	=*	=*	=*		=*	*=	*=	=*	=*	**	**	*=	*=	**	**			3	
48				ON		ON	ON	ON				ON			ON	ON			ON	ON		ON	ON	ON		I	1	0.473931E+04

Fig. 13 Branch and Bound Tree Nodes

PROCESS OPTIMISATION - DESIGN AND OPERATION: A PRACTICAL EXPERIENCE

D.H.Lott, M.L.Preston and D.Foster *

ICI has a long established involvement in the development and application of engineering software for the design and operation of chemical processes. 1967 saw the first real practical milestones of on-line process optimisation, as applied to one of its Olefines plants, and flowsheeting techniques, eg Network 67, applied to process design optimisation. This paper, twenty years on, describes in detail the current ICI approaches to process optimisation, firstly in the design stage using FLOWPACK II, and secondly during plant operation by the application of custom-built equation based simulations known as HAMBLE models.

INTRODUCTION

Process optimisation falls into two main activities:-

1. Optimising the design of new plants.

2. Optimising the operation of existing plants.

A third activity, evaluation of modifications to existing plant to increase the potential for making money is really a subset of the design activity.

Both the above problem areas have received attention within ICI since technical computing became a reality in the 1960's. By 1967 real benefits were being realised in the two areas. Rapidly evolving flowsheeting techniques and purpose built optimisation models were yielding significant savings in a number of new process designs (Ref 1).

In the north east of England a pioneering on-line optimiser was providing significant process optimisation benefits on one of ICI's Olefines plants. This proved to be the first in a long line of evolving and successful on-line optimisation systems within ICI.

In contrast the advent of ICI's general purpose flowsheeting packages FLOWPACK 1 and FLOWPACK II in the early 1970's initially removed the capability to efficiently optimise at the process design stage that had been present in the purpose built design models. Work on FLOWPACK II, initiated in 1982 and culminating in the current release described here, has allowed the incorporation of state-of-the-art optimisation techniques and their application to process flowsheets.

* ICI PLC, Engineering Department, Brunner House, Northwich.

OPTIMISING THE DESIGN OF NEW PLANTS

Here we are deciding on the plant items required to carry out a given process, how they should be connected together and what size they should be.

We are looking in the first instance at fundamentals such as whether the plant should be a batch or continuous. Then, if a reactor is required, what type, size and cost. As the reactor product can seldom be sold directly we must then look at a separation system. If there are un-reacted raw materials in the reactor product we will want to recover these and recycle them back to the beginning of the process. Choices of recycle system structure and methods of material recovery must be made.

As the process design develops in this way there are two types of decisions that must be made. First, the discrete decisions such as what operations are required? How should a particular processing step be carried out? Second, the trade-off decisions, how big should the purge be? In addition it may be necessary to investigate the trade-offs in several schemes in order to generate the information necessary to make the discrete decisions. ICI and its co-workers have detailed this approach elsewhere (Ref 2,3).

The ICI FLOWPACK II package (Ref 4) has been developed to handle this type of evolutionary design process.

FLOWPACK II

FLOWPACK II was launched in 1976 as the ICI steady state process simulation program after several years of flowsheet development. To the user, the executive falls into the sequential modular class (Ref 5). That is the unit operations exist as discrete modules and are evaluated in a sequential manner. Recycles are solved by identifying suitable tear streams and calculating through the flowsheet repeatedly with estimates of the torn streams until the calculated values are identical to their initial values. Each unit operation only needs to be capable of calculating the state of the output streams given the values of the input streams and the operating data. It is the job of the executive to present each unit operation with the necessary input data.

In common with other flowsheeting packages of this type FLOWPACK II exists as a pre-linked executable program with three basic phases.

First is the input program. Because the program is pre-linked containing a number of standard unit operations it is necessary to be able to create data driven models. That is, it must be possible to define flexibly the unit operations and structure of the flowsheet to be simulated.

Second, because the structure of the flowsheet is unknown until the input data has been read the order of solution cannot be embedded in the code. A network analysis phase is therefore required to sort out the most efficient calculation sequence.

Third, in the solution phase the unit operations must be evaluated in turn and the iterative calculations through the flowsheet carried out until convergence is achieved.

The great advantage of sequential modular flowsheeting packages is that they solve problems the way the engineer would solve the problem by hand. The engineer can readily understand what is going on and the cause of any problems is easily identified. Also, because the model is data driven, building a simulation model can be done in a sequence of convenient and natural steps where each step is a working simulation.

Although superficially a sequential modular system, FLOWPACK II does in fact fill a unique niche, part way between a sequential modular and an equation-based system (Ref 5). It has the ability to split up single unit operations into smaller sub-units connected together by internal streams, a facility known as sub-networking. The sub-networks can themselves be broken down, in principle as far as the individual equations defining the operation. Using this mechanism the executive includes only the minimum number of blocks necessary to determine balanced recycle streams, omitting all other parts of the unit operation from the iterative recycle calculations. This can greatly reduce the amount of computation necessary to solve the flowsheet and helps keep CPU times to a minimum. It is sub-networking that makes efficient optimisation a possibility.

Recycle Convergence

The default convergence algorithm used in FLOWPACK II is super linear. By calculating through the flowsheet initially in repeated substitution mode a matrix of derivatives (the inverse Jacobian) is generated. This is in fact a linear approximation to the complete process. The inverse Jacobian is then used to produce the next estimate of the recycle variables and the Jacobian is updated by a further pass through the flowsheet until convergence is achieved. Convergence using this approach is much faster than using the first order methods in most other flowsheeting packages and is similar to that used in simultaneous modular approaches (Ref 6).

Design

The traditional sequential modular programs compute in 'performance mode'. That is, for a given set of feed streams and operating data the flowrates and condition of all the streams in the process are calculated. This gives a useful but limited facility to the process engineer designing a new plant. He can obtain a converged heat and mass balance but cannot guarantee to achieve any specific conditions such as product purity or flowrate. It is therefore necessary to allow design constraints to be imposed on the flowsheet.

FLOWPACK II has been developed to accommodate this type of calculation. Any number of design constraints can be specified provided an equal number of process parameters are allowed to vary, thus maintaining the correct number degrees of freedom. In FLOWPACK II unit operations are defined using a MODEL instruction, design constraints are added using a CONTROL instruction. User defined names are given for the parameters that it is intended to vary and these are specified along with the design constraints to the CONTROL module. No other modifications need be made and the design calculation is a natural extension of the performance calculation.

One of the often quoted drawbacks of sequential modular packages is their inability to carry out design calculations efficiently. The normal method is to set up the design loop outside the normal performance loop. The flowsheet is then solved for an initial set of conditions in performance mode. The design loop checks to see if the constraints have been met, changes the variable values and rebalances the flowsheet until the design constraints are met. This can lead to excessive computing times because the performance of the flowsheet must be calculated several times on the way to finding the heat and mass balance that also meets the design constraints.

In FLOWPACK II design problems can be solved in two ways, the first as above being a nested or feasible path method, the other is the simultaneous or infeasible path method. When the simultaneous method for design problems is used the variable parameters, defined by the user, are added to the list of variables (flowrates, temperatures and pressures), defined by the recycle streams, and the design constraints are added as equality constraints to those defined by the tear streams. The whole problem is then solved in a single level of iteration. The choice between nested or simultaneous solution methods is made on grounds of computational efficiency.

Optimisation

Optimisation follows on naturally from performance and design calculations. If performance mode answers the question "what happens if?" and design mode "how is this done?", then optimisation answers the question "how is this done best?". In order to answer this question extra degrees of freedom and an objective function, to be maximised or minimised, must be specified to an OPTIMISE module.

As with design, optimisation in FLOWPACK II is implemented in two ways, feasible and infeasible path. For the feasible path mode the performance or design problem is solved for the initial conditions and the objective function evaluated. The parameters declared as the optimisation variables are then varied automatically and the flowsheet rebalanced until the desired maximum or minimum is achieved.

For the infeasible path method the whole flowsheeting problem is opened up to the Successive Quadratic Programming (SQP) algorithm of Powell (Ref 7). That is, the variables that define the component flowrates, temperatures and pressures in the selected recycle streams, plus the variables specified by the user, to define the design and optimisation problem, are all connected up to the optimisation routine. The optimisation problem is then solved subject to meeting the equality constraints, either implied by the necessity to balance the recycle streams, or to meet the specified design constraints (Ref 8). Inequality constraints can also be specified. Iterative calculations internal to unit operations or physical property routines are usually solved in a nested manner. However, the sub-networking facility does give the option for simultaneous solution without re-formulating the problem.

The approach is only viable using SQP for systems where the number of equality constraints remains small. When this becomes large the time taken to solve each QP becomes prohibitive unless the sparsity of the system is taken into account.

Equations

The objective function for optimisation is created in FLOWPACK II using the in-line EQUATION facility. This allows a Fortran type assignment to be made such that a user defined variable is set equal to some mathematical combination of other variables or stream variables. Values held in the vector representing a stream, such as the molar flowrate of a component or the temperature, can be referenced directly. Also, derived stream properties, such as total flowrate, density, bubble/dew points, enthalpy etc. can be referenced. The EQUATION facility affords great flexibility in flowsheet development. In addition to generating objective functions they can be used to calculate the input parameters to a unit operation based on the output from other units and/or streams.

For example the MULTI-SPLIT module will mix any number of streams and then split them into any number of output streams, with the fraction of each component to go into each output stream being specified. This module in combination with in-line EQUATIONS to calculate the split fractions can be used to represent many types of separation device. If the equations necessary to calculate the split fractions are simultaneous then these will be solved along with the rest of the flowsheet problem.

Functions

An extension of the in-line EQUATIONS facility is the FUNCTION facility. This is again very similar to a Fortran Statement Function. It allows equation structures that will be used several times to be set up and then used repeatedly by passing in different parameters. For example a FUNCTION could be defined to calculate Log Mean Temperature Difference (LMTD) around a heat exchanger. This could then be used several times simply by invoking the FUNCTION with different streams temperatures connected to it.

Case Study

A final feature of FLOWPACK II that is useful in optimisation studies is the CASE STUDY facility. With CASE STUDY a series of simulations can be set up within one run of FLOWPACK II. Any number of parameters can be defined as variables to be changed on a run to run basis. Their values can be set either explicitly or by setting a lower and an upper limit and specifying the number of runs to be carried out. Using CASE STUDY means that for a series of runs input and flowsheet analysis does not have to be rerun. Also, CPU time is reduced because each run can be set off from the previous converged solution. However, in optimisation studies the main benefit comes from the summary output facilities. These report the value of key user defined parameters, such as the objective function, as a function of the case study parameters.

The Use Of FLOWPACK II In Optimisation

The great benefit of carrying out process optimisation within FLOWPACK II is the way in which the model can be evolved in a natural progression with the engineer involved and in control at all stages. There should never be a time when the results of a simulation come as a great surprise. FLOWPACK II model generation is a building technique, the blocks needed to progress to the next stage are added to those of the previous one.

A typical scenario for building an optimisation model would be as follows.

1. Build up a performance mode model using the basic unit operations; mixers, splitters, reactors etc.

2. Solve the flowsheet in performance mode for a typical set of conditions.

3. Add any necessary design constraints using the CONTROL units and solve the design problem.

4. Use the in-line EQUATION facility to build up the desired objective function. This could be capital cost, operating cost, profit etc.

5. Use engineering judgement to decide on which parameters in the design may have an effect on the defined objective function and over which you have some flexibility to define the value.

6. Use the CASE STUDY facility to examine the effect of varying the parameters defined above on the objective function over a sensible range.

7. From the results of the studies in the previous step select the key variables.

8. Define the key variables to the OPTIMISE unit along with the ranges over which they can be varied. This will have been highlighted in the case study runs.

9. Run the optimisation problem and determine the optimal values for the selected key variables.

10. Carry out further case study runs in the area of the optimum to check its sensitivity to the key variables.

The objective function to be optimised can be any value that the engineer wishes to generate from the parameters in the flowsheet. The one most often used at the design stage is Economic Potential (Ref 2), this is defined as:-

EP = VALUE OF PRODUCTS - COST OF FEEDS -
COST OF UTILITIES - COST OF EQUIPMENT

In order for this expression to make sense all the costs must be expressed on the same basis and it is usually done on an annual one. The value of product streams, the costs of feeds and utilities are easily generated using in-line EQUATIONS. The total flowrate of a stream usually represented within the flowsheet in per hour terms is simply multiplied by the unit cost and the assumed number of operational hours per year. Calculation of the capital cost is a little more involved. For design optimisation to take place it must be possible to vary the size and hence the cost of the plant items. FLOWPACK II unit operation models have been modified, where necessary, to generate extra outputs of information, known as information streams, to facilitate this. For instance the short-cut distillation unit, which will carry out a distillation to specified recoveries, has addition output parameters for the number of theoretical plates and vapour rates. From these parameters the height and diameter of the column can be calculated and then by consideration of the operating pressure the cost calculated. The cost of heat exchangers can be calculated in a similar way. Calculate the log mean temperature difference, by use of either EQUATIONS or a FUNCTION and by combination with heat transfer coefficients calculate the area which in turn is used to calculate the cost.

The costing models used in design optimisation are usually of the form:-

COST = A*X**B

Where X is some characteristic parameter of the piece of equipment, such as the area of an heat exchanger or the volume of a reactor. The A and B coefficients for the correlation to give an annualized cost are determined off-line by consideration of the operating conditions and necessary specification of the plant item. They are then input into the simulation as user defined parameters. Once the approximate size and duty of an item is established the capital cost of the plant item at suitable spot points is generated using the ICI program NEWCOST. This is a factorial estimating package that calculates either, main plant item costs or, net installed costs, based on duty, size, material of construction etc. The spot costs generated in NEWCOST are then used to determine the A and B coefficients of the cost equation to be put into the FLOWPACK II model. In this way FLOWPACK II has available to it the variable cost of a plant item as a function of size and can carry out the trade-off optimisations between plant items ie. reactor size versus separation system size. Also, because raw material and utility costs are included in the objective function the raw material versus capital type trade-offs can also be optimised.

OPTIMISING THE OPERATION OF EXISTING PLANTS

Here the job is to optimise the economic performance of the plant during its lifetime against a background of a changing and unpredictable economic environment.

Steady-state, Dynamic And Discrete

At this time operational optimisation can be considered as falling into three categories:- steady-state, dynamic and discrete. The first is applicable to plants which are designed to run in a steady-state for long periods of time. Usually, these are large systems and the problem is to choose the best set of operating conditions from a wide continuous spectrum of possibilities. The majority of process optimisation problems fall into this category and they encompass the on-line systems developed by ICI for Olefines and Combined Heat and Power Stations. A more recent venture in this area has involved the amalgamation of large-scale steady-state optimisation with site-wide data gathering to provide a real-time based site optimisation facility. On the other hand, dynamic optimisation is relevant for processes like batch reactors where the changing state of the system is an important strategic element. Here the decision may be to derive an optimal temperature profile - a function rather than a point value and the objective may be to achieve minimum batch cycle time or perhaps minimum energy consumption. Experience to date has indicated that much can be saved in this area but progress is hampered by the absence of quantified kinetic mechanisms for the (usually very complex) reaction mechanisms of most batch processes. Finally, discrete optimisation or mixed integer non-linear programming concerns itself with choosing the best state for a system from a finite but large number of possibilities.

The operational problem of power station equipment selection is one area where ICI have implemented this type of optimisation but automatic flowsheet design at a realistic level could be amenable to the approach when better methods have been evolved. Once steady-state optimisation has been established on a plant or site where discrete options exist then often this is the next step in developing the full economic potential of the process.

The rest of this section on operational optimisation will be devoted to a more detailed discussion of some aspects of steady-state systems.

Some System Attributes

In providing a useful steady-state optimisation tool for the plant strategist the operations engineer is aware of several important considerations. The most important of these are that the system that he produces must be fast, robust and flexible. This is especially true of on-line optimisation where the computer needs to respond almost immediately to a disturbance which may have invalidated the current optimal set-points. For off-line use it must offer the investigating engineer the opportunity to scrutinize a wide range of related optimal scenarios quickly and reliably. He also recognises that the ultimate goal of the development is optimisation and not simply modelling. This is an important point because optimisation algorithms can impose mathematical restrictions (derivative continuity) on the modelling activity and if these are ignored then unreliable software and degraded performance can be the outcome. Although in theory equation solvers may require the same kind of restrictions to hold, in practice and when the restrictions are missing, they are far more robust than optimisation algorithms. For this reason it is often the case that existing models constructed simply for predictive use may need to be 'reconstituted' before they are suitable for optimisation.

Accuracy of the model predications is also important because, unless the model stands up reasonably well against actual plant performance, credibility will be effected and the system will not be used. Usually to achieve this for on-line systems the model may need to be supported by an on-line updating scheme based upon measured data from the process. In practice model complexity is limited by the requirement for fast execution but if the optimisation problem is formulated in the right way this limit can be very generous. As yet the most reliable way to achieve these demanding targets for a wide range of processes has been to generate 'custom-built' software with minimal computational overheads for each application. Because the system is specific to one plant this can usually be justified and in the case of on-line systems is almost obligatory.

Formulating Steady-state Problems

The preferred method of formulation is that of the 'simultaneous' approach. This is not in conflict with the use of a sequential modular system for design studies because the objectives are different. At one extreme models need to be configured and manipulated easily and quickly and computer execution time is secondary. At the other extreme the configuration is the priority. With the simultaneous method the iterative equations (g) which define the mathematical model are treated as equality constraints and augment the set of operational inequality constraints (h). Similarly, the internal variables of the model (x) are treated as extra optimisation variables by the algorithm and augment the normal set of decision variables (u). Symbolically, the problem now is ......

$$\min_{x,u} f(x,u) \qquad x = (x1, x2, \ldots xn)$$
$$\qquad\qquad\qquad u = (u1, u2, \ldots ur)$$

$$gi\ (x,u) = 0 \qquad i = 1,2, \ldots n$$
$$hj\ (x,u) > 0 \qquad j = 1,2, \ldots m$$

Expressing the problem in this way has many advantages. Foremost among these is the elimination of unnecessary model convergence at intermediate points. Time-consuming nested iterative loops which might appear in a sequential modular formulation are eliminated and treated simply as extra constraints relating the necessary variables. This feature is particularly valuable for modelling highly integrated plants where it is not possible to represent the economic performance adequately without modelling the interactions across the entire system. Consequently, the computation of the problem functions is fast and because there is no iteration the numerical accuracy of these functions can be very near machine precision. This is important if optimiser gradient information is to be evaluated by differencing methods although the possibility of an exact gradient calculation is now more likely than when iteration is present. The disadvantage is an obvious increase in problem size, especially when n is large and the difficulty of 'interpreting' intermediate results in terms of the process after a computational failure. With the right solution technique, however, large values of n need not be a hindrance.

Solution Methods

Nowadays the numerical solution of constrained optimisation problems of the above type is best achieved by using a Successive Quadratic Programming routine such as Powell's 1977 Variable Metric Algorithm (Ref 7). These methods treat non-linear equality constraints very efficiently and are thus very suited to solving problems with the above formulation. However, when n becomes large the methods need modification in accordance with an idea published by Westerberg et al (Ref 9). In essence, this idea again eliminates the x variables from the problem at each major iteration of the SQP algorithm and also allows any significant sparsity or structure of the mathematical model to be used to reduce the computation. The result is superior numerical performance which is particularly impressive when n is very large.

When Is Steady-state Optimisation Worthwhile?

Experience has shown that the benefits of applying this technology on suitable plants can improve variable gross-margin by 1-2%. To be suitable the process must offer a sufficiently complex strategic problem where the solution is not obvious and is perhaps subject to change because of uncontrolled disturbances outside the system. The complexity of the problem may come about because there are many free decision variables that need to be set or alternatively the process itself may be complicated because it is highly interactive or produces a wide range of differently valued products. Large numbers of potential process inequality constraints also contribute because different sets may be invoked under different combinations of product and feedstock values. All processes are of course subject to 'change' even if it is only in the economics but more often than not new feedstocks, drifts in internal efficiencies and the rescheduling of equipment create a continually changing problem. The change element is a particularly important factor for on-line systems where the objective is best described as the pursuit of a 'moving optimum'.

DISCUSSION

The practical use of a general purpose flowsheeting tool, FLOWPACK II, for use in design optimisation and custom-built models for operational optimisation has been described. A question which must be asked is, could the model created at the design stage be used as the operational optimiser? To answer this question it is necessary to look at the features of the design optimiser and check them against the requirements of the operations optimiser.

When FLOWPACK II is used in optimisation mode the models used are from the library of unit operations. The job of these units is to calculate the output streams from a given set of input streams and operating conditions at steady state. For optimisation, gradient information showing how the residual of each variable is affected by changes in each of the other variables is required. To provide this information a single pass through the flowsheet must be made for each of the variables connected to the optimiser. Based on the gradient information the optimiser will determine the direction in which to move each variable and by how much in order to meet the equality constraints and simultaneously optimise the objective function. A pass through the flowsheet, on a MicroVAX II, may take 1 minute for a large flowsheet with complex unit operations and rigorous thermodynamic models. If there is a recycle stream with 10 components plus 10 additional degrees of freedom defined by the user it would take 20 minutes to establish a set of gradients. After taking an optimisation step gradients must be re-established. It may take typically 7 steps to find the optimum so the total CPU time for the optimisation would be around 140 minutes. Although, for a single component steam system, were the physical properties can be calculated very quickly from steam tables the CPU time might be down to 10 minutes.

In practice 140 minutes is not a problem for a design optimisation, provided it works reliably. It is the final in a series of evolving runs and one in which large savings in terms of capital, raw material and operating costs may be identified. In the context of a design project, even if the optimisation took 24 hours and identified a more profitable plant it would be very worthwhile. What is far more important than the CPU time is the flexibility of the model building system so that different options can be included quickly and then optimised.

In contrast, the first criterion that must be met by an operational optimiser is the elapsed time in which the results are required. If the FLOWPACK II model developed for design purposes is not capable of meeting this then a dedicated model would have to be written.

The second thing to be considered is the use to which the optimiser is being put. In design optimisation we are looking at getting the balance of the plant right in terms of, the relative size of plant items, the capital versus operating cost trade-offs, and that the plant design is flexible and robust at full load. The accuracy to which the profit at the optimum is calculated is not that important. As described FLOWPACK II gets derivative information by perturbation of models that may have iteration within them. Provided they do not lead to premature termination at a non-optimal point these are adequate for the type of trade-offs being investigated at the design stage. Whereas, in operational optimisation we may be looking to maximise the profit function by 1-2% in a situation where the size of the plant items and the capital cost has been fixed.

Further, in writing very fast models, the on-line optimiser always has the advantage that: (a) Plant data are available so that immediate feedback on the adequacy of simplified formulations is available. The design models must always use the most rigorous formulation available because there is no information on how the model is to be used or extrapolated. (b) The on-line model can be geared to the appropriate physical properties for the particular system under study. The design optimiser model must be written to accommodate even as yet unforseen physical property correlations and to guard against phases that may, in principle, occur.

FLOWPACK II design optimisers could be used as operational optimisers if the objective function and variable values at the optimum were calculated to sufficient accuracy and the elapsed time was acceptable. If this was not the case it would be possible to do two things. One, speed up the simulation by writing specific unit operations to replace the ones from the general library. These models would probably include fast local thermodynamic models. The advent of very fast general purpose dual-level models (Ref 10) shows some promise of providing a facility that can be tuned either against rigorous models or plant data. This development could bridge the gap between the two, but the approach has yet to be evaluated in practice. Two, further develop the unit operation models to calculate derivative information in addition to the normal outputs. This would both reduce the time needed to generate derivatives and improve their accuracy. This is an attractive strategy as it would be possible to stop model development as soon as the required CPU and accuracy had been reached. However, the development effort to move from using the current models to writing specific models for individual units would not be insignificant.

CONCLUSIONS

There are two types of process optimisation, design and operation. ICI has tools specifically developed for each of these applications. It would be preferable to use the same tool for both applications but the design models may not have the performance required of operational optimisers.

It should be possible to improve the performance of design optimisation models by extra development work but this has not been attempted in ICI to date. All on-line optimisers currently installed in ICI are dedicated models.

Acknowledgements

The authors wish to thank all their colleagues who have contributed to the development of the software systems described, and also ICI PLC for permission to publish this paper.

References

1. Andrew, SM (1969), Trans.Inst.Chem.Engrs, 47,79.

2. Douglas, JM,(1985) "A Hierarchical Decision Procedure for Process Synthesis" AIChE Journal,31,3,p 353.

3. Douglas, JM, and Woodcock, DC, (1985) "A New Cost Diagram Approach For Preliminary Process Design" Inst.Chem.Symp.Ser,92,403.

4. Perris, FAP, "FLOWPACK II, A Third Generation of Flowsheeting System" Computer Aided Process Plant Design - (Leesley, ed), Gulf 1982.

5. Rosen, EM, "Steady State Chemical Process Simulation: A State-of-the-Art Review" Computer Applications to Chemical Engineering - (Squires and Reklaitis, ed) ACS Symposium Series 124.

6. Umeda, T, and Nishio, M, "Comparison Between Sequential and Simultaneous Approaches in Process Simulation" Ind.Eng.Chem.Process Des.Dev.,Vol 11,2 (1972).

7. Powell, MJD, "A Fast Algorithm for Non-linearly Constrained Optimisation Calculations", first presented at 1977 Dundee Conference on Numerical Analysis.

8. Biegler, LT, and Hughes, RR, "Infeasible Path Optimisation with Sequential Modular Simulators", AIChE Journal,28,6,p 994 (1982).

9. Lock, Edahl and Westerberg, "An Improved Successive Quadratic Programming Optimisation Algorithm for Chemical Engineering Systems", Carniege-Mellon University, June 1982.

10. Johns, WR, and Vadhwana, V, "A Dual-Level Flowsheeting System", Symposium Series No. 92.

REVAMP OF HIGH PRESSURE DISTILLATION COLUMNS USING RANDOM PACKINGS

* K. ROBINSON

Following the development work on new random packings and distributor technology it is shown that high pressure distillation columns can be re-vamped using random packings. Successful re-vamps in refineries and petrochemical plants are discussed and the details of an olefine plant demethaniser re-vamp are presented

INTRODUCTION

Although packings were almost certainly the first items of equipment to be utilised for gas liquid contacting in the process of distillation, trays have dominated the scene in "modern" large scale plants. The reason is almost certainly one concerning the reliability and re-producibility of designs.

In the mid 1960's it was unusual to find a packed distillation column larger than 1m in diameter. Design data and techniques were virtually unknown and rules of thumb proliferated. Varied performance and lack of reproducibility added to this reluctance to build packed columns.

On the other hand the bubblecap tray had been developed and although design techniques were somewhat crude it was shown to give reasonable and generally reproducible efficiency over a wide range of operating conditions. The natural step forward was therefore to better trays thus leading to the development of sophisticated design techniques being developed by Fractionation Research Inc. and others for sieve and valve tray columns.

Pressure to develop large packed columns came intially in the field of acid gas removal columns for the "new" large ammonia plants. These 1000 te/day plants of the late 1950's changed from using trayed amine absorption systems to packed carbonate systems. The liquid loads were very high and had a tendency to foam thus making tray columns uneconomic. The packed columns had diameters in the range of 4500mm to 6000mm and had total packing heights in the region of 24m.

* Norton Chemical Process Products, Stoke-on-Trent, UK.

As process engineering and awareness of energy conservation developed it became clear that the low pressure drop of packings when compared to that of trays gave considerable scope when carrying out distillation under vacuum conditions. The first major process unit where packings were installed was in the refinery vacuum column. Here lower operating pressures in the flash zone give higher yields and based on the quantity processed in these huge towers (up to 13m in diameter) very great savings can be achieved. These units were initially packed using grid packings which were later superceded by random packing and occasionally ordered packings.

The benefits of the low pressure drops from packings in other vacuum columns was also recognised and in the mid 1970's some designers built large units (greater than 2m diameter) with varying degrees of success. It was not until the early eighties that it was recognised that the solution to the problem of predictability and reliability was in the design of the liquid and vapour distribution and re-distribution systems. Since this time equipment has been developed and data banks on packing efficiency have been established to the point that the choice of a separation device can be made based on an economic analysis of the capital and operating requirements for equipment to produce a specification product.

The above history relates to vacuum distillation only but is equally relevant to atmospheric and low pressure processes. High pressure operation has additional problems relating to high specific fluid loadings and to the physical properties of the type of materials normally distilled under pressure.

1. High Pressure Distillation Processes

High pressure distillation is required for the separation of materials not normally liquid at atmospheric pressure. In many cases cryogenic conditions apply also.

The major industrial processes using high pressure separation are associated with the separation of C1 - C4 hydrocarbons.

Refineries generate a number of gas streams containing these low molecular weight hydrocarbons principally from cracking and reforming operations. The principle steps (See Fig. 1) involve two stage absorption using absorption oil and a stripper for stripping the C2 and H2S content of the rich absorption oil. The next step is to separate the C4 and lower molecular weight hydrocarbons from the heavier absorption oil fractions. The C3's and C4's are then separated each going on to other separations and processes. Table I shows brief details of the columns involved in this process.

TABLE I

Pressure	Bar	Theo. Stage	Actual Trays	Typical Diameter
1st Stage Abs.	14	8	24	2400
2nd Stage Abs.	14	4	20	1200
Stripper	14	12	20	3400
De-butaniser	11	30	40	2800/3600
C3/C4 Splitter	14	25	30	2500

Petrochemical plants have large high pressure columns associated with the production of high purity ethylene and propylene. The first towers in these distillation trains are also cryogenic operating at -100 Deg. C. The separation train (Fig. 2) is on paper a very simple one with the full range of low molecular weight hydrocarbons through to a light gasoline fraction. A small amount of hydrogen is also present. The first column, the demethaniser separates methane and hydrogen overhead the bottoms passes to the de-ethaniser which separates the C2 fraction, the subsequent two columns, the depropaniser and de-butaniser take the C3 and C4 fractions overhead. The C2 stream from the de-ethaniser is further separated to make the maximum yield of high purity ethylene and similarly the C3 stream is separated to yield high purity propylene. Table 2 summarises details of the columns involved in this process.

TABLE 2

	Pressure BAR	Theo. Stage	Actual Trays	Typical Diameter
De-Methaniser	31	35	56	1500/2500
De-Ethaniser	25	35	52	2000/2600
C2 Splitter	19	90	110	3500
De-Propaniser	8	29	44	2000
De Butaniser	5	26	40	1700
C3 Splitter	20	120	150	2500

Another long established high pressure separation is the separation of ethane from the "dry" gas of the U.S. gas fields. Ref. 1 describes this cryogenic process. Ethane separated in this manner has long been the major feedstock for the U.S. ethylene industry. Separation of light hydrocarbons from "wet" gas is very similar to that described above for refinery gas and hence is another major user of high pressure distillation.

2. The Requirement for Re-Vamping

Most of the units described in Section 1 will have been built using valve or sieve trays with the possible exception of some of the "dry" gas field units which are often small in diameter and may possibly use packing. The trays would be designed to operate in the region of 80% of flood which is the typical design point for minimising column diameter and hence capital cost. Small increases in plant rate soon hit a bottleneck. More sophisticated trays can generally gain 10% or so but the final limit is soon reached.

Table 3 shows typical flows and system properties for an ethylene plant de-ethaniser.

TABLE 3

Olefine Plant De-ethaniser - Typical Conditions

Position	Pressure BAR	Gas Flow Kg/hr	Gas Density Kg/m3	Liquid Flow	Liquid Density	Surface Tension dyne/cm	Viscosity C.P.
Tray 1 (Bottom)	29.5	124028	71.1	145185	405	2.0	0.06
Tray 53 (Top)	29.5	104717	57.82	44025	403	3.5	0.053

The most notable points are the high vapour densities, low liquid densities and low surface tensions. This particular example was an actual re-vamp case after the re-construction.

The diameter at the top was 2050mm and at the bottom 2650mm. On this diameter the specific liquid loadings were 33m3/m2hr and 65m3/m2hr respectively. These are very high figures when compared with the figures experienced in atmospheric and low pressure distillation. A normal tray design for these flows would give diameters of 2250mm in the top section and 2850mm in the bottom section. A particular point about these tray designs is the area occupied by downcomers and underdownflow areas namely 22% of the column cross section in the top section and 33% in the high liquid load lower section. Hence only 67% of the column cross section is available for vapour flow. Two further complications in downcomer design, additional to the high liquid load are firstly the time required for separation of the dense vapour from the low density liquid and secondly the pressure drop when measured in height of actual liquid is high thus causing a much higher back-up in the downcomer.

3. Random Packings in High Pressure Distillation

Some of the first random packed high pressure distillation columns were put into service around 15 years ago in the gas field de-methanisers in the US. These units separate ethane from the natural gas under high pressure cryogenic conditions.

The typical method of designing packed columns at this time was to calculate the diameter for a pressure drop of 40mm water gauge per meter of packed height. By trial and error it was discovered that these units when using 25mm Pall Rings reached maximum capacity at 28mm w.g./m and that the design point for efficient stable operation should be in the region of 20mm w.g./m. It is sufficient to note that this figure equates to approximately 45mm liquid/m packing at the liquid density for the operating conditions. It is clear therefore that the physical properties also have an effect on the packing capacity.

The recovery of ethane was very low, generally in the 50 - 70% region and the number of theoretical stages required from the column was less than 10 thus any efficiency shortfall could be countered by the provision of more packed height.

Although the above gave some degree of experience of distillation at high pressure further attempts in more exacting distillations, where large numbers of theoretical stages were required to meet high purity product specifications, came to failure. It should also be said that similar results were experienced with similar high purity distillations at all pressures. However, in the late 1970's economic pressures to use random packings catalysed Norton Company and later others including FRI into carrying out some basic research into the capacity and efficiency of packed towers.

3.1. Capacity

Due to the launch of a new generation high capacity packing, the IMTP family, Norton carried out a test programme intended to generate a capacity correlation for packings as reliable as those used for trays. It was hoped that packed columns could then be designed at a high percentage of their maximum capacity, a figure akin to the 80% of flood used for trays.

The test programme utilised laboratory data from a 380mm diameter column, field data for high vacuum distillation, with a large number of stages, from a 914mm diameter column and most important for its relevance to this paper high pressure data from a C3 splitter. The C3 splitter also had a large number of theoretical stages, was 1100/1400mm in diameter and obtained data at pressures up to 29 BAR.

The final correlation has been described by Strigle and others (refs. 2, 3, 4). It is shown in Fig 3.

The logic for the correlation is a very simple one. Packing efficiency (or HETP) is generally constant over a wide range of vapour rates as defined by Cs where $Cs = V\sqrt{\rho v/(\rho l-\rho v)}$

Fig. 4 shows such a curve. The horizontal part of the curve would be the design HETP and the extension of this at point A is defined as the maximum capacity based on the fact that the design HETP has been exceeded. Fig. 3 is a plot of this maximum capacity point versus rate and physical property factors. Other physical properties such as surface tension and viscosity have been shown above to have an effect on capacity. Fig. 3 is applicable for surface tensions in the range of 18 - 20 dynes/cm and viscosities around 0.2 c.p. For properties widely outside this region such as encountered in high pressure distillation further physical property factors have to be applied. Fig. 5 shows part of the curve corrected for operating conditions for a C3 splitter. The major change is due to the low surface tension in the high pressure system.

Also shown on Fig. 5 is a plot of the capacity of a sieve tray also at C3 splitter conditions. This indicates the possibilities of re-vamping a trayed column for a similar capacity and more theoretical stages or for a similar number of theoretical stages at a higher capacity. The options depending on the selection of packing size.

3.2. Efficiency

The early part of the IMTP field test programme under the vacuum conditions revealed that good liquid distribution was the key to obtaining high efficiencies from packed towers. It was also shown to be the key to reproducibility.

Fig. 6 illustrates the results of some work carried out on distribution.

The plot shows the number of theoretical stages generated in a column versus the wetted area of the column. This particular plot was for a column initially having no support ring and distribution holes right up to the wall and then with progressively larger support rings. A similar performance is observed due to any interuption of an even distribution pattern whether this be due to gas risers or structural beams.

Similar problems are observed even with a perfect distribution pattern if for some reason the quantity of liquid varies from one part of the column to another. Variations in liquid flow may be due to distribution hole tolerances, varying pressure heads due to levelness of plate type distributors, kinetic factors in pipe type distributors or even liquid feed pipes onto plate type distributors.

When considering the above the explanation for the apparent random behavior of the packed columns of 20 years ago is very clear. Packing efficiency is constant, poor liquid (or vapour) distribution changes the slope of the operating line and hence the separation achieved in each part of the tower is different. We, of course, measure the average compositions and calculate the efficiency of the packing. Hence variability of distribution effects the average efficiency of the column. However, using the same distributor and the same packing but in one case operating close to minimum reflux ratio and another a significant factor from minimum can give very different product qualities and hence average column efficiencies. Hence giving the idea that packed column are not reproducible.

Distributors designed taking account of the above factors and typically as illustrated in Fig. 7 were utilised in the high pressure test programme using the C3 system. Capacity and efficiency data were obtained over a range of reflux ratios and product purities. The results from this test programme are shown on Fig. 8 plotted onto a section of the capacity curves as shown in Fig. 3. It should be noted that the test data has been adjusted, by the physical property factors for the various operating conditions, back to the standard conditions of Fig. 3. Further field data from high pressure units is now incorporated into the Fig. 3 curves. The high pressure data essentially fixes the high X end of the curves.

Packing efficiency data was collected from the early C3 runs and having solved the liquid distribution problem the data was reproducible and HETP's were found to be very low, especially when compared to the rule of thumb figures used in designing the early gas field demethanisers.

Typical figures for close boiling hydrocarbons under high pressure conditions are in the region of 400 - 500mm for the No. 25 IMTP. Higher figures are found in units separating C3's from a mixed feedstock such as a depropaniser or C4's in a gasoline stabiliser. The higher HETP's in such systems can at least be partly explained by the higher relative volatility and large ratio of the slope of the equilibrium to operating line as discussed by Rukovena and Koshy (Ref. 5).

4. Re-Vamping of High Pressure Units

4.1. Following the development of design data and the evolution of equipment as described above Norton has successfully re-vamped a wide range of columns in the refining and petrochemical industries.

In refining the principle and recurring revamps have been in the low molecular weight gas recovery units. All columns as shown in Table I have been re-vamped on many occasions. Other one off units have also been re-vamped, a notable one being a de-isohexaniser. This column separates a mixed hydrocarbon stream between nC6 and iC6. The particular re-vamp was of a column 3000 mm/2500 mm diameter with 95 sieve trays achieving 64 theoretical stages. The revamp was to achieve 75 theoretical stages. This was achieved by 38 m of No. 40 IMTP in 3 separate beds.

The re-vamp of high pressure columns in Petrochemicals has been principally in Olefine plants where at least one of each of the columns shown in Table II, with the exception of the de-propaniser, have been re-vamped. More than 10 de-methanisers have been re-vamped.

4.2. Re-Vamps of an Ethylene Plant De-Methaniser

The reason for this project was very simple - more capacity at the same product quality.

The de-methaniser is a very complex column essentially separating hydrogen and methane from a mixed hydrocarbon feed stream ranging from methane to a light gasoline fraction. It has four feed streams, (all flashing) and requires a minimum ethylene loss overhead (generally less than 0.5%) and minimum methane in the bottoms. The operating pressure is 31.5 Bar and has a range of temperatures from - 97 Deg. C at the top and + 4 Deg. C in the bottom.

Simulation of the column indicated that the existing column with 56 actual trays installed would be required to generate 53 theoretical stages under the new operating conditions. The column layout used is shown in Fig. 9 together with the number of stages required from each bed. The heights of the beds and types of packing used were dictated by many factors namely:

Theoretical Stages Required
Hydraulic Load in the particular section
Location of vessel manways
Location of existing feed points

Back calculations indicate that the duty in some sections was easier than in others but the general order of HETP's used was 490mm for No. 25 IMTP in the upper section to 560mm for No. 40 IMTP in the lower section. These figures were relevant to the equilibrium set used for the simulation and obtained from a similar but simpler column in the same service using the same data set.

Physically the replacement of trays by packings consisted of removing the existing trays and removal of any sloping bolting bars, all other tray fittings remained in the vessel. It was not possible to weld to the vessel wall and hence all the new equipment had to be of a fully bolted construction.

The packing supports, liquid distributors, redistributors and flashing liquid distributors were constructed onto existing bolting bars.

The column was commissioned along with another one in the same train. The Norton guaranteed capacity was 113% of nameplate compared with 105% being achieved by the trays. The eventual operating rate achieved but restricted by other equipment was 119% of nameplate. Fig. 10 is a before and after summary under similar operating conditions. The cracked gas shown is upstream of a sequence of 4 flash vessels the liquid stream from each comprising one of the four feeds to the column.

FIG. 10

		Trays	Packing
Cracked Gas Compositon			
(Mole %)	H2	24.55	25.35
	CH4	24.44	21.64
	CH6	10.13	11.88
	C2H4	31.35	32.22
	C3+	9.53	8.91
Demethaniser Top Product			
(Mole %)	H2	6.62	6.55
	CH4	92.70	92.56
	C2H4	0.64	0.84
Demethaniser Bottom Product			
(Mole %)	CH4	0.08	0.06
	C2H6	20.05	22.45
	C2H4	61.12	60.74
	C3+	18.75	16.75
Max. Capacity obtained as % of plant design		105	119

5. Conclusions

It is concluded that random packings can be designed to give reliable performance regardless of diameter, number of stages or operating pressure. Economics will, however, be the final arbitor as to the choice of contacting device to be used. The most likely split will be for ordered packings for very low pressures, ordered or random packings for medium vacuum to atomospheric pressure. Trays for new equipment for atmospheric to high pressure and random packings (plus possibly ordered packings) for atmospheric to high pressure re-vamps.

LIST OF SYMBOLS

Cs	Vapour Capacity Factor	$= V\sqrt{\frac{\rho v}{\rho l - \rho v}}$
L	Liquid Load kg/hr m2	
V	Vapour Load kg/hr m2	
X	Flow Parameter	$= \frac{L}{V}\sqrt{\frac{\rho v}{\rho l}}$
ρl	Liquid Density kg/m3	
ρv	Vapour Density kg/m3	

REFERENCES

1.	Klehm J. J. and Singletory J. E.	Hydrocarbon Processing	Aug. 1971
2.	Dolan M. J. and Strigle R. F.	C.E.P.	Nov. 1980
3.	Dolan M. J. and Strigle R. F.	Canadian Process Equipment and Control News	Oct. 1983
4.	Strigle R. F.	C.E.P.	Apr. 1985
5.	Koshy T. D. and Rukovena F.	Hydrocarbon Processing	May 1986

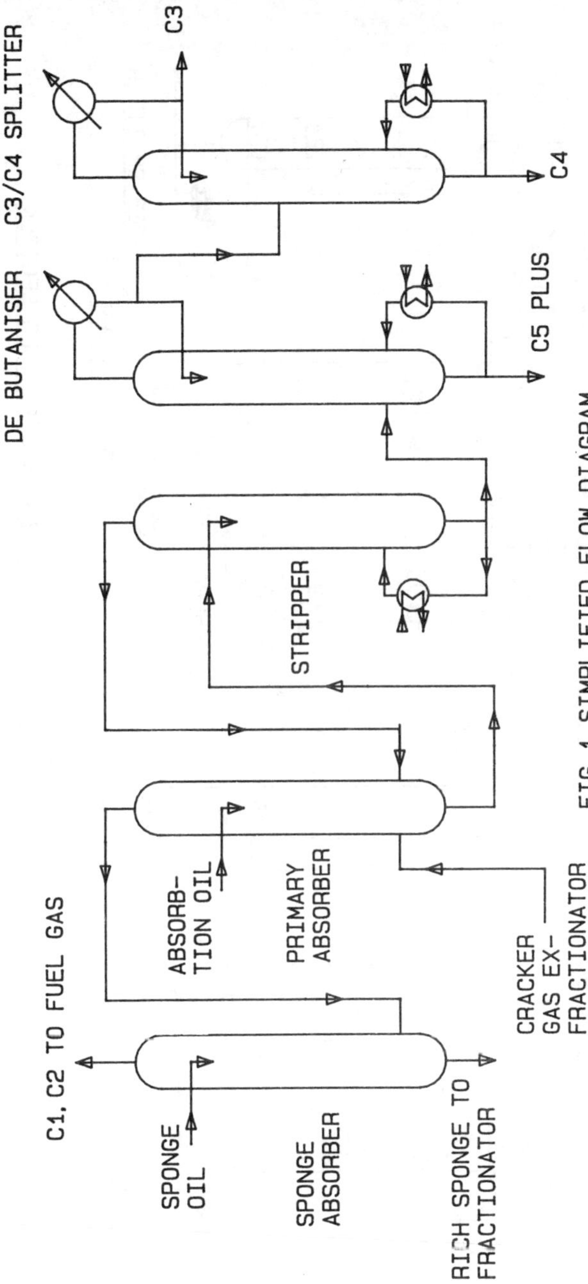

FIG.1 SIMPLIFIED FLOW DIAGRAM
FOR REFINERY GAS PLANT

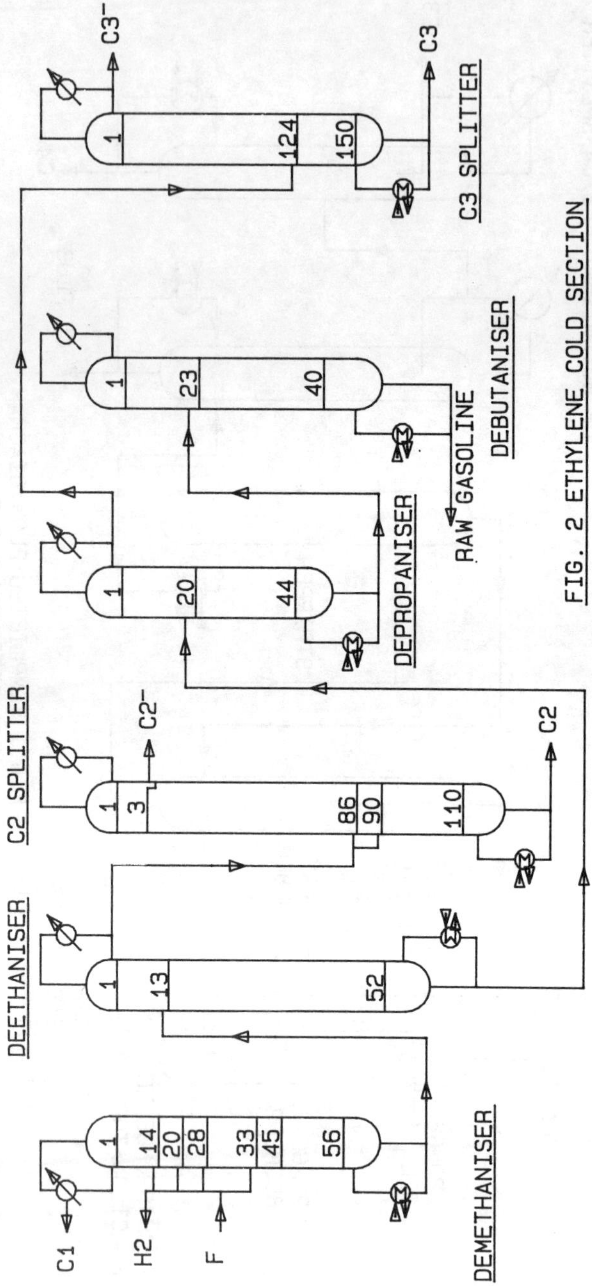

FIG. 2 ETHYLENE COLD SECTION

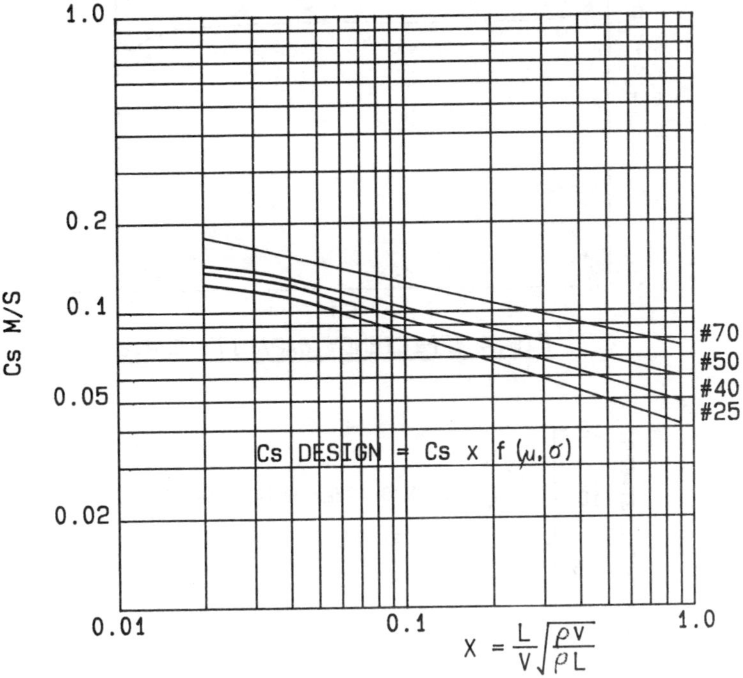

FIG.3 MAXIMUM OPERATIONAL CAPACITY FOR IMTP PACKING

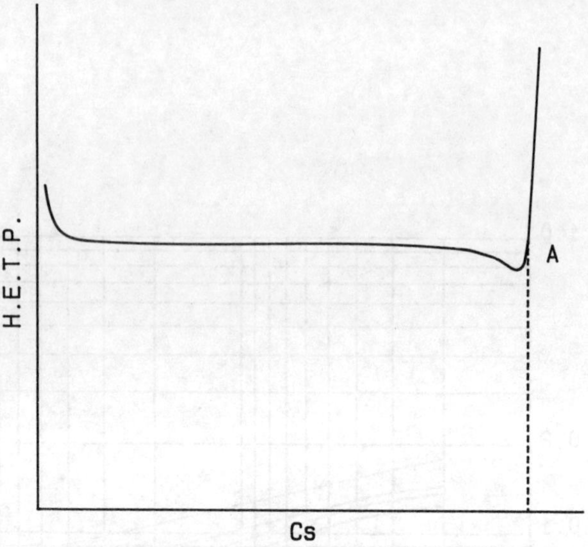

FIG.4 EFFICIENCY CAPACITY CURVE

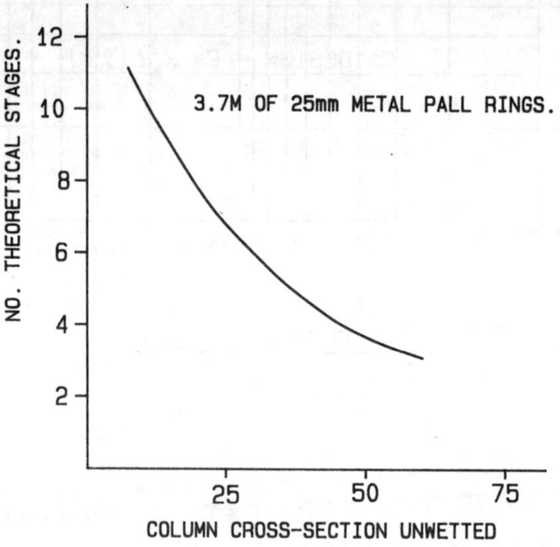

FIG.6 EFFECT OF MALDISTRIBUTION.

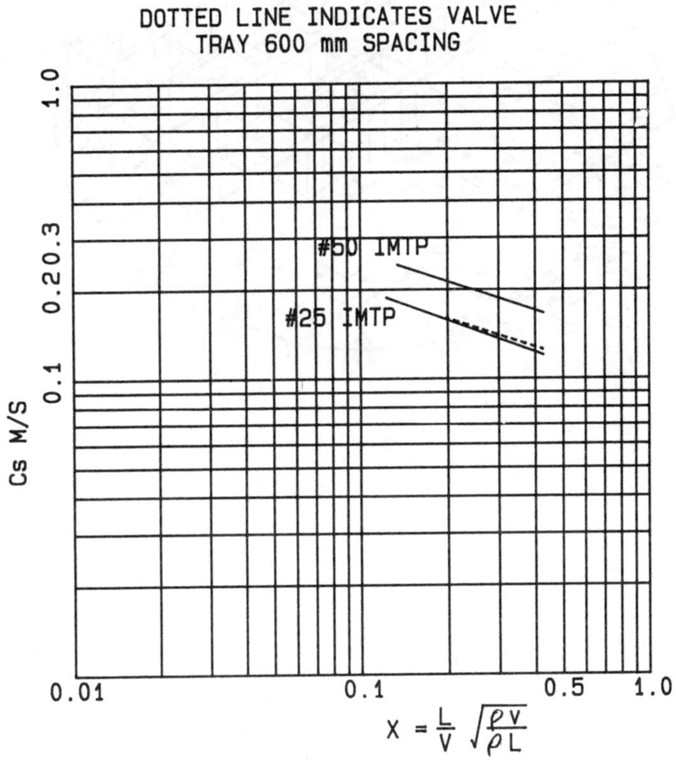

FIG.5 IMTP PACKING & TRAY CAPACITY
AT C3 SPLITTER CONDITIONS

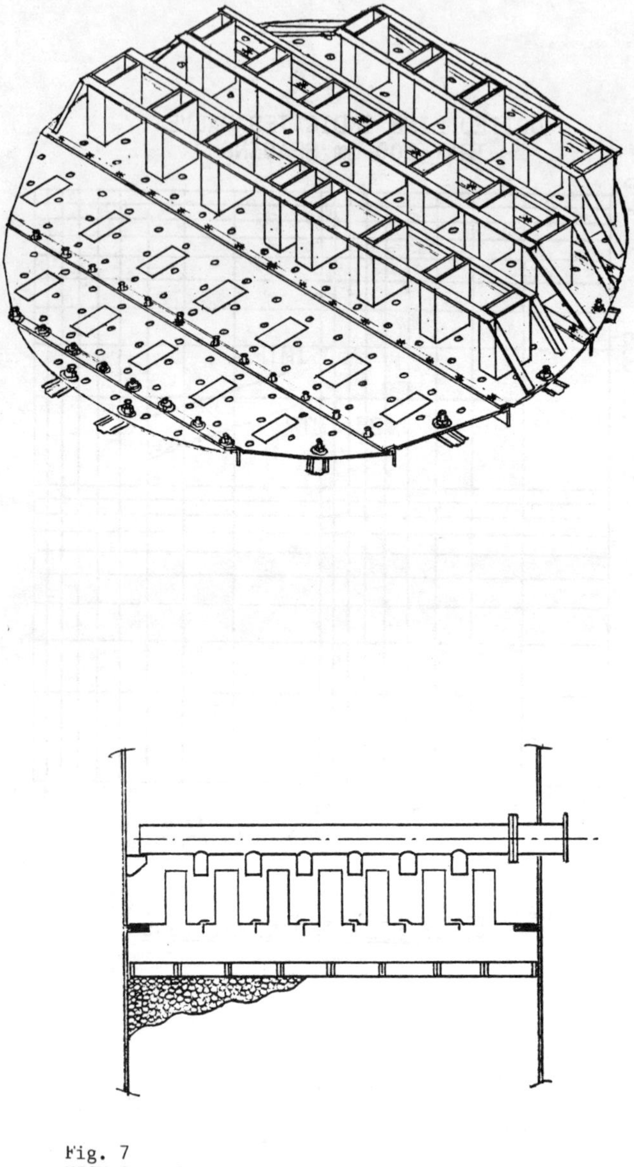

Fig. 7
HIGH QUALITY DISTRIBUTOR

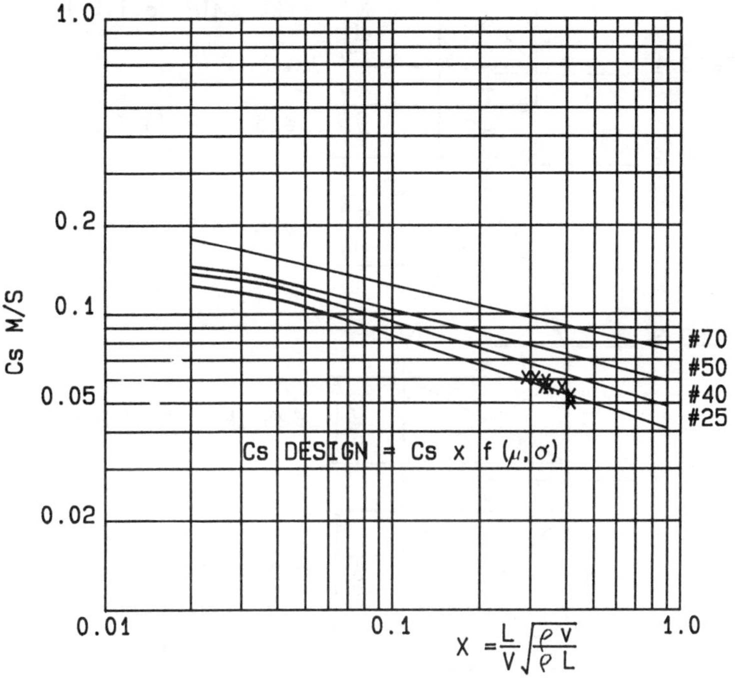

FIG.8 MAXIMUM OPERATIONAL CAPACITY FOR IMTP PACKING

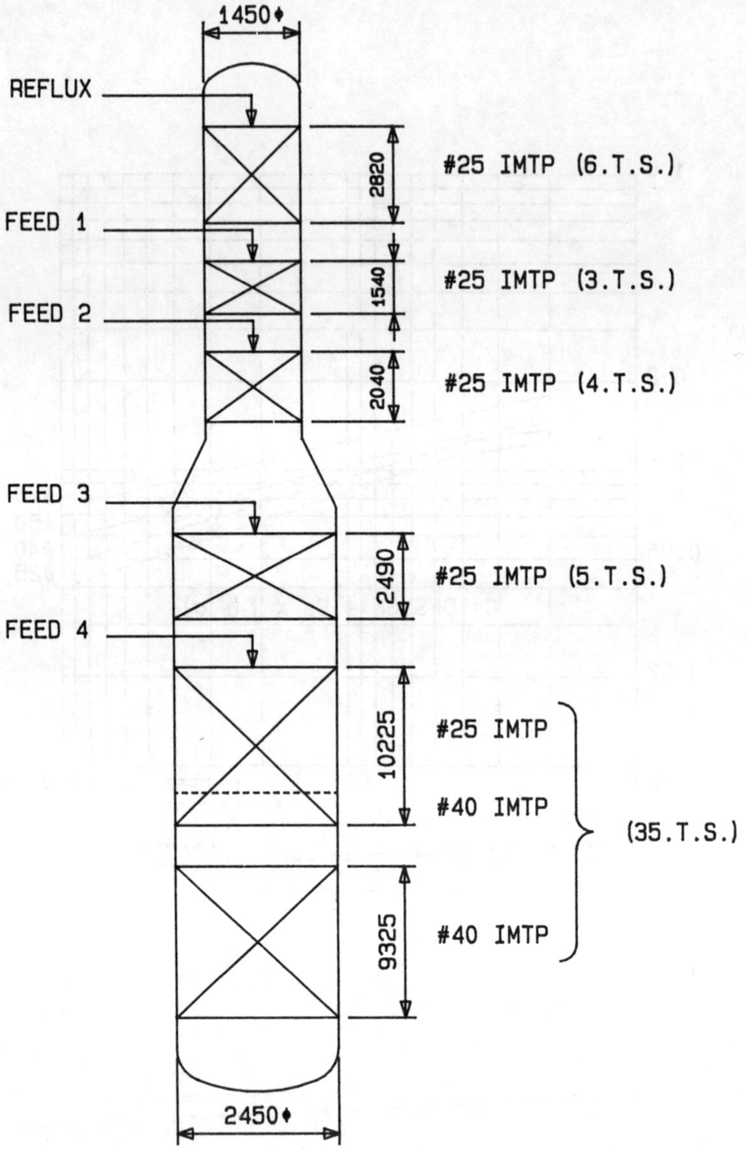

FIG.9 LAYOUT OF REVAMPED DEMETHANISER

OPTIMISATION OF PHOSPHATE GRINDING

Y. VINCENT* and S.B. SMOULT**

Recently developped techniques in process control allow relatively unstable and complex operations to be run at their optimum point. The computer programmes used to drive the operations are put together specifically for each application using the latest methods in analysis of sensitive variables, data validation and modellisation. The approach used has been successfull in several different applications, including grinding of talc and phosphate, drying, filtration and granulation/dryer systems.

INTRODUCTION

Many user objectives can be achieved simultaneously or separately by optimizing production of grinding mills.

Examples are :

- increased unit capacity
- reduced cost price per ton of ground matter
- no violations of technical constraints :
  - . electrical and mechanical limits, etc
  - . type of output grade, particle size, etc.

One of the most difficult problems to resolve when implementing an optimal control system in a grinding unit is the continuous measurement of particle sizes. The latest granulometers used in continuous processes measure only small quantities, therefore the problem becomes one of taking representative samples from the air/particle stream.

The entire programme of studies spanned several years at a crushing and grinding unit in the TALCS de LUZENAC plant in the Ariège department of FRANCE. Six units at this plant are currently optimised, as is a phosphate crushing unit at the PEC RHIN plant at Ottmarsheim in the Haut-Rhin department.

* **FMA** , NANCY, FRANCE
** **PEC-ENGINEERING** , CERGY-PONTOISE, FRANCE

VERTICAL GRINDERS

The operating principle is shown on figure 1.

The mill is fed by a vane feeder. The raw material is then ground by rollers and blown upwards by a column of air originating from a circular chamber at the base of the mill. Air enters the mill at a tangent, causing an upwards moving vortex to blast up the grinding chamber, churning the raw material in the process.

An initial separation of the particles takes place at the grinding stage, as only those particles which are sufficiently small relative to air speed can be carried upwards.

The increased section size of the separator reduces the air throughput speed. In the resulting drop in pressure, the biggest particles fall back into the mill, as they can no longer be kept suspended.

This form of separation is static.

In contrast, dynamic separation occurs by centrifugal action through a system of horizontal disks which support selector blades. In this case, the largest particles of raw material are the first to reach the separator walls. As the air flow here is weaker or nil, these particles fall back into the mill.

A cyclone is used to separate the coarse solids from the air and a bag filter removes the fines. The mill is also vented via the bag filter.

OPTIMISATION PROCEDURE

General

We have followed the usual sequence of steps when designing any production system :

- Identification of all the appropriate state and control variables

- Instrumentation of mill.

- Programme of system behaviour trials and execution of measurement runs.

- Validation of designed model(s). Localized controls.

- Determining the optimum control procedure.

Working principles

a) Analog and digital data inputs :

The computer is constantly aware of the state of the crushing circuit via an interface made up of standard industrial electronic cards.

There are some variables common to all types of mill :

- humidity
- air temperature
- air flow
- power consumption by fan motor
- power consumption by mill
- selector speed
- Delta P over mill
- Delta P over filter
- feed rate

Others are only measured if this is considered necessary after an initial study :

- Position of fresh air flap
- Position of flap in air flow circuit
- Position of air back-flow flap
- Pressure of grinders
- Rotating speed of grinders
- Negative pressure in front of mill, ...

Additional data inputs can serve to control levels in the raw material or mill output silos or to start or stop the mill.

b) Analog or digital data outputs :

By experience it is also possible to identify which variables to control or regulate :

- Delta P over the mill
- Power consumption of mill motor
- Selector speed, ...

The software is in any case adapted for all existing types of mill controls (0 - 10V, 20 mA, etc...).

Digital output sockets are available for the automatic control of silos and alarms and for actuating or switching-off circuits, etc...

c) <u>Optimum calculation algorithm</u> :

This particular module of the software is designed to calculate the commands to be sent. These calculations are done on the basis of the measured data and the applicable constraints.
The module falls into two parts :

- The first section estimates the mill output particle size by means of both a mathematical model to determine proper particle size and the measured variables. It then proceeds with a comparison between "estimated" and "controlled" particle size.

- If the variance is greater than a particular threshold value, the settings are recalculated :

* allowing for	: the mill status, particle size and electric power constraints, the previously calculated variance
* in order to arrive at	: a maximum possible flow-rate, conditions in which the mill operates in a stable manner.

The mathematical models needed for determining the proper settings vary according to individual units. Consequently, it is necessary to conduct a series of measurements to obtain a maximum amount of data, which is then processed using statistical methods to arrive at a model.

Before any models or control algorithms are used on site they are first checked for consistency and accuracy using simulation software.

Description of software

The software is designed to handle two separate tasks, always in active mode :

- Operator dialogue task : in this way, any operating parameters can be changed at any moment, a block diagram of mill operation with constantly updated measurements can be displayed ; a printout can be obtained of operating reports, data files, etc... This task is done via a menu, the first part of which is accessible to everyone. The second part of the menu requires a password before a requested option can be activated.

- Control-regulation task : this portion of the software deals with data inputting, calculation and automatic control in the framework of the operating parameters defined in the first task. Communication between this task and the interface cards is by a serial line which carries analog and digital data inputting and output commands.

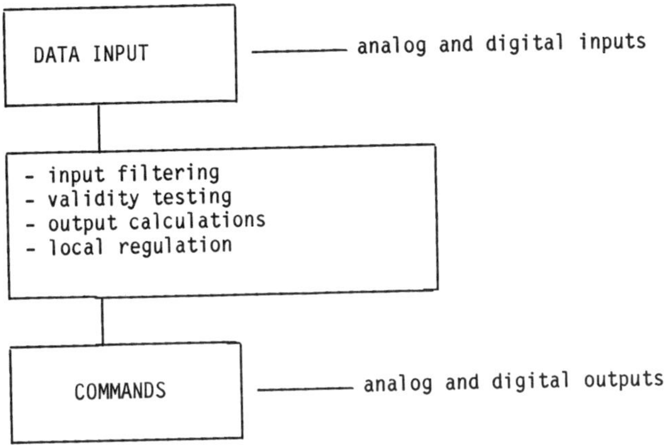

DIAGRAM OF CONTROL-REGULATION TASK

Implementation

A) Preliminary study

The initial phase entails work partly done on site and partly in the office in order to ascertain :

- the potential throughput and energy savings

- the necessary investment to achieve optimal functioning

- the pay-back time for a grinding mill operating on an optimized control basis

The preliminary study is summarised in a report covering the above mentioned points and it takes 2 to 3 weeks, including 1 on site.

B) Optimisation study and optimized control

Once the mill output circuit has been properly equipped with transducers, actuators, interface cabinet and microcomputer and the objectives have been clearly defined, the study proper can commence.
It involves the following stages :

- series of measurements : 3 weeks
- identification and mathematical modelling : 3 weeks
- fitting local regulating devices : 2 weeks
- finalizing optimum controls : 1 week
- test run + training of personel : 1 week

RESULTS

The increase in throughput at the TALCS de LUZENAC plant is 40 %, with an accompanying 10 % increase in electric power consumption.

These results are based on production reports over a period of two years.

The increase in throughput at the PEC RHIN plant is 68 %, with an increase in electric power consumption of 8.5 %.

OTHER PROCESS APPLICATIONS

Fertilizer granulation line (planetary granulator)

The major items of equipment are :

- a mixer
- a planetary granulator
- a rotary drier

Preliminary studies showed that there was interference between recycled material (fines on oversize) and the primary product. The key to limiting recycle is the regulation of the moisture content at the drier outlet. The unit has been optimized and is running in a fully satisfactory manner.

Fertilizer granulation line (spherodizer)

The process sections are :

- a wet section in which the phosphate rock attack, the ammonisation and the potash addition is carried out.
  The product is a slurry.

- A dry section. The slurry is injected into the spherodizer which granulates and dries the slurry.

Studies are in progress to optimize the dry section. The main parameters are :

- the moisture content and the flow of the slurry

- flow of hot gas through the spherodizer and the temperature drop

- heating capacity

- change of humidity of slurry and air in the spherodizer

- solid material flow (belt weighing devices to give recycle ratio, over and undersize, production rate)

- pressure in the spherodizer

The primary concern at this stage of the study is to lower the recycle ratio so as to lower the energy consumption per ton of product and to increase production.

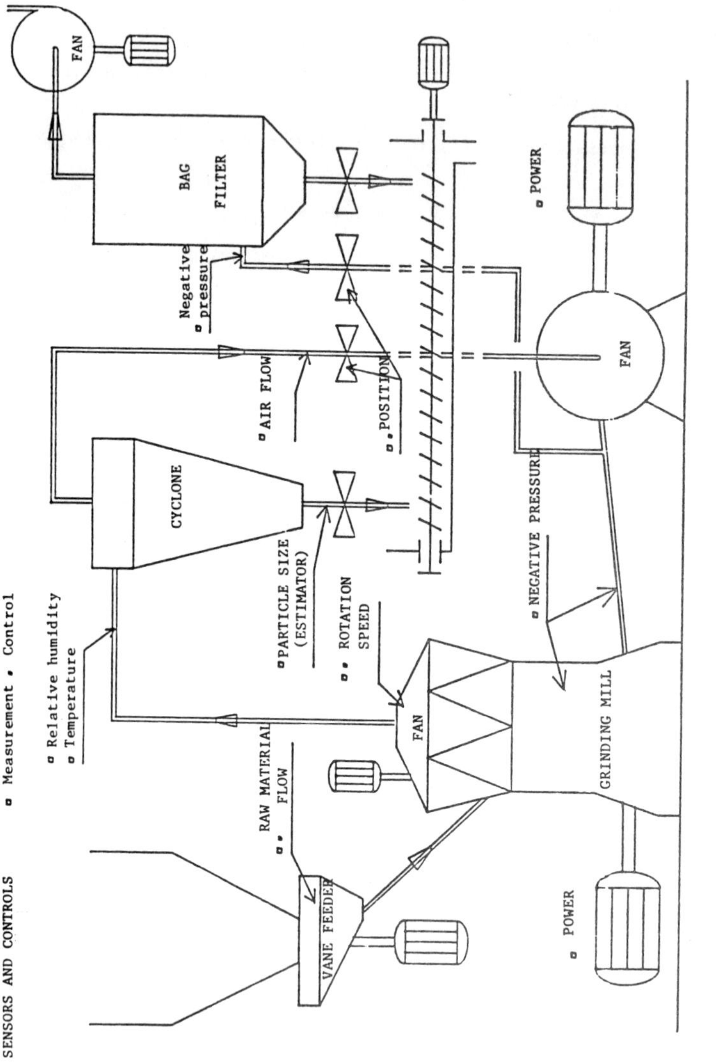

CRUSHING AND GRINDING UNIT

FIGURE NO.1

FIGURE N°2

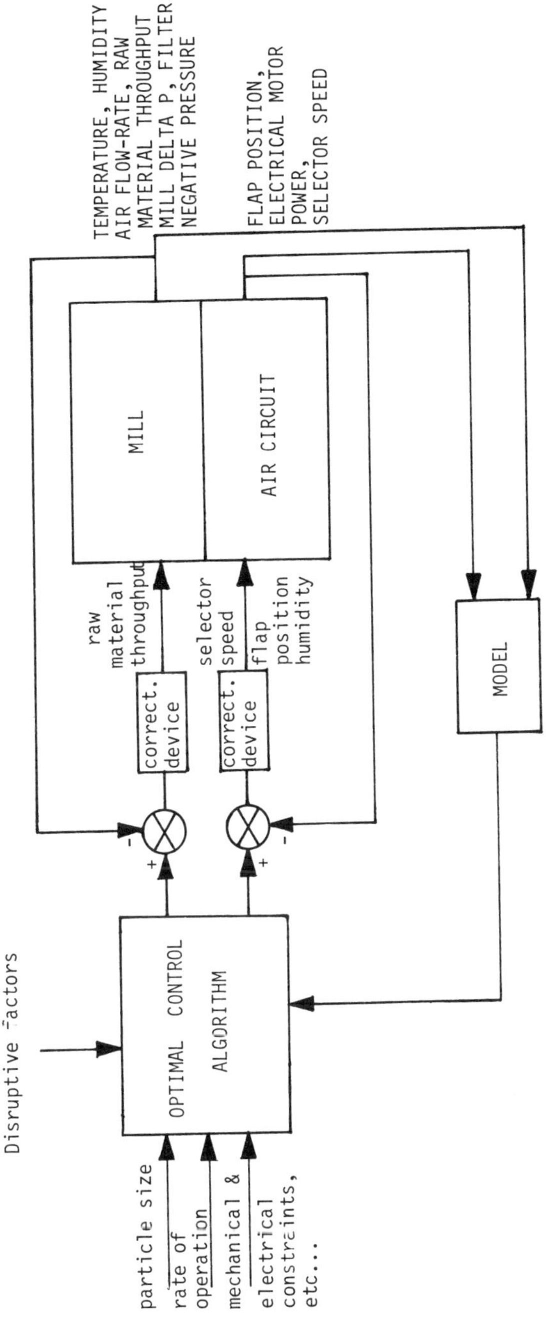

SCHEMATIC CONTROL DIAGRAM OF CRUCHING AND GRINDING UNIT

INDEX